Practical Handbook of
WAREHOUSING

Third Edition

Practical Handbook of
WAREHOUSING

Kenneth B. Ackerman

VAN NOSTRAND REINHOLD
New York

Cover photograph courtesy of
Trammel Crow Distribution Corporation

Copyright © 1990 by Van Nostrand Reinhold

Library of Congress Catalog Number 90-12471
ISBN 0-442-00557-1

All rights reserved. No part of this work covered by the copyright hereon may be reproduced or used in any form by any means—graphic, electronic, or mechanical, including photocopying, recording, taping, or information storage and retrieval systems—without written permission of the publisher.

Printed in the United States of America

Van Nostrand Reinhold
115 Fifth Avenue
New York, New York 10003

Van Nostrand Reinhold International Company Limited
11 New Fetter Lane
London EC4P 4EE, England

Van Nostrand Reinhold
480 La Trobe Street
Melbourne, Victoria 3000, Australia

Nelson Canada
1120 Birchmount Road
Scarborough, Ontario M1K 5G4, Canada

16 15 14 13 12 11 10 9 8 7 6 5 4 3 2 1

Produced by AAH, Seven Fountains, VA 22652

Introduction

This is the third edition of a book first published in 1983. It contains two new chapters. One of these deals with the growing use of "just-in-time" techniques and their effect on warehousing. A second new chapter covers specialized warehousing and three types of special storage – temperature control, protection of hazardous materials and "fulfillment" warehousing. Major changes in both materials handling and information processing and transmitting have caused appropriate changes in the contents of this book.

For over two decades, the author has been involved in writing about the warehousing industry. Earlier writings were oriented toward users and operators of public warehouses. The goal of this early writing was to develop a better understanding between the third party warehouse operator and his consumer, through a common appreciation of technical problems.

Since that time, the number of information sources on warehousing has increased. A professional society, the Warehousing Education and Research Council did not even exist when we first started writing about the industry. Today it provides an important source of information in the field.

Both private and third-party warehousing have grown in volume. More importantly, public awareness of warehousing as an industry in itself has increased.

This book is intended to serve both the private and third-party warehouse operator. It is designed primarily as a handbook for the warehousing practitioner, but it also serves as a guide for students, being based on the experience of the author as both an operating warehouseman and as a management consultant. It draws heavily upon our writings as editor and publisher of *WAREHOUSING FORUM*, as well as our past service as editor of *Warehousing and Physical Distribution Productivity Report*.

It is designed as a management reference for anyone who is involved in operating, using, constructing or trading in industrial warehouses.

Acknowledgements

Dozens of people made this book possible; they are credited in the footnotes wherever possible. The book could not have been prepared without the generosity of people who were willing to share their ideas and writings with me. Special thanks are offered to Warren Blanding of Marketing Publications, Inc. for permission to use extensive writings which appeared first in his publication, *Warehousing and Physical Distribution Productivity Report*.

William J. Ransom of Ransom and Associates reviewed the book and made significant contributions to this edition.

Professor Joseph Cavinato reviewed the second edition and made some excellent suggestions for change, which were followed in the preparation of this edition.

This is a successor to two earlier books, one published in 1977 and titled *Warehousing*, the other published in 1972 and titled *Understanding Today's Distribution Center*. The 1972 work was co-authored by R. W. Gardner of Distribution Centers, Inc. and Lee P. Thomas, consultant. Susan Aumiller and Cathy Avenido assisted the author in word processing and organization of the material. Jeannette Faelchle checked proofs and handled the hopeless task of keeping the author organized. The late Richard M. Fox, advertising consultant, made many recommendations for improving the writing.

To all these people we are most grateful.

Contents

Background of the Warehouse Industry

1   **WAREHOUSING:**
     ORIGINS, HISTORY AND DEVELOPMENT 3
     History 3
     Warehousing in America 4
     Role of the Independent Warehouseman 5
     Financial Security Aspects 6
     Role in Mass Production 7
     From Terminal to Distribution Center 7
     The Emergence of Business Logistics 9
     The Rise of Professional Associations 10
     Trade Associations 10

2   **WAREHOUSING IN TODAY'S ECONOMY** 13
     The Business Environment 13
     Freight Consolidation 16
     Real Estate as an Investment 17
     Warehouse Locations Shift 17

3   **THE FUNCTIONS OF WAREHOUSING** 19
     Six Functions of Warehousing 19
     Three Warehousing Alternatives 23
     Cost Structure 26
     A Contract to Control Risks and Costs 26
     Changing Values of Warehousing 28
     Space 29
     Equipment 29

	People 29
4	**WAREHOUSING ALTERNATIVES.**
	THE MAKE OR BUY DECISION 31
	Corporate Financial Policy 32
	Labor Relations Policies 34
	Strategic Decisions 35
	The Final Decision 35
	Public vs. Contract 36
	The Economics of Third Party Warehousing 37
	Bonus Incentives 38
	Other Factors 38

The Elements of Warehouse Management

5	**COMMUNICATIONS** .43
	Hard Copy Versus Voice 44
	Public Warehousing Forms 44
	Damages 47
	Real-Time Inventory Control 48
	Circulation of Stock Status Information 49
	Effective Order Picking 50
	Faulty Communications Equal Wasted Effort 50
	Complaints 50
	Standard Procedures 51
	Uniform Communications—A New Development 51
	Uniformity for Warehousing 52
	System Priorities 54
6	**PACKAGING AND IDENTIFICATION** 55
	Functions of a Package 55
	How the Package Affects Warehousing 56
	Effect on Stacking 57
	Packaging for Unitization 60
	Identification 60
	Container Handling 61
	Package Design 62

Efficient Packaging—A Joint Responsibility 64
A Series of Compromises 64

7 TRANSPORTATION 65
Location 65
Effects of Transport Modes 67
Freight Rates 68
Transit Time 68
Private Fleets 69
Warehouse Freight Consolidation 69
Reducing Transport Costs in the Deregulation Era 70
Reducing Freight Costs 71
Transportation Brokerage—A Useful Tool 75
The Second Industrial Revolution
and Its Effect on Transportation 79

8 ACCOUNTABILITY 83
Title 83
Inventory Responsibility 86
Product Liability 88
Insurance 88
Summary of Claims Procedures 89

**9 STARTING-UP OR MOVING
A WAREHOUSE OPERATION** 91
Initial Planning 91
Cost Considerations 92
Transportation 92
Taxation 93
Insurance on Goods in Storage 93
To Move or Not to Move 94
Timing Can Be Critical 95
Estimating the Cost of Moving 96
How to Load the Trucks 96
Cost of Transferring the Load 96
How Long Will It Take? 98

Deciding to Continue Services
or Suspend Operations for the Move 99
What to Move 99
Customer Communications 100
Other Communications 101
Opening the Relocated Warehouse 101
Planning the Opening of the Distribution Center 107
Writing a Warehouse Operating Manual 107

10 WAREHOUSE PERFORMANCE AUDITS **113**
Tracking Accuracy 115
Account Profitability 116
Qualitative Performance Ratings 117
Other Performance Guides 117

Real Estate Aspects of Warehousing

11 FINDING THE RIGHT LOCATION **123**
Reasons for Site Seeking 123
Location Theories 123
Outside Advisers 124
Requirements Definition 126
Comparisons From Published Sources 132
The Final Selection Process 132

**12 BUILDING, BUYING, LEASING,
   OR REHABILITATING?** **135**
Acquisition 135
A Worksheet for the Buy/Lease Decision 136
Assumptions 139
The Rehabilitation Alternative 140
Rejuvenating Warehouse Roofs 141
Rejuvenating Warehouse Floors 144

13 CONSTRUCTION **147**
Test Borings 148
Government 148

Storm Water Management 148
Utilities 149
Layout Design 149
The Contract 150
Risk Management 151
Roofs 151
Fire Protection Systems 152
Heating 153
Roads 153
Docks 154
Wear and Tear 155
Floors 156
Rails 156
Illumination 157
Walls 157
Landscape 158

14 PLANNING FOR FUTURE USES 159
Design Structure for Versatility 159
Changes in Basic Function 160
Internal and External Changes 162

Planning Warehouse Operations

15 SPACE PLANNING 167
Sizing the Warehouse 167
Storage Requirements 169
Interaction of Storage and Handling Systems 169
Stock Location Systems—Fixed versus Floating Slots . . 170
Family Groupings 170
A Locator Address System 171
Cube Utilization 172
Lot Size 173
Reducing Aisle Losses 173
Reducing Number of Aisles 174
Managing Your Space 174

16 PLANNING FOR PEOPLE AND EQUIPMENT 177
The Process of Making Flow Charts 177
The Best Sequence? 180
'What If' Questions 182
Personnel 182
The Critical Role of the Supervisor 183
Planning for Equipment Use 186

17 CONTINGENCY PLANNING 189
Strategies 190
Trigger Points 190
Planning for the Impact of Strikes
—Your Own Firm's as Well as Others' 191
Strike Against Suppliers 192
Strikes Against Customers 192
Trucking Strikes 192
Charting the Planning Process 193
Planning Steps 193

18 POSTPONEMENT 197
Background of Postponement 197
Types of Postponement 198
Postponement of Commitment 198
Postponement of Title Passage 199
Postponement of Branding 200
Postponement of Consumer Packaging 200
Postponement of Final Assembly 201
Postponement of Blending 201
Increased Warehouse Efficiency 201
Potential Challenges 203
Postponement in the Warehouse 204

19 STRATEGIC PLANNING 205
What Is It? 205
How to Start 206
Corporate Strategy as a Morale Builder........209
Production Changes 210

Strategic Thinking in the Public Warehouse 210
Financial Changes 211
Planning Warehouse Networks 212

20 SELECTING A PUBLIC WAREHOUSE **215**
The Preparatory Phase 216
The Investigative Phase 216
Evaluating Quality 217
The Decision 224

Protecting the Warehouse Operation

21 PREVENTING CASUALTY LOSSES **227**
Types of Casualty Losses 227
Controlling Fire Risks 228
Hazard Ratings for Stored Merchandise 231
Wind Storm Losses 233
Flood and Leakage 234
Mass Theft 234
Vandalism 236
Plant-Emergency Organizations 236
New Developments in Fire Protection 237
The Importance of Management 238

22 'MYSTERIOUS DISAPPEARANCE' **241**
Electronic Detectors and Alarms 241
What About False Alarms? 242
Door Protection Alarms 243
Grounds Security 244
Seals 244
Storage Rack 245
Restricted Access 245
Marketability of Product 246
Hiring Honest People 246
Using Supervision for Security 247
Procedures to Promote Security 247
Customer Pick-ups and Returns 248

    Security Audits 248

23 SAFETY, SANITATION AND HOUSEKEEPING 251
    Improving Safety in Lift Truck Operations 251
    Action Steps to Improve Safety 252
    Sanitation 254
    Housekeeping 255
    Damage Spreads 257
    Personal Appearance—Part of Good Housekeeping . . . 257
    Effect on Employee Morale 258
    Effects on Customer Service 259
    A Source of Shame 259
    Impact on Safety 260
    A Housekeeping Checklist 260

24 CONTRACTS AND LIABILITY 265
    The Changing Role of the Warehouseman 266
    The Insurance Problem 267
    Customer's Contracts 267
    Liability Limitation 268
    A Meeting of the Minds 268

25 VERIFICATION OF INVENTORIES
    AND CYCLE COUNTING 271
    Purpose of Inventories 271
    Inventory Carrying Cost 272
    Benefits of Improvement 272
    Problem Areas 273
    How Can a 'Physical' Be Improved? 273
    Incentives 274
    Managing the Job 275
    Involve the Whole Team 275
    Getting Organized 275
    Advance Preparation 275
    The Importance of Cutoff 276
    Zones 276
    Arithmetic Aids 277

Controlling the Count 277
Reconciliation 277
Eliminating Inventory Errors by Cycle Counting 278
What Does the Future Hold? 284

The Human Element

26 ORIENTATION AND TRAINING 287
Training 290
Some Training Examples 290
Retraining for New Equipment 291
Films and Video Tape 295
Operator's Guide 296
Mentoring 296
Record Keeping 297
Goals of Training 297

27 LABOR RELATIONS 301
Creating a Union-Free Environment in the Warehouse . . 302
An Example 303
Public Warehousing Alternative 304
Public Warehouse Strikes 305
Using Subcontractor Warehouse Services 305
Transportation Strikes 305
Inventory Hedging 306
Turning from Union to Union-Free 306
Pre-Employment Screening 307
Employee Complaints 310
The Role of the Supervisor 310
Arbitration Standards Pertaining to Discharge 311
The Essence of Union-Free Management 312
Summary 313

28 MOTIVATION 315
Output Criteria 315
Feedback 316
Skills 318

Balance of Consequences 318
Participatory Circles 319
Developing a High-Performance Work Culture 321
A Case Example of Participative Management 321
Warehouse Incentives 322
The Importance of Listening 323

29 MAINTAINING AND IMPROVING PERFORMANCE . . 325
Development 326
Discipline 326
Peer Review—A New Approach to Discipline 326
New Personnel 329
Pre-Employment Investigation 329
The Interview 330
Drug and Substance Abuse in the Warehouse 332
Probation 335
Personnel Management 335
The Labor Contract 336
Wage Reviews 336
Job-Performance Appraisal 336
Attitude Surveys 338
Promotion From Within 340
Pride in the Company 340

Productivity and Quality Control

30 MAKING WAREHOUSING MORE EFFICIENT 343
1. Improve Forecasting Accuracy 345
2. Free Labor Bottlenecks 345
3. Reduce Item Handling 345
4. Improve the Packaging 346
5. Smooth the Flow Variation 346
6. Set Improvement Targets 347
High-Priority Goals 347
7. Reduce Distances Traveled 348
8. Increase the Size of Units Handled 349

9. Seek Round-Trip Opportunities 349
　　　10. Improve Cube Utilization 350
　　　Technology to Achieve Increased Efficiency 351

31　MEASURING PRODUCTIVITY **353**
　　　What to Measure? 353
　　　Historical Standards 354
　　　Predetermined Standards 356
　　　Observation Techniques 356
　　　Stopwatch Standards 356
　　　Multiple Regression Analysis 356
　　　Uses of Regression Analysis 358
　　　Quality Measurements 359
　　　Common Methods of Measuring Productivity 360
　　　Guidelines for Measurement Systems 360

32　SCHEDULING WAREHOUSE OPERATIONS **363**
　　　Eliminating Delay 363
　　　Truck Dock Schedules 364
　　　Backlog 365
　　　Scheduling for Peak Demand 365
　　　Real Time Information and Scheduling 366
　　　Stock Locations and Scheduling 366
　　　Aids to Scheduling 367

33　CUSTOMER SERVICE— ITS ROLE IN WAREHOUSING　369
　　　1. Is This Warehouse Necessary? 370
　　　2. Phone Orders and Toll-free Numbers 371
　　　3. Consolidations 371
　　　4. Free Shipping 373
　　　5. Classify Orders by Size 374
　　　6. Backhaul 374
　　　7. Accuracy in the Warehouse 375
　　　8. Keep Communications Channels Open 376
　　　9. Seasonality and Sales Contests 377
　　　The Customers Pay the Bills 378
　　　Summary 379

34   IMPROVING ASSET UTILIZATION 381
    Space Utilization 381
    Energy Utilization 381
    Equipment Utilization 384
    The Equipment Utilization Ratio 385
    Improving Performance 386
    Lift Truck Rebuild Program 387
    Key Points of the Rebuild Program 388
    The Value of the Program 389
    Inventory Performance 390
    The Management Factor 390
    Why Measure? 391
    Danger Signals 391
    Inventory Turnover and Distribution Analysis 393
    Case History 394
    An Action Plan—And Potential Benefits 399
    Summary 400

35   'JUST-IN-TIME' 401
    JIT in Manufacturing 402
    JIT in Service Support 402
    JIT As a Merchandising Strategy 403
    JIT in Warehousing 406

36   IDENTIFYING AND CONTROLLING COSTS 409
    Three Cost Centers 409
    Space Utilization 410
    The Effect of Product Turns 411
    Other Performance Specifications 411
    Measuring Productivity 414
    A Cost Calculation Example 415
    The Throughput Rate Concept 418
    Adjustments 418
    Summary 420

37   MANAGEMENT PRODUCTIVITY 421
    Defining Management 421

Why Warehousing Is Different 422
The Effective Manager 422
Measurement Tools for the Manager 424
The Critical Tasks of a Manager 425
The Manager's Hardest Job 427
The Buck Stops Here 429
How Good Are Your Warehouse Supervisors? 430
The Supervisor Promoted from Within 431
The Importance of Clarifying Expectations 432
Putting It All Together 433

38 REDUCING ERRORS 435
The Value of Order Checking 435
Order Picking Errors 436
Order Taking Errors 437
A Good Item Location System 437
Clear Item Identification 438
Clear Description of Quantity Required 439
'Good' Order Picking Document 440
Order Identification 440
The Order-Filling Form 440
Other Factors to Consider 443
Multi-Line Order Filling 444
Personnel Factors 444
Identification With Work 445
Posting Error Rates 445
Using Case Labels 446
'Picking Rhythm' 446
Summary 447

Handling of Materials

39 RECEIVING AND SHIPPING 451
Receiving 452
Shipping 453
Dealing with Shippers Load and Count (SL&C) 455

Controlling Damage Claims 459

40 SPECIALIZED WAREHOUSING **461**
  Temperature Controlled Warehousing:
  The Essential Differences 462
  Hazardous Materials Warehousing 466
  State and Local Building Inspectors 468
  Fulfillment Warehousing 470
  Summary 472

41 ORDER PICKING **475**
  Quality in Order Picking 475
  Order-Picking Forms 476
  Inventory Relationship to Order Picking 476
  Order Picking Systems 477
  Fixed Versus Floating Slots 478
  Order Picking Methods 478
  Discrete Versus Batch Picking 479
  Zone System of Picking 480
  Designing Your Order Picking System 481
  Modeling Order Picking 482
  Safety 482
  Hardware 482
  Summary 483

42 EQUIPMENT: STORAGE AND LIVE STORAGE **485**
  Defining the Job 485
  Improving Storage with Racks 487
  Reducing Number of Aisles 488
  Other Types of Pallet Rack 490
  Live Storage—Gravity Flow Rack 491
  When to Use Flow Racks 492
  What Flow Racks Will Not Do for You 492
  Using the Carousel 493
  Stock Layout in a Carousel 494
  Justifying the Carousel 495
  Conveyor Systems 495

Types of Conveyors 496
The AGVS Alternative 498
Summary 498

43 FORKLIFTS AND OTHER MOBILE EQUIPMENT 499
Choosing a Lift Truck 499
Lift Truck Attachments 502
Order Picking Trucks 502
Justifying Narrow-Aisle Equipment 503
Strategic Considerations 507
Automatic Guided Vehicle Systems 507
System Integrators 511
What to Look For in Mobile Equipment 511

44 APPROACHING AUTOMATED HANDLING 513
Strategic Planning 515
Ergonomics 515
Robotics 516
Conventional Warehouse System Vs. High Rise 517
The Building 518
Storage Equipment 519
Materials Handling Equipment 520
Back-up Systems and Maintenance 520
Security 521
Computer Requirements 521
Coping with Change 522
The Future of Automated Systems 523
Conclusions 524

45 UNIT LOADS 525
Pallets 526
Parts 527
Types of Pallets 527
Types of Wood Employed 528
Performance Specifications 529
Box Pallets or Pallet Containers 529
Slip Sheets 529

Performance Specifications 530
　　　Building and Handling Unit Loads 531
　　　Loading Unit Loads on Trailer or Rail Cars 531
　　　Equipment to Handle Unit Loads 532
　　　A Case Example of Slip-sheet Handling 532
　　　The Potential Is Great 534

46 DEALING WITH DAMAGE 535
　　　Causes of Damage in Physical Distribution 536
　　　Causes of External Damage 536
　　　Disposal of Carrier Damage 537
　　　Concealed Damage 537
　　　Warehouse Damage 539
　　　Reducing Warehouse Damage 540
　　　Recooperage and Repair 541
　　　Protecting Damaged Merchandise 542
　　　Storage of Damaged Products 542
　　　Loss and Damage Guidelines 543
　　　Damage Control Ratios 544
　　　Damaged Loads Ratio 544
　　　Inventory Shrinkage Ratio 545
　　　A Management Checklist for Damage Control 546

Handling of Information

47 CLERICAL PROCEDURES 549
　　　Customer Service Standards 550
　　　Location of Clerical Center 550
　　　Office Environment 551
　　　Organizing the Clerical Function 551
　　　Interaction with Marketing 552
　　　Manual and Electronic Data Processing 553
　　　Control of Automation 553
　　　Measuring Clerical Costs 554
　　　Management Information 554

48 COMPUTER HARDWARE 555
Micro, Mini, Mainframe 557
Computer Jargon 557
Portable Data Collection Terminals 558
Purchasing Hardware 559
Hardware Development 559
Applications in Public Warehousing 561

49 SOFTWARE FOR WAREHOUSING OPERATIONS . . . 567
Warehouse Floor 568
Customer Service 570
Management Information 571
Transportation 571
Electronic Data Interchange (EDI) 572
Cost Analysis Programs 573
Program Costs 574
The Requirements Definition 574
The Role of the User 574
Finding Software Sources 575
Evaluation 578
Demonstration 579
Flexibility and Control 582
The Vendor 582
Training 582

50 ELECTRONIC IDENTIFICATION 585
Bar Codes—What They Are and How They Work 585
Scanners 587
Contact and Non-Contact Readers 589
Printing Bar Codes 590
Using Bar Coding in a Warehouse 591
Benefits 591
Using the System 593
The Future of Electronic Identification 594

Appendices
Appendix 1
UNIFORM COMMERCIAL CODE 595
Appendix 2
AWA BOOKLET OF
DEPOSITOR-WAREHOUSEMAN AGREEMENTS 609

INDEX 629
ABOUT THE AUTHOR 639

Dedication

This book is dedicated to Sheldon B. Ackerman, retired chairman and founder of Distribution Centers, Inc

Ultimately he provided the environment which made this work possible.

Part 1

BACKGROUND OF THE WAREHOUSING INDUSTRY

1

WAREHOUSING: ORIGINS, HISTORY AND DEVELOPMENT

The value of "warehousing" of some kind has been known from our earliest history. Since primeval times, living creatures have stored food in periods of plenty to provide nourishment when little is available.

Today, warehousing involves a great deal more than the stockpiling of foodstuffs. As civilization has developed, warehousing has become broader, more diverse, more complex.

Basically, merchandise warehousing is the storage of commodities and products for profit. It evolved from ancient granaries to the storage of manufactured goods in shipping, trading, and production.

History

The book of Genesis in the Bible tells how Joseph became a hero in Egypt by showing the people of that country how to warehouse agricultural surpluses for famine years. Joseph's granaries became emergency sources of food during crop failures. Chapter 14 of Genesis tells of Joseph's enterprise in ancient Egypt:

> ... *And during the seven years of plenty, when the earth bore ample crops, he collected all the produce of the rich years over Egypt and stored the grain within the towns ... There was a famine in every country, but there was food everywhere in Egypt. And when all Egypt itself grew famished ... Joseph opened all the granaries and sold grain.*

As early transportation systems were developed, warehouses became terminal points for the major land and sea trade routes. By the Middle Ages, terminal warehousing was well established in such major trade route centers as Venice. Thus, the use of warehousing became related to transportation as well as the storage of food.

Because of its strategic location between Europe and the Orient, Venice developed into the busiest commercial warehousing center of its time. The Venetians were the first known users of warehouse receipts, which they traded as negotiable instruments. The bankers of the early province of Lombardy were among the first to employ warehouse receipts as collateral for loans. Unlike the Biblical storehouses that were the property of a king or government, Venetian warehouses were operated for profit by groups of merchants known as trade guilds.

Warehousing in America

North America's settlement and early growth occurred during an era when ocean commerce was dominant. The first American warehouses were transit sheds located at the docks; the birth of the railroad industry changed this role. Railroad companies established depot warehouses for consolidating and distributing rail freight.

While commercial warehousing of agricultural commodities can be traced back to the earliest days of the American colonies, the emergence of general merchandise warehousing seems to have occurred shortly before the Civil War. An 1840 business directory for the city of New York listed 17 entries under "storage," and similar directories for Baltimore, New Orleans and San Francisco began to show similar entries at about the same time. Some of the first general merchandise warehousemen were former forwarding merchants driven out of business by the railroads.[1]

In the early days of railroading, the use of freight cars as ware-

---

[1] David Mitch: "Public Warehousing," Working Paper, Department of Economics, University of Maryland Baltimore County. A shorter version of this paper will be published in (David O. Whitten, editor) *Handbook of American Business History,* vol. II (Westport, Conn., Greenwood Press, forthcoming).

houses on wheels was common. During the grain harvest season, every available empty freight car, as well as empty freight depots, were utilized to store the crop. As the railroad industry began to experience car shortages, a separation of the transportation function from warehousing and storage became necessary. Storage charges and car demurrage charges were assessed in order to discourage the use of railroad cars and freight houses for warehousing.

Early railroad practices favored large shippers. The history of American railroading is filled with examples of favoritism. Warehousing was sometimes offered as a free service for major customers.

In the middle of the 19th Century, customs bonded warehouses were authorized to store goods exempted from payment of customs duty. Congress attempted in 1861 to abolish customs bonded warehouses, but that attempt failed and bonded warehousing has continued to the present day. The bonding of warehouses was extended to include distilled spirits and other commodities which carried special taxes.[2]

Role of the Independent Warehouseman

Independent warehousemen in the U.S. organized the American Warehousemen's Association (AWA) in 1891. One of their first activities was to seek legislation that would prohibit rail carriers from providing warehousing as a customer accommodation. The AWA was only six years old when it filed a complaint against 52 railroads to eliminate the free use of space in freight depots by favored shippers. The Interstate Commerce Commission, formed primarily for the purpose of ending the railroads' discriminatory practices, favored the position of the warehousemen and limited "free time" storage at the depot to four days. The Hepburn Act of 1906 ended freight-depot warehousing by the railroads. This legislation defined "storage and handling of property transported" as part of the railroading function and required the application of published tariff rates for all such services furnished by railroads.

---

2   Ibid.

While the warehousing industry would have grown without the ICC and the Hepburn Act, there is no doubt that public warehousing grew more rapidly after railroad activity in that industry was brought under government control. The AWA frequently sought legal review when transportation companies were discovered using warehousing as a "loss leader" in marketing of freight services.

In 1980, Congress passed legislation that removed much of the regulation existing for all modes of common carriage, including railroads. Shortly after this legislation was passed, a change of administration in Washington was marked by substantial reduction in government involvement in all kinds of industry, including transportation.

Financial Security Aspects

The warehousing industry has long been closely linked with banking and insurance. The use of warehouse receipts as negotiable commercial paper originated in medieval Venice. These receipts were used as collateral when money was loaned against commodities stored in warehouses. The banker, facing the risk of casualty loss of the commodity held in the warehouse, sought insurance on the inventory. In turn, the underwriter and the banker often sought to reduce the commercial risk by requiring that a third party in the business of warehousing guard and secure the merchandise.

The earliest years of the American Warehousemen's Association were marked by concerns of responsible warehouse operators about "fly-by-night, irresponsible warehousemen" entering the business. AWA stated as its objective the "lifting from warehousing some of the stigma that has hitherto been attached to that business by the misdoings of irresponsible so-called warehousemen." One of AWA's goals was to achieve a degree of respectability which would allow warehouse receipts issued by AWA members to be bankable as collateral all over the country.[3]

In some parts of the world, such as Mexico and Colombia, third

---

3   Quotes are from proceedings of American Warehousemen's Association, 1903.

ORIGINS, HISTORY, DEVELOPMENT

party warehousing is performed by the banks, and it is regulated by government banking laws. Typically, a Colombian coffee grower who needs to borrow money will place the harvested crop in a bank warehouse where it is retained until the loan is paid. In effect, the bank warehouse functions as a pawn broker.

In the United States a similar practice is known as field warehousing. A financial institution secures the inventory by storing it apart so that it can be held as collateral. In some cases the field warehouse inventory is not removed from the borrower's premises—it is fenced off and placed in the custody of an independent party.

Role in Mass Production

The industrial revolution brought about a conversion of craft shops into factories. One result was mass production. To balance production against demand, warehousing may be used as the safety valve. When goods are produced to meet a sales forecast, they must often be held in storage awaiting demand. Thus, mass production created new uses for warehousing. At first, this warehousing was done primarily at the factory. As distribution patterns developed, warehousing moved closer to market areas. Private and public warehousing services were used, both close to the factory and in the market place.

From Terminal to Distribution Center

In the early years of public warehousing in the United States, nearly all companies in the industry were family-owned and managed. Typically, they started as an outgrowth of a delivery business. Warehousing was offered as a supplemental service to transportation, with the warehouse a part of the drayage terminal. The word "terminal" still exists in the company names of some firms in the warehousing industry. Terminal warehouses were located in the center of the city, usually close to the railroad depot and the wholesale market district. As demands for storage space increased and land values rose, multi-story buildings were erected to provide more storage space on minimum amounts of land. A broad range of manufactured products, foodstuffs, and household goods was stored at these terminal warehouses.

Private warehouse development followed a similar pattern, locating close to the railroad depots or at the factory site.

Until World War II, the most common materials-handling methods in warehouses involved the use of hand trucks. Cases were hand-stacked using pyramid or step face stacking for heights of more than eight feet. However, most buildings were designed to allow stacking of only 8-to-12 feet high.

The mechanized equipment that became common during World War II created major changes in the warehousing business. The principal factor in this change was mass production of the forklift truck and the wooden pallet. This equipment allowed the practical pile height of merchandise to be doubled or tripled. Further, merchandise could be efficiently hauled over a much wider area of floor space.

Two factors combined to contribute to the obsolescence of the multi-story warehouse: increasing availability of forklift trucks, and increases in cost of warehouse labor. The war years caused a rapid rise in labor costs. Multi-story warehouses were gradually replaced by single story ones on relatively large land areas. Since large tracts of land were usually not available in downtown locations, warehouse sites moved from the railroad terminal area to the outskirts of the community. Ceiling heights were dictated by the capabilities of the forklift truck, generally ranging from 20-to-30 feet.

While the forklift truck did the most to revolutionize warehousing after World War II, other improvements in materials handling also changed the industry. These include specialized lift trucks, conveying equipment, stacker cranes, and unitized loads.

Most recently, substantial reductions in the cost of computers and processing and transmitting information have created significant changes in the design and use of all kinds of warehousing. While the cost of materials handling has steadily risen, the cost of information handling has decreased. Therefore, warehouse operators seek ways of using timely information to reduce or avoid materials handling costs.

The Emergence of Business Logistics

Through much of American business history, warehousing was considered a function of the traffic or transportation department. During World War II, the concept of logistics emerged in American military planning as a key to battlefield success. The U.S. characterized itself as "the arsenal of democracy" and American military forces were noted for their ability to maintain supply lines that provided necessary supplies of food, fuel and ammunition for rapidly moving armies. Some historians claim that superior logistics was the most important American contribution to victory in World War II.

After the war, business strategists gradually recognized that similar principles should be applied in private industry. Peter Drucker, a management theorist, started writing on this subject about a decade after the end of World War II. When Mr. Drucker described business logistics as the last great frontier for major cost reduction in American manufacturing, an increasing number of managers recognized the advantages of organizing their companies with a physical distribution or logistics department.

Physical distribution or logistics was actually a synthesis or integration of several activities that had previously been fragmented in most companies. The term physical distribution management has been defined as "the integration of two or more activities for the purpose of planning, implementing and controlling the efficient flow of raw materials, in-process inventory and finished goods from point of origin to point of consumption."[4]

From the beginning, warehousing was recognized as an integral part of physical distribution or business logistics. However, the rise of the concept of logistics caused warehousing services to be analyzed by marketing and production executives, in addition to the traffic and transportation managers who had been primarily concerned with warehousing in the past.

---

[4] Adapted from a definition of business logistics published by the Council of Logistics Management.

The Rise of Professional Associations

Some of the credit for spreading the concept of business logistics or physical distribution must go to the establishment of a professional organization in 1963 which is now known as Council of Logistics Management (CLM). Unlike many of the organizations that preceded it, the Council (originally known as the National Council of Physical Distribution Management) was established as a professional society rather than a trade association. Its founders included professors and consultants, as well as managers. The group emphasized education and idea exchange, and its membership policy was designed to embrace anyone with a sincere interest in the field.

About 15 years later, a similar organization was established for the field of warehousing. The Warehousing Education and Research Council (WERC) is a professional society with by-laws and policies similar to those of CLM. The prime difference between the two is that WERC restricts its agenda to warehousing. Like CLM, WERC created a new environment in warehousing by becoming the first professional society in this field to open its membership to any interested party, regardless of occupation. WERC provides a meeting ground for both private and public warehousemen, as well as educators, consultants and other managers involved in this field.

Trade Associations

Other organizations provide useful educational input for warehouse operators under the format of trade associations. Unlike professional societies, trade associations provide admission by company rather than by individual, and most are restrictive in their membership admission policy.

The American Warehousemen's Association continues to be a vital force in research and education concerning warehousing, though its membership is restricted to firms engaged in the public warehousing business. In recent years, AWA has made an increasing amount of its research and training courses available to non-members engaged in private warehousing.

Several transportation associations, such as the National Industrial Transportation League, admit warehousing executives as members and provide an interface between warehousing and transportation.

Countless other trade associations for manufacturers of grocery products, pharmaceuticals and chemicals include warehousing and distribution as part of their committee functions. Wholesaler trade associations also have placed considerable emphasis on research in warehousing.

As a result, the interested student of warehousing has many sources of information—both from professional societies and trade associations.

2

WAREHOUSING IN TODAY'S ECONOMY

The Business Environment

The contemporary economic climate is marked by rapid change, which of course affects warehousing. Perhaps the most troublesome change is the high cost of capital. Expensive and scarce capital can affect size and location of inventories, and can certainly affect decisions regarding construction of new warehouse facilities. When capital is short, the creative financial officer will generate new money by cutting inventories, by rescheduling production, or by persuading his supplier or customer to take some of the stocks he is now carrying.

Another way to reduce inventory costs is to simplify the assortment of products that is offered in the market place. This can significantly affect warehousing operations by reducing the number of stock-keeping units. Because both storage and handling costs tend to rise geometrically as an inventory becomes more complex, reducing the number of stock-keeping units will substantially reduce warehousing costs. If inventories are simplified and reduced, last year's full warehouse may only be half-full this year. Since most warehouse costs are fixed, the allocated cost per stored item will nearly double! Under such conditions, the warehouse user may consider closing some of his locations and consolidating inventories into the remaining ones.

There are at least four aspects of increased corporate concern with material flow:[1]

- Transportation deregulation,
- The rise of global supply chains,
- New technology,
- A drive to improve asset productivity.

In the late '70's, the transportation system in the United States was essentially deregulated. This new environment offered a whole new range of options to the shipper to control inbound and outbound flow. It allowed transportation cost, which is the largest logistics cost component for most firms, to become a negotiable cost as opposed to the fixed-cost nature of transportation expense under a regulated environment. The old "green eyeshade" traffic manager was skilled at seeking the best price in the printed tariff. In contrast, today's logistics manager seeks competitive advantage by developing innovative transportation networks and relationships.

A second factor was the evolution of global supply chains for many firms. With the fluctuating value of the dollar, imports of raw materials and finished goods increased, and a greater number of assemblies and subassemblies were purchased overseas. One result was a recognition of the complexity of managing a global logistics network. U.S. managers acquired a new set of management skills to control global material flow networks and thus to maintain a competitive advantage in both domestic and overseas markets.

A third factor which conditioned this evolution of logistics management was the appearance of significant new technology. Traditionally, businesses buffered supply and demand uncertainty with inventory deployment. Production snags, service complaints, or weak markets were doctored if not cured by establishing new warehouse stocks. New information technology has allowed managers to buffer this uncertainty with information rather than inventory. Management during the '70's and '80's discovered that investments in computers,

---

[1] The next five paragraphs were taken from an article by Bernard J. LaLonde, Sept. 1988.

electronic linkages with customer and vendors, bar coding, and overnight air freight offer an excellent tradeoff advantage when compared with inventory investment.

A fourth factor is an overall corporate drive for improved asset productivity. Inventory assets make up a significant share of the current assets of most firms. Senior management's concern with asset productivity has been both pervasive and competitive. It has extended to the deployment of inventory assets and the role of effective logistics management in improving asset productivity. The ability to create more customer value with less inventory investment has become a competitive goal for many firms.

Consider the following when planning your future warehousing programs:

1. Should you have more warehouse locations, or less?
2. Should these locations be private warehouses or public?
3. If they are private, should the real estate be owned or leased?
4. Should the location, configuration or design of those warehouses be changed?

What are some of the pressures involved? High-priced money means continued pressures from corporate financial departments to reduce inventory. As long as the price of money is high compared to other commodities, this is not likely to change. You can manage your business with a smaller amount of inventory when you have a smaller number of stocking locations. Therefore, high-priced money will probably bring about the reduction of the number of warehouses.

There are some other factors which let you do more with less inventory. The most important of these is a good communications system—an information system which relies on electronics rather than mail, and real time inventory control so you know how many cases of item XYZ you have in your warehouse this morning—including your deduction for the shipment which just left the dock 30 minutes ago and the addition for the load to be unloaded this afternoon.

If you are using private trucks today to replace the boxcars of a few years ago, you have also probably expanded the area which you can dependably serve from a given distribution point.

On the other hand, there are trends which could cause you to increase the number of stocking locations or at least to move them. One of these trends is the popularity of the concept of just in time inventory deployment which requires schedules as precise as the hour in which a shipment will be received (See Chapter 35).

In the decade of the 1990's, leading edge firms will use logistical competency to gain competitive advantage. This use of operating competency is frequently referred to as strategic logistics. It offers a way to gain an advantage in the market place. Traditional methods of product differentiation (advertising campaigns, etc.) are subject to rapid duplication by competitors. In contrast, logistical competency is created when your firm can deliver products to the correct place as required at the lowest possible cost.

Integrated logistics is efficiency-driven. Strategic logistics is customer-driven. Strategic logistics is concerned with effectiveness. It takes system design a step beyond integrated logistics by acknowledging that superior performance is equally as, or more important than minimum cost. A key point about strategic logistics is that it represents value-added performance. The creative part of logistics is the design and operation of competitively attractive value-added services.

Strategic logistics emphasizes the more effective use of off-balance sheet assets. The current trend toward smaller, leaner corporate structures renders third party alliances a very attractive capitalization alternative. Key capital resources such as transportation vehicles, warehouse space and materials handling equipment can be obtained from third party sources. In addition, scarce human resources can be shared.[2]

Freight Consolidation

Before deregulation, a key marketing strategy of the public warehouse industry was freight consolidation—the ability of the

---

[2] The last three paragraphs taken from an address by Donald J. Bowersox at the Annual Conference of WERC, May 1988.

warehouse operator to merge the shipments of several manufacturers shipping to the same destination, providing a saving for all of them. Today, the deregulation of transportation has made the concept of freight consolidation irrelevant.

Freight consolidation was based on the assumption that common carrier prices were regulated by tariffs, and the only way to achieve a lower price was to move a greater volume on the same shipment. In this deregulated environment, most of the advantages of consolidation can be achieved by individual shippers who have the clout to bargain for rates. Thus, for the major shippers, or the aggressive buyer, warehouse freight consolidation has little appeal.

Real Estate as an Investment

Some companies which have elected to use private warehousing were richly rewarded by the escalation of the real estate values of their warehouse buildings. For three decades after World War II, the escalation of value of industrial real estate was even greater than the escalation of the cost of living. There is evidence that this inflation is over, and the chances for big profits in warehousing real estate may be much slimmer today.

Warehouse Locations Shift

Location patterns obviously follow the market and population shifts. Many manufacturers now need more locations in the sunbelt and fewer in the east. Removal of some of the old East Coast railroad bottlenecks which required numerous locations will accentuate this trend.

Some warehousing points have been made redundant by the demise of storage-in-transit as well as the lowered priority of cheap labor.

A warehouse location in rural Nebraska selected in the 1960's was probably chosen because of rail traffic considerations or because the price and availability of warehouse labor was favorable. In today's labor market, you might find a better and cheaper supply of warehouse workers in Pittsburgh than in the middle of the corn belt. Today, electronic communications, back-haul availability, and access to independent truckers are more important than either railroad or labor considerations.

3

THE FUNCTIONS OF WAREHOUSING

In order to evaluate the use of warehousing in your business, it is essential to understand ways in which warehousing functions to add value to products. Essentially, warehousing provides time and place utility for any product.

Six Functions of Warehousing

Ultimately, the use of warehousing in commercial activity is related to its function in the business cycle. The following common functions of warehousing will be considered:

- Stockpiling
- Product mixing
- Production logistics
- Consolidation
- Distribution
- Customer service

Stockpiling is the use of the warehouse as a reservoir to handle production overflow. Such overflow reservoirs are needed in two situations—one involves seasonal production and level demand, the other arises from level production and seasonal demand. For example, the canner of tomato products builds a warehouse inventory at harvest time, while customer demand for the product is fairly level through the year. For the toy manufacturer, the demand comes at certain seasons or holidays, but the manufacturer may need to stockpile

PRACTICAL HANDBOOK OF WAREHOUSING

in order to accommodate seasonal demand. In either case, the warehouse is the reservoir used to balance supply and demand.

A manufacturer that has product-oriented factories in different locations also has the opportunity to use a product mixing warehouse to combine the items in the entire line. For example, one major appliance manufacturer has factories in several communities, with each factory producing a distinct line of products. In order to satisfy customers who wish to order full carloads or truckloads containing a mixture of the entire line, warehousing points are selected at locations that permit economical mixing of the product.

Some manufacturers need to stockpile semi-finished products, and this function is called *production logistics*. One automobile manufacturer runs its engine plants on a relatively level production schedule, while assembly of finished vehicles fluctuates with demand. A storage bank of finished engines is accumulated during those times when the engine plants are running at higher speed than assembly lines.

One use of warehousing in production logistics is the principle of "just in time." "Just in time" requires either closely coordinated

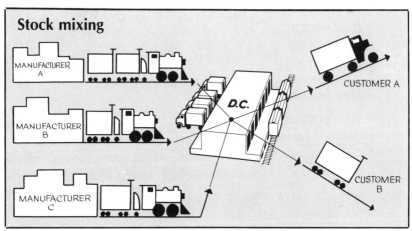

Stock mixing aids a marketing firm that buys from different manufacturers. The full line of products is maintained at the distribution center for shipment to customers.

FUNCTION OF WAREHOUSING

manufacturing or well organized warehousing. It also requires precise loading of vehicles so that, for example, the blue auto seat is unloaded in precisely the right sequence to meet the designated blue automobile on the assembly line.

Consolidation is the use of warehousing to gather goods that are to be shipped to final destination. Warehousing costs are justified by savings in outbound shipping costs achieved through volume loads. In one case, a fast-food company uses consolidation warehouses to serve clusters of retail stores, thereby reducing costs and frequency of small shipments to the stores. Suppliers of the food company are instructed to place volume loads of their products in these consolidation centers. This enables the fast-food company to cut its transportation costs by moving its supplies closer to its food-serving outlets. At the same time, the food retailer reduces inventory costs by arranging for its suppliers to retain title to these inventories until they are shipped from the consolidation center.

Distribution is the reverse of consolidation. Like consolidation, it is justified primarily by the freight savings achieved in higher volume shipments. Distribution involves the push of finished products

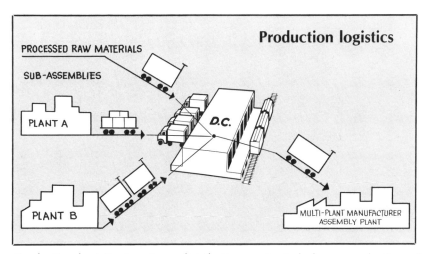

Production logistics is using a distribution center to balance production of sub-assemblies with finished unit production.

PRACTICAL HANDBOOK OF WAREHOUSING

by the manufacturer to the market, whereas consolidation involves the pull of supplies by the customer.

In one case, a pet food manufacturer uses distribution warehouses to position full carloads of products at locations convenient to customers. Fifteen distribution centers at strategic locations allow the pet food canner to achieve overnight service to major customers in the continental U.S.

Both consolidation and distribution provide service improvements by positioning merchandise at a convenient location. Both involve cost tradeoffs that balance warehousing expense against transportation savings. Both provide improved time and place utility for inventories.

At times, customer service improvement is the only motive for establishing an inventory. The five warehousing functions considered earlier all relate to production, marketing and transportation costs. Yet, at times a warehouse stock is justified only by the demands of the customer, which may be far from frivolous.

A southern manufacturer of toilet seats persuaded a chemical supplier to place a warehouse stock of raw plastics in Cincinnati. The

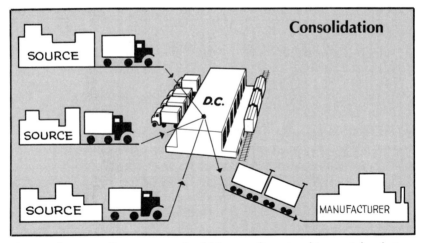

A manufacturer directs several of his suppliers to ship part loads to a consolidation point where they are combined to make economical full load shipments to the manufacturer's plant.

FUNCTION OF WAREHOUSING

motive for this request was the need for backhaul freight for the manufacturer's private truck fleet. By establishing a raw material inventory in Cincinnati, the manufacturer is able to move finished goods into Ohio and then return his trucks with backhaul cargoes of plastics. The supplier positioned this inventory in Ohio to accommodate the manufacturer.

Three Warehousing Alternatives

The three types of distribution centers are differentiated by the extent of user control. These are the private warehouse, the public warehouse, and the contract warehouse. Many companies use a combination of the three methods.

The private warehouse is operated by the user, and it offers the advantage of total control. If the user is distributing pharmaceutical products in which the penalty for a shipping error could be loss of life, such total control is often the only acceptable alternative. Where storage volume is large and handling volume is constant, the private warehouse is often the most economical means of handling the job. Yet, the private warehouse carries the burden of fixed costs, total exposure

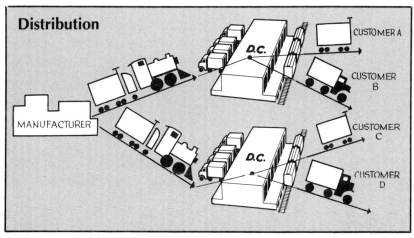

Distribution takes place when a manufacturer sends full loads to distribution centers, each serving a separate market area. Partial load shipments are made from the center to customers.

PRACTICAL HANDBOOK OF WAREHOUSING

Table 3-1   Contract Warehousing Alternatives

	Warehouse space	Handling labor and material handling tools	Clerical/office	Supervision provided (includes care, custody, and control of product)	Does Bailment usually exist?
1.	Real estate lease	Charged for by the unit	Included in handling rate or charged separate	Provided by operator	Yes
2.	Real estate lease	Charged for by the unit	Included in handling rate or charged separate	Not provided by operator	No
3.	Real estate lease	Charged for by time expended	Charged for by time expended	Charged for by time expended	Yes
4.	Real estate lease	Charged for by time expended	Charged for by time expended	Not provided by operator	No
5.	Space rental	Charged for by the unit	Included in handling rate or charged separate	Provided by operator	Yes
6.	Space rental	Charged for by the unit	Included in handling rate or charged separate	Not provided by operator	No
7.	Space rental	Charged for by time expended	Charged for by time expended	Charged for by time expended	Yes
8.	Space rental	Charged for by time expended	Charged for by time expended	Not provided by operator	No
9.	Charged for by the unit	Charged for by the unit	Included in handling rate or charged separate	Provided by operator	Yes
10.	Charged for by the unit	Charged for by the unit	Included in handling rate or charged separate	Not provided by operator	No
11.	Charged for by the unit	Charged for by time expended	Charged for by time expended	Charged for by time expended	Yes
12.	Charged for by the unit	Charged for by time expended	Charged for by time expended	Not provided by operator	No
13.	Others	Charged by the unit	Included in handling rate or charged separate	Not provided by operator	No
14.	Operate depositor's space	Charged for by the time expended	Charged for by the time expended	Not provided by operator	No

Source: Roger W. Carlson, "What is Contract Warehousing," published by WERC, Sarasota FL '83.

in the event of labor disruptions, and requires management. Though labor costs can be controlled through layoffs of unnecessary workers, many warehouse operators are reluctant to release skilled people who may be needed in the next few days or weeks. During a strike, the union can picket and impede shipments from a private warehouse. Warehousing clearly has its own management skills, and the private warehouse operator must attract and maintain such skills in his management team.

The public warehouseman is an independent contractor who offers services to more than one user. The public operator does not own the merchandise that is stored, and usually the warehouse company is independent from the firms owning the inventory. By serving a number of customers, the warehouseman is able to balance the variations of inventory, or of handling the workload, and, therefore, develops relatively level demands for storage space and personnel. As an independent contractor, the public warehouseman is immune to involvement in the labor disputes of any clients.

Contract warehousing is a combination of public and private services. Unlike the public warehouse, which offers a month-to-month agreement, the contract warehouse usually has a long-term arrangement. This contract may be used to govern supplemental warehousing services such as packaging, assembly or other extraordinary activities. In such cases, the contract provides an element of stability in procuring services.

As an example, a machine manufacturer needed a parts supply depot but was unable to come up with the specifications necessary for public warehouse rates. It was desirable to have the same workers handle this merchandise on a regular basis. A long-term lease was developed for the space in a public warehouse, and workers and lift trucks were offered on a fixed fee per week. In this instance, the warehousing contract included elements of both public and private warehousing.

PRACTICAL HANDBOOK OF WAREHOUSING

Cost Structure

The three figures shown here illustrate the cost structures found in third party warehousing.

All warehousing costs are a mixture of fixed and variable costs. The fixed costs in a warehousing operation account for more than half of the overall costs in that operation. Figure 3.1 shows how fixed and variable costs vary when volume increases.

To these costs the third-party operator must add a profit. However, in making rates, the public warehouseman must decide what level of unit throughput will be used to establish the rate.

The warehouse *user* may have negotiated a rate to pay $1 per unit to move merchandise through the warehouse. But what happens to the operator if activity is reduced? Figure 3.2 illustrates what happens. The vertical line in the middle of the chart shows the projected level of warehouse throughput. At that level, the warehouse contractor makes the profit which was planned. When throughput falls below that level, profit rapidly drops and moves to a loss. As throughput moves to the lower left corner of the chart, the warehouseman is unable to recover fixed costs. In contrast, when the throughput level moves ahead of projection, the gain turns from planned profit to windfall profit (which the operator keeps to protect against the rainy day when volume falls below projections). Much of the risk in traditional public warehousing is connected with the losses when throughput is below projection, which are balanced by windfall profits when volumes are ahead of predicted levels.

A Contract to Control Risks and Costs

When both parties to the contract understand the impact of the fixed and variable costs just described, and when both determine that they are going to have a long-term contractual relationship, they then have an opportunity to reduce these risks. A long-term relationship may also give them a chance to look at capital investments aimed at increasing handling efficiency, investments which might be too risky in a typical warehousing arrangement.

Figure 3.3 suggests a rate structure which controls these risks.

FUNCTION OF WAREHOUSING

Cost Structures in Third Party Warehousing

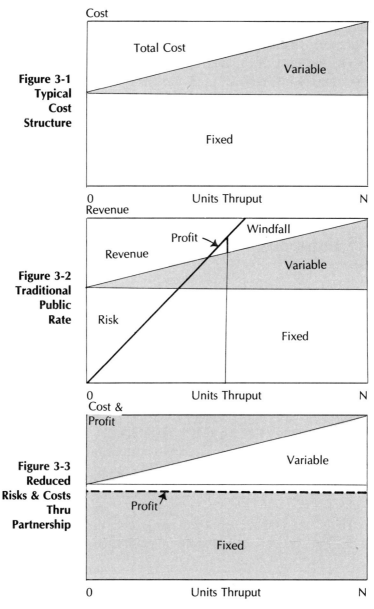

Figure 3-1 Typical Cost Structure

Figure 3-2 Traditional Public Rate

Figure 3-3 Reduced Risks & Costs Thru Partnership

Instead of a fixed rate per-unit, the user pays a level minimum fee which is designed to cover fixed costs and profit for the operator. That fee is represented by the solid horizontal line—a cost of $1,000 per month. In addition to this monthly fee, the user will pay a variable rate per-unit when throughput levels exceed the projection represented by the left vertical axis, in this case 1,000 units. If fewer than 1,000 units are moved, the minimum charge of $1,000 still applies. When volume doubles, fixed costs and profits stay the same, and the variable cost of moving the extra units is applied.

In choosing among the three warehousing alternatives, the user must measure the degree of risk involved in the decision. One may determine that the risk of public warehousing is unacceptable because of high penalties for service failure. However, this risk may be moderated by choosing the contract warehousing alternative with the user's own supervision to increase control over the warehousing function. A third alternative is to control the risk by having a private warehouse.[1]

Changing Values of Warehousing

In the current business environment, the ways in which warehousing can add value have been changed by structural changes in business logistics.

Today's logistics manager may be dealing with a system that manages tradeoffs among purchasing, manufacturing, transportation and warehousing. For example, purchasing decisions can be altered to provide a balanced two-way trucking route. Purchasing and distribution managers may work together to position warehousing in a location that helps to create this balance. When multi-national operations are developed, the warehousing job becomes much more complex.

To the extent that capital remains scarce, the logistics manager will continue to work at reducing inventory. Slow movers will be

---

1 From an article by W.G. Sheehan, Distribution Centers, Inc.

purged from the product line, and every effort will be made to speed turnover in warehouses. Getting the most out of inventory assets has become a key job for the logistics manager.

Space

Warehouse space is a commodity. Like any commodity, its price can show great volatility with changes in demand. At times, this commodity can become so scarce that the acquisition of additional space seems nearly impossible. Scarcity or high cost of space can be alleviated through changes in use or specifications of the other two warehousing components—people and equipment. On the other hand, when space is relatively cheap, people and equipment are used in a totally different fashion.

Attitudes towards warehouse space are the product of cultural influences in different parts of the world. In countries long accustomed to crowded living conditions, warehouses are designed to utilize space to the utmost. In contrast, part of the American frontier tradition is a general disregard for the value of space.

Equipment

Equipment consists of materials handling devices—racks, conveyors and all of the hardware and software used to make a warehouse function. Some equipment is especially designed to save space, although it may also require greater amounts of labor. Obviously, different kinds of equipment will be used when space is cheap, when there are low ceilings, or when there is an unusual environment such as refrigeration.

The useful life of most warehousing equipment is substantially shorter than the useful life of the space. On the other hand, warehouse equipment depreciates in value, while warehouse buildings often increase in value over time.

People

The most critical component in warehousing is people. The personal performance of warehouse workers often makes the difference between high- and low-quality warehousing. By comparison, the variations in quality of buildings and equipment are relatively small.

It is possible to design warehousing systems that waste space and equipment, but reduce the cost of labor. An example would be a warehouse in which all freight was stacked just one pallet high and could be handled quickly, but space would be grossly wasted.

While most warehouse productivity studies have emphasized the better use of people, the cost of labor since World War II has not escalated as rapidly as the cost of new space. Therefore, productivity improvement systems should consider all three components of warehousing, not just labor.

We will consider space, equipment, and people, the three components of warehousing in much greater detail in the remainder of this book. At all times, consider the tradeoffs among the major components, and that judicious employment of any one of them can affect the performance or cost of the other two.

4

WAREHOUSING ALTERNATIVES. THE MAKE OR BUY DECISION

The make-or-buy decision was first used by management to determine whether a certain component should be manufactured in-house or purchased from an outside vendor. The fact that a similar decision can be made in selecting warehouses is sometimes overlooked. Most users of warehouse services also are able to use their own people to get the job done. To this extent, the decision whether to hire a warehouse operator or run the warehouse yourself is the same classical "make-or-buy" problem.

If the answer to the make-or-buy question is to buy from an outside supplier, the next consideration is whether that outside source should be public warehousing or contract warehousing.

An important part of the make-or-buy question in warehousing is the organization philosophy of the user company. In some firms, an increase in size is a sign of achievement. In a bureaucratic organization, the more people and assets a manager supervises, the more successful that manager is perceived to be. When a company's philosophy follows this pattern, the distribution manager is not motivated to purchase outside warehousing services.

On the other hand, some companies reward managers for reducing personnel and/or assets. This is increasingly true in today's economy. As a result, the decision whether to do your own warehousing or to buy from others is being reexamined by many distribution executives. At least three factors that should be considered when answering the make-or-buy question: corporate financial policy, labor relations, and corporate strategy.

Corporate Financial Policy

Financial strategy has a major influence on the make-or-buy question. The most common measure of corporate success is the ratio of net profit to sales—the "bottom line" represented by net profit figures. A more sophisticated financial strategy suggests that the most important success barometer is not profits, but return on invested capital.

In an exercise known as the "DuPont Model" (named after the chemical company of the same name), managers have a tool to measure progress in improving return on assets (ROA) or return on invested capital (ROIC).

In the model shown as Figure 4-1, gross margin is the difference between sales and the cost of goods sold, and all expenses are combined to create a total expense figure. Operating profit is the difference between gross margin and total expenses, and operating profit is divided by gross sales to develop the traditional percentage of profit margin. This is the normal means of keeping score found in almost every American business.

However, the rest of the model is not so commonplace. Current assets are calculated by totaling inventory, accounts receivable and other current assets. Fixed assets are usually made up of real estate, plant and equipment. The total sales figure is divided by the asset total to create a capital turnover. Capital turnover is the ratio of total assets to sales revenue. The final ROA number is determined by multiplying the profit margin by capital turnover.

The distribution manager can use this model in several ways. Traditional profit margins can be improved by making decisions in physical distribution that will improve sales and profits. In addition, one can enhance capital turnover by being an effective inventory manager, by reducing inventory or improving its turnover. A warehouse manager may reduce fixed assets by reducing investments in ware-

THE MAKE OR BUY DECISION

houses. Any or all of these steps will reduce total assets, thus increasing capital turnover and usually improving return on assets.

How will this affect the make-or-buy decision? If you can control warehousing assets through the means of a short-term lease, one which has terms that would avoid the necessity of capitalizing that lease under current accounting standards, you can acquire a warehousing "asset" without having to declare the value of that asset on the DuPont Model. Similar strategies can be followed with materials

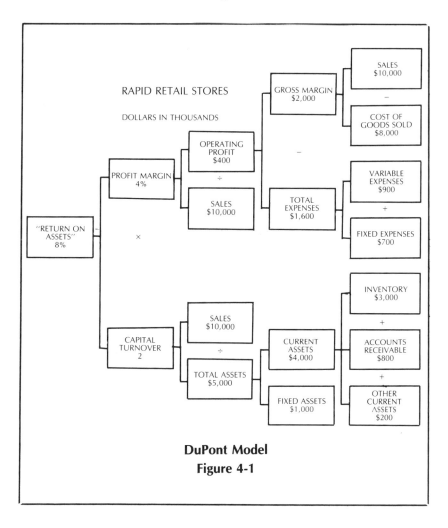

DuPont Model
Figure 4-1

handling equipment, transportation equipment, and other assets used in physical distribution.

Labor Relations Policies

Labor relations policies may influence the warehousing decision. For example, some manufacturers who accept unionization in production plants will strongly resist unionization of the distribution system. This happens when management views the distribution system as a safety valve to be used when production is disrupted by labor troubles.

For whatever reason, the distribution manager who intends to stay non-union is likely to make a different decision than one who is unconcerned about unions. In some cases, the user may promote non-union status best by controlling the distribution function.

In one non-union chain of retail stores, a relatively luxurious environment within the distribution center was provided because management felt it was a valuable "fringe benefit" for the workers. In that situation, buying outside warehousing services could not be considered because no warehouse contractor could duplicate the environment offered by the company in its own facility.

In other cases, the user may choose to buy warehousing services from an outside non-union contractor, recognizing that the contract need not be renewed if the labor situation deteriorates.

In still other cases, the user may select public warehousing, recognizing that National Labor Relations Board and Court decisions[1] have held that a public warehouse cannot be picketed by a union that has a dispute with one of its customers unless that customer's employees are stationed on the warehouse premises.

In examining the use of outside warehouses as a strike hedge, it's important to note that the outside contractor may also be the target of a strike. Therefore, it is essential to research the labor relations his-

---

1 Squillacote v. Lodges 516 and 2067. IAM, 459 F. Supp. 371 (E. D. Wisc., 1978).

tory of the outside contractor, as well as the current status of that contractor with respect to unions and labor contract expiration dates.

Strategic Decisions

Perhaps the most important influence on the make-or-buy decision is the overall business strategy of your company. Strategic planners typically consider what business activities the firm should be in. Does that list of activities include warehousing? Is real estate owned or leased? Some strategists may conclude that the company shouldn't be in warehousing; others may decide that developing management skills in warehousing is a good strategy for your company. Real estate gains realized through selection of good warehouse properties may be an important source of future profits.

A typical corporate strategy might be to rely on service partnerships.[2] Such partnerships offer several benefits. One is the cost reduction generated from specialization. A second is that the external specialist has the potential to spread or share risk. New business may be attracted as a result of a successful relationship. A fourth benefit of partnerships is shared creativity. A service partnership is relational rather than transactional, and the agreement which creates that partnership is not necessarily formal.

Nearly every corporation is in more than one business, and some enter certain businesses by accident rather than by choice. The corporate strategist typically explores whether these peripheral business activities—such as warehousing and distribution—should be continued or abolished.

The Final Decision

The decision to choose between internally managed warehousing or an outside supplier is based on corporate strategy and management goals.

The decision to use private warehousing may be made for bu-

---

[2] Taken from an address by Donald J. Bowersox at the Annual Conference of WERC, May 1988.

reaucratic reasons, or could be made to develop internal warehousing management expertise. A decision to use an outside contractor could be made solely to improve return on assets, even if the direct expenses are higher. The same decision may be reached because better warehouse management is available outside the company than inside, or because of economies of scale, or the ability to consolidate freight with other shippers in a public warehouse. And the decision to go outside might be made to provide a contingency capability of continuing distribution operations in the face of labor disputes, fire, severe weather or other disasters. The decision whether to make or buy warehousing services probably won't be made on a purely financial basis.

Public vs. Contract

Assume that the first question, whether to make or buy, was answered by deciding to use an outside contractor. The second question is whether the arrangement should be public or contract warehousing.

Public warehousing is a full-service system in which the contractor issues a warehouse receipt for the goods, and also provides space, equipment and labor for materials handling and clerical operations. Charges are usually based on the unit of product handled, sometimes expressed on a per-case basis, other times on a price per-hundredweight or ton.

Other than the payment of a modest minimum fee, there are no fixed charges or long-term agreements in public warehousing, and the cost of the service usually varies directly with the volume of products moving through the facility.

In a contract warehouse, the user signs an agreement with the operator for a specific period of time, frequently for a number of years and always longer than the 30-day public warehousing arrangement. Firms using contract warehousing typically do not need the flexibility allowed with the 30-day cancellation agreement used in public warehousing. With contract warehousing, the user absorbs some of the contractor's risk by committing for fixed amounts of space and labor for a period of time. As a result, the user receives a lower price reflecting the reduced risk. The buyer trades a measure of flexibility for a measure of cost control.

In "What Is Contract of Warehousing?" Roger W. Carlson[3] notes that contract warehousing agreements tend to vary in three major areas: (1) Who retains inventory responsibility (also known as bailment); (2) how space is reserved, whether on a month-to-month contract, or a lease, or purchased by the user, or not purchased at all; and (3) whether handling is charged by the case, the hundredweight or the man-hour. Clerical services will have a similar set of variables.

Under contract arrangements, some users may buy handling services without purchasing space, by requesting the warehouse contractor to operate a building the user already owns or leases. Conversely, the user may buy space without buying labor services, or buy space and handling services without buying clerical services.

The Economics of Third Party Warehousing

In public warehousing, consider the economics from the operator's point-of-view.[4] Today in the United States, space alone may cost $4.00 a square foot to rent. (This price includes utilities and maintenance.) At these prices, space doesn't have to be vacant very long before it begins to erode any profit that the warehouseman may have made on it when it was full.

In fact, four to eight weeks of vacancy will erode every bit of profit the warehouse operator might have made on that space during the previous ten to eleven months. Under such conditions, it is not easy for the warehouseman to speculate with construction of new buildings in the absence of any customer commitment.

Through contract warehousing, the relationship between the warehouse operator and the owner of the goods can be formalized and

---

3   Carlson, Roger W., "What Is Contract Warehousing?"; an article published by WERC, Sarasota, FL, Feb., 1983.
4   Sheehan, William G., Contract Warehousing The Advantages of Having Additional Control of Your Third Party Warehouse. Volume 20 No. 8, Warehousing and Physical Distribution Productivity Report, copyright Marketing Publications, Inc. Silver Spring, MD.

lengthened, assuring both sides that their primary concerns will be met. The operator no longer has to worry about a client pulling out unexpectedly, leaving him with empty space. No longer are both parties operating month-to-month risking possible changes from both sides; instead they are operating under a long-term contract which can be written to satisfy both parties.

The more units are shipped through the warehouse, the more total dollars are spent. However the per-unit throughput rate comes down. This happens because the great bulk of costs are fixed. Since the variable costs as a percentage of total are relatively low, as volume goes up, the total cost per unit goes down.

Bonus Incentives

In choosing between public or contract warehousing, be sure to consider incentives for performance. The public warehouse operator has a built-in incentive to improve his internal profits, and to keep his rates competitive in order not to lose the business to rival firms.

This incentive may be reduced in a contract arrangement because of a fixed price agreement. However, the contract for warehousing services can be structured to provide incentive through a productivity improvement sharing formula. Under this type of formula, the user and contractor may agree, for example, that an acceptable materials handling standard is 5,000 pounds per man-hour. When performance climbs above this standard, 30 percent of that improvement might be paid to the contractor as a productivity bonus.

One problem is that incentives based on standards cannot always be accurate in a diversified warehouse operation, simply because a change in product mix often will create different working conditions. On the other hand, incentives for performance usually will improve productivity and should be investigated as one way of improving productivity in contract warehousing.

Other Factors

Under certain economic or market conditions, it may be difficult to find a public warehouse operator willing to bid on an outside operation. Not long ago, many public warehouses passed the break-even point when occupancy rose above 50 percent. Now, the break-even point has moved considerably higher, and the cost of carrying empty space must be compensated by a proportionately higher price for services sold.

A potential advantage of contract warehousing is a reduction of the adversary relationship between the supplier and customer. By decreasing the frequency of bidding contests, the parties are more likely to function effectively as a team.

In real estate, location has long been considered the most important criterion for valuation of the property. Since the distribution facility is normally a piece of real estate, the decision of where to locate that facility may be the most important one made by corporate planners. Selecting a poor location could make the difference between success or failure in the overall distribution program.

One of the major factors is the corporate customer service policy. If the goal is to provide better customer service than the competition, the firm may choose to have many warehouses located throughout the market. On the other hand, a strategic decision to offer the lowest prices in the industry would probably result in a decision to serve the entire market from a single distribution center located next to the production plant.

Another factor is whether the distribution facility functions as a distribution center, a consolidation center, or both. If distribution is the primary function, the location of major customers is the key. If consolidation is the main function, then location of the sources of inbound materials becomes important.

Incentive financing offered by state, county or city governments is another factor that must be considered. These incentives may be in the form of industrial revenue bonds, which are free of local or federal income taxes to the investor and thus carry an interest rate somewhat below the market. At times when interest rates are extraordinarily high, the spread between industrial revenue bond rates and conventional mortgage rates may be quite significant. Similar financial incentives may be offered in the form of tax abatement for an inner-city location.

Many of the factors mentioned above are beyond the control of the logistics manager. If you cannot control these factors, you can at least be knowledgeable about them. In addition, you can make decisions that hedge the effect of any changes by anticipating that they will take place.

The best way to know whether your private warehouse is truly productive is to go through the make-or-buy decision. Whether you are running your own warehouse or hiring a warehouse service, have you asked yourself why?

Ultimately, warehousing is a tradeoff, and the value of the trade must be periodically measured.

Warehouses exist because they should create certain economies. If these are not measured periodically, you cannot be sure the economies still exist.

Part II

ELEMENTS OF WAREHOUSE
MANAGEMENT

5

COMMUNICATIONS

Because the warehouse is part of a larger logistics system, communications to and from it assume critical importance. Sometimes the warehouse serves as a buffer between manufacturer and customer, sometimes between manufacturer and supplier and sometimes between wholesaler and retailer. It is always a place where accurate and timely communication is vital.

In essence, the most critical communications with a warehouse are those designed to answer five questions:

1. What has been or is to be received?
2. What has been or is to be shipped?
3. On what date and at what time was the customer's shipment delivered?
4. What is the correct amount of money due to each common carrier for transportation services?
5. What remains in inventory?

There are four general groups of people who will need answers to these and related questions: the operators of the warehouse, outside users of the warehouse, suppliers of transportation and the ultimate consignees of shipments from the warehouse.

Non-routine problems also must be handled in the communications procedure. These include quarantine holds on inventory, credit clearance, damage and emergency shipping procedures.

The warehouse manager must frequently accept communications in a variety of forms, some of which may be easy to handle, some not. This chapter will focus on public warehouse communications. Because of the speed of computer development, methods of communicating are changing so rapidly that any attempt to describe them becomes obsolete even before the ink is dry. However, certain principles are more permanent.

Hard Copy Versus Voice

The most common fast communication in business is the telephone. However, transmitting numerical data or other detailed instructions can be risky, since there is no written record of what was said. In some businesses, tape recordings of telephone conversations are retained in order to provide a backup record of the conversation in the event of a later dispute. Misunderstanding or transposition of numbers by telephone is not unusual, and the responsibility for such errors may be difficult to resolve. When the telephone is the primary means of message transmission, the users must be willing to accept the risk of errors. Written confirmation should be a required supplement to follow all telephone communications.

Mail is the oldest form of hard copy transmission. However, mail is slow and increasingly uncertain. When mail is the primary order transmitting medium, extra inventory may be needed to compensate for delivery time lag. When inventory is valuable, the cost of faster electronic means of transmitting hard copy is often justified.

An intermediate step between mail and electronics is the use of courier or parcel delivery service. Such services are frequently faster and more reliable than mail, but they also are more expensive.

There is a wide variety of equipment that can transmit hard copy over telephone lines, or occasionally by radio. This includes electronic transmission of computer data that can be stored by a variety of methods for later printing or typing. The facsimile or "Fax" machine is perhaps the fastest growing means of sending hard copy over telephone.

Public Warehousing Forms

The warehousing industry uses several forms, which, while not strictly standardized, have only slight variances as they are employed throughout the country. Some forms will contain preprinted legal clauses, such as those customarily shown on a common carrier bill of lading. Exhibits 5-1 through 5-3 show sample public warehouse forms provided by the American Warehousemen's Association.

Receiving goods at the warehouse involves three different

AMERICAN WAREHOUSE COMPANY
STREET ADDRESS • CITY & AMERICA 00000
TELEPHONE: (312) – 123-4567

ORIGINAL
NON-NEGOTIABLE WAREHOUSE RECEIPT

AMERICAN WAREHOUSE COMPANY claims a lien for all lawful charges for storage and preservation of the goods; also for all lawful claims for money advanced, interest, insurance, transportation, labor, weighing, coopering and other charges and expenses in relation to such goods, and for the balance on any other accounts that may be due. The property covered by this receipt has NOT been insured by this Company for the benefit of the depositor against fire or any other casualty.

DOCUMENT NUMBER
DATE
CUSTOMER NUMBER
CUSTOMER ORDER NO.
WAREHOUSE NO.

RECEIVED FROM

THIS IS TO CERTIFY THAT WE HAVE RECEIVED the goods listed hereon in apparent good order, except as noted herein (contents, condition and quality unknown), SUBJECT TO ALL TERMS AND CONDITIONS INCLUDING LIMITATION OF LIABILITY HEREIN AND ON THE REVERSE HEREOF. Such property to be delivered to THE DEPOSITOR upon the payment of all storage, handling and other charges. Advances have been made and liability incurred on these goods as follows:

FOR ACCOUNT OF

DELIVERING CARRIER | CARRIER NUMBER | PREPAID/COLLECT | SHIPPERS NUMBER

QUANTITY	SAID TO BE OR CONTAIN (CUSTOMER ITEM NO., WAREHOUSE ITEM NO., LOT NUMBER, DESCRIPTION, ETC.)	WEIGHT	RATE CODE	STORAGE RATE / HANDLING RATE	DAMAGE & EXCEPTIONS
	TOTALS				

NO DELIVERY WILL BE MADE ON THIS RECEIPT EXCEPT ON WRITTEN ORDER.

AMERICAN WAREHOUSE COMPANY
BY
AUTHORIZED SIGNATURE

FORM 2 3/70

Exhibit 5-1—Warehouse Receipt

forms: the shipper's advance notice, the warehouse receiving tally and a non-negotiable warehouse receipt. The receipt normally has some clauses describing the responsibilities of the public warehouse.

The purpose of the advance notice is to give the shipper a listing

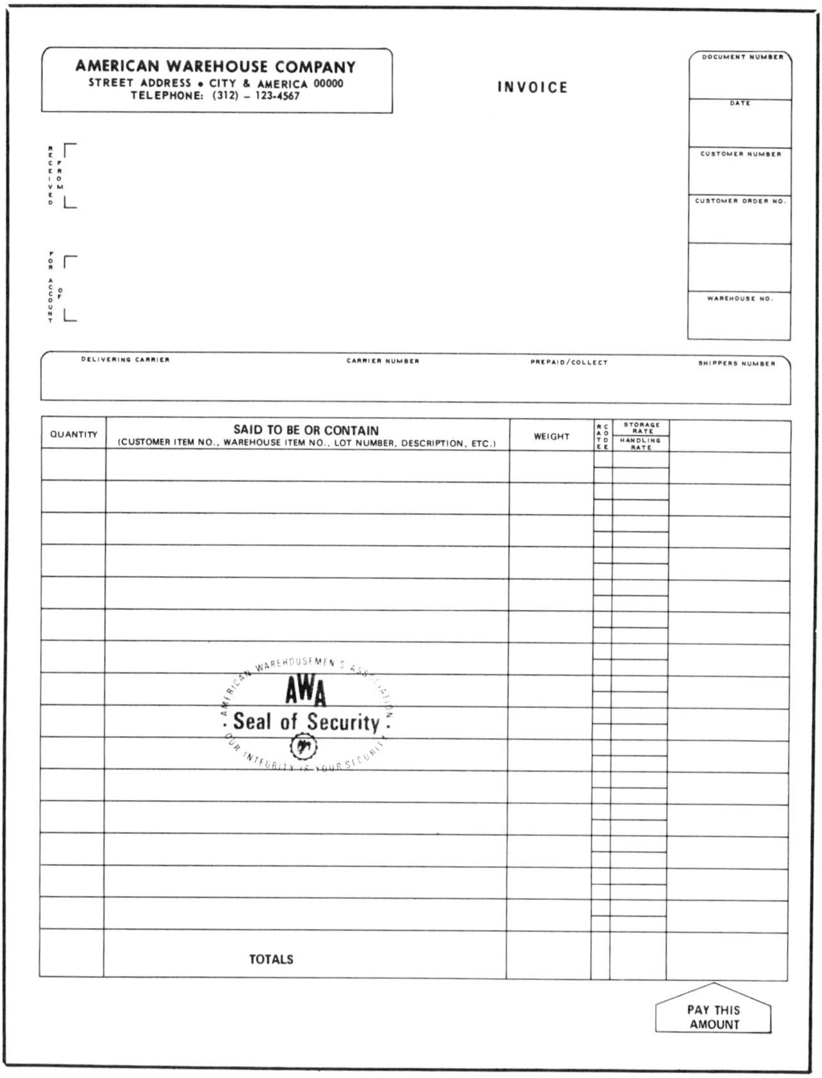
AMERICAN WAREHOUSE COMPANY
STREET ADDRESS • CITY & AMERICA 00000
TELEPHONE: (312) – 123-4567

INVOICE

DOCUMENT NUMBER
DATE
CUSTOMER NUMBER
CUSTOMER ORDER NO.
WAREHOUSE NO.

RECEIVED FROM

FOR ACCOUNT OF

DELIVERING CARRIER	CARRIER NUMBER	PREPAID/COLLECT	SHIPPERS NUMBER

QUANTITY	SAID TO BE OR CONTAIN (CUSTOMER ITEM NO., WAREHOUSE ITEM NO., LOT NUMBER, DESCRIPTION, ETC.)	WEIGHT	RATE CODE	STORAGE RATE / HANDLING RATE	
	TOTALS				

PAY THIS AMOUNT

Exhibit 5-2—Invoice

of goods shipped *before* the inbound vehicle gets to the warehouse. The advance notice also identifies the trailer or box car number on which the product was shipped. Careful use of advance notice information can provide opportunities for savings in materials handling. For example, if goods contained on the inbound shipment are urgently needed to fill orders, those critical items can be kept on the dock for immediate shipment, thus avoiding a wasted round-trip to a storage slot.

A receiving tally is the warehouse receiver's independent listing of the goods that came off the inbound vehicle. This tally should be prepared "blind." The warehouseman filling it out should not know what the shipment is *supposed* to contain. The tally can then be compared with the advance notice, with an immediate recheck to resolve any discrepancies.

The purpose of the non-negotiable warehouse receipt is to provide legal certification that the goods listed on it are now in the custody of the public warehouse.

Public warehouses usually provide a stock-status form at the end of each month that also includes inventory levels. Some of these reports will show a beginning balance, a summary of receipts and shipments, and an ending balance. This provides a means of checking the public warehouse records against the user's records.

When common carriers are used, the bill of lading is a contract between shipper and carrier. It provides proof that the merchandise was transferred from shipper to carrier, and that the carrier has assumed responsibility for the cargo until it can prove it was delivered.

Damages

Sometimes a common carrier shipment to a warehouse will either have damaged freight or a discrepancy in the count. Such exceptions are reported on a report known as the "OS&D," which stands for *over, short and damaged*. This document, prepared by the warehouse operator, is the basis for settlement of a claim between the shipper and the common carrier.

PRACTICAL HANDBOOK OF WAREHOUSING

Real-Time Inventory Control

Before electronic information transmission was common, the question of when a replenishment shipment had actually arrived at a

Exhibit 5-3—"OS&D" Report

warehouse was not easy to determine. Some users applied a standard time allowance, making an assumption that common carrier shipments from the manufacturing plant would arrive at the center within "X" days. Such a system made no allowance for transportation variations. If the replenishment shipment was delayed, the inventory did not accurately reflect goods physically in the warehouse.

Today's information systems should allow all inventory to be added or subtracted at the exact time when goods arrive at the warehouse or leave it. Particularly when inventory is costly, precision in the adding and subtracting of inventory can improve profits.

Circulation of Stock Status Information

The purpose of a stock-status report is to allow the warehouse operator and the user to compare notes on the quantity of merchandise available in inventory. If user and operator do not agree on balances, they must find the source of error. Third parties, such as the user's customers, may be anxious to see stock-status reports. But wide circulation of such information can cause confusion or even embarrassment. Sometimes inventory is available, but held in reserve for an important customer, and revealing that this stock is "set aside" could cause customer relations problems.

The stock reporting system should have a means of reflecting errors that are identified but not yet adjusted. By adding a variance column to the stock-status report, the user is informed of the existence of these unresolved errors. Typical of such mistakes is an inventory cross-over, in which a shortage of one model is balanced by an overage of another. This happens because the wrong item was shipped. The variance column provides an interim report that an error exists. When a physical inventory adjustment is made, the variance disappears.

Communications regarding warehouse releases or shipments must be precisely controlled. The warehouse user must determine which parties have the power to release merchandise, and how and when credit authorization will be handled. Some warehouse users have salesmen or brokers, and the power of these people to make warehouse releases must be carefully documented.

The bill of lading provides proof that goods moved from warehouse to common carrier. However, information about when and in what condition goods are received by the consignee can be obtained only through the carrier's freight bill.

Effective Order Picking

The format of order pick sheets affords a labor-saving opportunity in order selection. Ability to sequence line items by location and to process orders for zone picking can greatly improve the effectiveness of the warehouse order pickers. If goods are arranged in a fixed location, the order-picking document should list merchandise in the same order in which it is located on the line.

Faulty Communications Equal Wasted Effort

One of the most common communications breakdowns in warehousing is confusion as to whether an order is in cases or units. Some orders may call for full cases and individual packages from broken cases. If the order is not clear as to whether cases or packages are involved, there can be substantial errors.

Another symptom of poor communication is error in transcription of orders. With today's information handling techniques, manual copying of information should be prevented, thus reducing opportunities to make copying errors.

Complaints

Any busy operator will inevitably experience complaints. However, complaints should be looked upon as an opportunity for improvement. When they are handled properly, a faulty procedure may be identified. Logging complaints can be a means of comparing the relative efficiency of various distribution centers. But, to do this fairly, the number of complaints must be related to the number of transactions.

Standard Procedures

Both the user and the operator of the warehouse should provide written procedures to describe how each job is to be performed. No warehouse is too small to have written procedures, and the presence of such procedures provides a basis for training new people and quality control over existing operations. Misunderstandings about the best way to get a job done usually occur when there's no agreed procedure. Written procedures frequently no longer exist in mature companies where employees have been on the job for many years and are performing their jobs from memory. Unwritten procedures frequently become erratic.

One procedure typical of grocery product warehousing uses "tie and high" instructions. This is a written pallet pattern to be used when goods are received at the warehouse. This procedure applies only for non-unitized shipments of cases, products that would typically be palletized by the delivering truck driver. In the absence of specific instructions, the driver may use an ineffective storage pattern on the pallet, or even worse, he may use a pattern that makes it easy to conceal a product shortage by leaving voids in the center of loaded pallets. Precise description of stacking pallet patterns is essential when non-unitized case goods are received.

Uniform Communications—A New Development[1]

The problem in communications between warehouse operator and user is frequently one of format. Unless everyone involved in a transaction formats the information in the same way, the information cannot be used on preprinted business forms.

Historically, the grocery products industry has been a leader in standardization of distribution systems. Five groups within the grocery industry—chain retailers, food brokers, manufacturers, transportation companies and public warehouses—are all potential bene-

---

1 This subchapter is based on information provided by William Arbuckle at Quaker Oats Company in Chicago and Craig T. Hall at TLC Warehousing Services, in Holland, MI.

ficiaries of improved communications. But when each party uses a different format, the manufacturers must write and maintain a separate list of computer programs for each format received.

The transportation industry was the first to recognize the problem of different formats. In 1976, the Transportation Data Coordinating Committee was formed in Washington, D.C., with the express purpose of developing Electronic Data Interchange Standards (EDI). TDCC developed standard formats for EDI messages used by the railroads, trucking, ocean and airline industries.

TDCC added message standards for the grocery industry and the warehousing industry. All these industries share a common data dictionary, which is maintained by TDCC, and they all use the same TDCC concepts.

The grocery industry started to search for a way to achieve uniformity in communications between retailers, brokers, and manufacturers. Six organizations joined in the effort: GMA (Grocery Manufacturers of America); FMI (Food Marketing Institute); CFDA (Cooperative Food Distributors Association); NARGUS (National Association of Retail Grocers in the US); NAWGA (National American Wholesale Grocers Association); NFBA (National Food Brokers Association).

They formed the Uniform Communications Standards committee for the grocery industry (UCS). The committee is composed of representatives from retail chain stores, food brokers, and manufacturers. There are over 50 companies represented.

The uniform communication system (UCS) is an effort to standardize a format for computer-to-computer communications between grocery manufacturers, distributors, and ultimately public warehouses and common carriers. As the program has developed, there are six message types which have been reduced to a standard format: purchase order; invoice; shipment notification; receiving notification; acceptance/rejection advice; price change announcements.

Uniformity for Warehousing

A set of standard messages for the warehousing industry was

also written by the Warehouse Information Network Standards (WINS) committee.

WINS was supported by the American Warehousemen's Association (AWA) and the International Association of Refrigerated Warehouses (IARW) and approved by TDCC (Transportation Data Coordinating Committee). UCS (Uniform Communications Standard) had been initiated by the grocery products industry. In 1988, the Uniform Code Council (UCC) became the administrative agency for both WINS and UCS. This organization established an office at Dayton, Ohio and has published user guides for information systems. Today, most of the effort is referred to as Electronic Data Interchange (EDI).

The EDI committee has developed standards for six types of messages: warehouse shipping order; warehouse inventory status report; warehouse activity report; stock transfer receipt advice; warehouse shipping advice; acceptance/rejection advice.

Warehouse operators should realize many benefits when the uniform system is in place. These include the following:

1. Reduction of Cost

—Less clerical effort to enter orders. EDI will greatly reduce the manual effort required to process an order. The manual edit, corrections, and order review will take less time than the manual effort required to enter an order by re-keying the entire order.

—Eliminate manual effort to key shipping information. Warehouses are presently required to key shipping information into the manufacturer's terminal. With EDI, the computer can transmit it. Cost of warehouse transmission equipment will be partially offset by this savings.

—More efficient labor scheduling. Greater lead time created by direct entry will allow more time to effectively schedule picking and loading labor.

2. Improved Accuracy

—Clerical errors are decreased. There is less manual entry, both on order and shipping advices.

—Reduced picking errors. Consistent picking lists and bills of lading forms will reduce errors in both picking and traffic processing.

3. Improved Service to Manufacturers

—More freight consolidation. Greater lead time allows more time to consolidate freight, thus greater freight savings.

—Better labor scheduling. Again, greater lead time means better scheduling of labor, better and faster picking and loading.

—Faster and more accurate reporting. Direct transmission of shipping and stock replenishment receipts will be faster and more accurate. This allows the manufacturer to update his inventory sooner.

System Priorities

For most users of warehouses, on-time delivery is the first priority. Therefore the most important purpose of EDI is to ensure that the correct shipment was delivered to the consignee on time.

It is not surprising that EDI's widest use is as a means of checking and confirming on-time delivery. Perhaps the second most common use is to control freight payments. As industry grows more accustomed to uniform communications, the uses of EDI will be extended.

6

PACKAGING AND IDENTIFICATION

This subject is dealt with from the perspective of the warehouse operator, who presumably is not a packaging specialist and thus is not empowered to make changes in packaging and identification.

Warehousemen in the U.S. receive, store, handle and ship a broad range of products.[1] All products (with the exception of some bulk commodities) are packaged in some way. A food packaging system may consist of a primary package that touches the product, an intermediate or secondary package such as a folding carton showing the advertising message, and a distribution container that identifies and carries multiples of the secondary package. The warehouseman receives, stores and handles the distribution container but must deal with it according to the size and strength of the total package.

Functions of a Package

A packaging system has four functions: (1) protection, (2) containment, (3) information and (4) utility.

The package should provide protection against the common hazards of warehousing and distribution. These include stacking compression, shock and vibration in transit and handling, protection against temperature extremes and changes, moisture, and infestation. The containment quality of a package is its ability to resist leakage and spilling, and containment methods will vary with the type of prod-

---

[1] The packaging section of this chapter was adapted from an article by Paul L. Peoples, School of Packaging, Michigan State University, East Lansing, MI.

uct. The information on packaging may consist of advertising messages, as well as a listing of ingredients, For the warehouseman, the most important information is that which enables the materials handler to identify correctly the package. Utility of the package itself involves its convenience of opening and closing. In warehousing, utility is usually the ease and efficiency of handling the unit in storage.

The manufacturer naturally seeks a minimum cost for these functions, but physical distribution people must be concerned about the normal stresses involved in warehousing and transportation of the product. Clearly, packaging represents a compromise between the minimum cost desired by the manufacturer and the minimum strength needed to protect the product as it moves through the various channels of transportation and warehousing.

Ideally, the package designer provides sufficient packaging so that the package, when coupled with the inherent strength and characteristics of the product, is equal to its environment at every stage. Less than adequate packaging leads to damage. Excessive packaging leads to excessive costs. The variables in this equation are, of course, constantly changing and are seldom precisely known. The transportation environment varies in different regions of the country. It also varies between modes, and even between different trucks and highways. The storage environment and other potential hazards can also vary. The use of new packaging materials can cause major variations in package performance. The net effect is that even the best of packages can become a poor performer due to unforeseen changes in its environment.

How the Package Affects Warehousing

Most warehousing operations deal with the distribution package, and the strength and dimensions of this package have a direct effect on handling and storage efficiencies. The weight of the individual unit affects handling and order picking efficiency. The compressive strength of the package and product affect the stacking heights and thereby the storage efficiency of the system. The height of the package affects the utilization of individual rack spaces; its length and width influence the stacking pattern and the effective use of space on the pallet, as well as the use of overlapping packages to make firm pallet loads.

Quite frequently, the dimensions and performance of the package-product combination are accepted without challenge by the warehouse operator. It is assumed (falsely in many cases) that changes are not possible. In reality, changes are possible and can yield large benefits in warehouse utilization and efficiency. Very minor dimensional changes of the distribution package can yield large increases in rack efficiency, pallet cube utilization and use of trailers. If the distribution carton itself cannot be altered, the sizes of the interior package should be investigated. Minor reductions in height that do not affect product weights and volume, or changes in height with compensating increases in diameter, could permit the required changes in the distribution container. The need for such an investigation will be readily apparent if the cubic space occupied by cases of the product is compared with the cube required to store a pallet load of the product. In a similar manner, the number of units within the distribution package also can be changed to affect container dimensions and improve its storage.

The sizing of a container for interior packaging to achieve the most efficient pallet pattern, height and good trailer utilization involves a multitude of variables. Computer programs have been developed to analyze palletization, case count and arrangement, package sizing, and truck/rail car loading.

Effect on Stacking

In recent years, the shift from metal and glass containers to plastic containers for many liquid products has resulted in significant decreases in stacking height for these products. The metal and glass containers provided the necessary strength to withstand four-pallet-high stacking. The thinner and weaker walls of the plastic container often reduce maximum stacking height to two pallets. Although the use of dividers and heavier container materials can improve plastic container stacking heights slightly, they often will not permit the stacking heights of glass and metal containers. Warehousemen can look forward to further increase in the use of plastic containers for food and liquid products.

In addition, many food products now in cans and jars will soon be marketed in packages formed from composites of plastic film, foil and paperboard. It is quite probable that stack heights for these commodities will be reduced. The sidewalls of these containers are flexible and bulging containers will be more prevalent. The prudent warehouseman concerned with maximum utilization of warehouse space should be looking at alternative storage methods to include drive-in and drive-through racks, as well as other methods for improving warehouse cube utilization.

Most distribution containers in the U.S. are constructed from corrugated board. The strength of the total packaging system is a combination of the product and the container. Some products, by their nature, are inherently load-sustaining. Examples are metal cans, glass jars and industrial products with rigid cushioning, blocking and bracing systems. Products in non-rigid container, such as paper products and loose food products, cannot contribute to the stacking and compressive strength of the container/product system. Accordingly, the stacking capabilities of such systems are totally dependent on the strength of the fiberboard container.

By its nature, fiberboard is a hygroscopic (water-absorbent) material. As moisture is absorbed by the fibers, they swell and the bond between them begins to weaken. At a temperature of 73°F. and relative humidity of 50 percent, fiberboard has a moisture content of about 8 percent. If the temperature remains constant and the relative humidity is increased to 90 percent, the moisture content of the fiberboard increases to 20 percent. This increase in moisture content lowers the compression strength of fiberboard box by nearly 50 percent. Another characteristic of corrugated fiberboard is its tendency to creep and lose strength when subjected to a steady load, as in a storage stack. The rate at which this creep and strength loss occurs is based on many factors, including the support provided by the product, and stacking misalignment.

PACKAGING AND IDENTIFICATION

In many warehouses, carton clamps (see Figure 6-1) are used to move individual layers of packaged products or full unitized loads. Many package/product combinations can be handled by this method efficiently and without damage. However, when the product is dense and has no inherent stacking strength (bags of dried peas, beans, noodles), the corrugated container that must support the entire stacking load can be damaged by use of clamps. The clamping operation sometimes creases the sides and corners of the container, collapsing the corrugated wall that provides the major stacking strength of the container.

For best use of space with minimum risk of stack failure, the warehouseman should work closely with available packaging specialists to verify stacking capabilities under various temperature, humidity and storage duration situations.

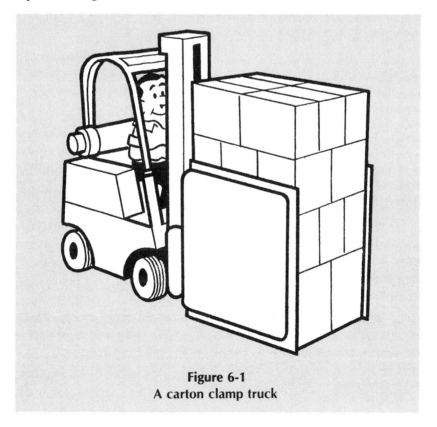

Figure 6-1
A carton clamp truck

Packaging for Unitization

Packaging can save labor at the distribution center through unitization, which is simply the consolidating of smaller boxes into a larger single unit in order to reduce the number of pieces handled. There are several ways to unitize packages, the most common being palletization. A pallet load of small packages may be glued together with non-slip adhesive. Such an adhesive has a high shear (side-to-side) strength and low tensile strength. This prevents the packages from sliding off the pallet, while enabling them to be easily removed by lifting.

Pallet loads also can be unitized with shrink film. This is a plastic film that is wrapped around the load and then heated to shrink and form a tight outer shell around the pallet, offering protection as well as load integrity. Strapping or tape also may be used to stabilize a unit load. A disadvantage of unitization occurs when one of the cartons in the unit is damaged, and the entire load must be rehandled.

More information about unit loads will be found in Chapter 45.

Identification

There are three hidden costs resulting from poor package identification:

1. Labor cost involved in extra searching during order selection.
2. Cost of correcting shipping errors resulting from identification problems.
3. Customer dissatisfaction created by shipping errors.

One cause of poor identification is package advertising that obscures the identification code. Another is complicated marking systems, which are difficult to read, understand or recall. Some numbering systems are conducive to numerical transposition. An example is quantity-oriented coding, such as 12/24 ounce and 24/12 ounce. These numbers describe the quantity and size of the inner packages, but are easily reversed or transposed.

The packaging department, in conjunction with marketing, determines the system used for identifying products—often without considering the problems of order-selection in the warehouse. Small

PACKAGING AND IDENTIFICATION

numbers, combinations of alpha-numeric identifiers, poor lighting and poor color contrast will inevitably lead to mistakes in order picking and shipping errors. Warehouse operators should insist upon product identification systems that are legible in the warehouse storage environment and use only the minimum number of letter and number combinations necessary for identification.

When the product moves in less than truckload quantities, the carriers normally require that each container be stenciled with the appropriate address. Container printing should allow space for the stencil or label.

Effective identification has certain characteristics. One is the use of large and bold marks with adequate background contrast. Another is marking on two or more sides of the container. A third is the use of color.

Increasingly, package identification systems are equipped with bar-code printing designed for reading by scanners. More about scanner technology will be found in Chapter 50.

Container Handling

Many products are picked and shipped in less than pallet quantities. Order picking and handling costs are typically developed around the cost-per-container. Each of the units within the container carries a portion of that handling cost. By increasing the number of units within each container, the handling cost per unit can be dramatically reduced as can the material costs for containers.

By tradition, many products are sold at the retail and wholesale level in some multiple of six, 12 and 10. This is often perpetuated by sales order blanks that identify only those multiples. But when the wholesaler or retailer is purchasing in quantities of 96, 100, or several hundred, there is no need to package the product in smaller multiples. The use of master cartons that approximate the quantity typically purchased, or the most economic order quantity, can reduce a number of case handlings and thereby the handling cost per unit.

Package Design

Product packaging should be designed with consideration for all of the transportation and handling operations involved. If the package is designed to save on handling and transportation costs as well as provide protection, it may be different from one designed solely to minimize costs. A tradeoff calculation will demonstrate whether additional packaging expense is justified.

Increased packaging costs may be offset by savings in handling or transportation. Unitized handling with a carton-clamp truck can provide substantial labor savings in the warehouse, but the strength of the package must be sufficient to protect the product during carton-clamp handling. The extra protection built into the package to allow clamp handling also will provide a margin of product protection throughout the distribution cycle.

There are also cost tradeoffs in improved storage capability. Most warehouses built since World War II have a clear pile height of 20 feet or more. Unfortunately, there are many products that cannot withstand a stack height of 20 feet so the overhead space is likely to be wasted. While the addition of steel storage racks will improve space utilization, it is costly to equip a warehouse with racks. Moreover, the storage height gained by racking is frequently offset by the lateral space that is lost for aisles. For this reason, storage racks are not always an economical investment. Furthermore, insurance underwriters have expressed concern as to whether normal fire protection systems can control fires in storage racks.

Any distribution manager with products now limited to less than 20 feet of free standing pile height would be wise to evaluate packaging that would allow increased stacking height.

Package design must also afford protection from damage, so the designer must understand the causes of damage in physical distribution. In storage, carton compression is a potential problem in the stacking of merchandise. When a stack falls in a warehouse, the causes of the collapse should be studied. In many geographic areas, humidity is the most serious environmental problem because most packages will be weakened by high moisture absorption.

PACKAGING AND IDENTIFICATION

The packaging engineer also must consider risks of mishandling of his product throughout the distribution system. He must know what happens to the package and to the product when it is abused. For example, one appliance manufacturer discovered that a cartoned washing machine could be dropped from a height of 15 feet; if it landed squarely on the floor, the package would show no evidence of damage. Yet, inside the package, the machine itself was a total loss.

The responsibility for such concealed damage usually cannot be determined. Frequently the damage is not discovered until the product reaches the consumer. Sometimes it is not discovered until the product is actually put to use. An ideal package is one designed to reveal *all* damage, thus eliminating the problem of assigning responsibility for concealed damage.

A certain amount of wear and tear on all packages is inevitable, since some are handled many times during the distribution process. Normal transportation risks also must be considered. Users of rail transportation should be aware of the effect rail car switching impacts at various speeds have on their products. Impact registers can be used to check the performance of railroad crews, but it is not realistic to assume that any rail car will be handled without a certain number of impacts at speeds in excess of six miles per hour. While some box cars have cushion underframes designed to absorb the shock of normal switching impacts, most box car users cannot be assured they will be able to obtain this equipment. Therefore, the packaging planner must design a carton that can withstand most rail impacts.

Other kinds of transportation damage must also be expected. One is abrasion, or the rubbing of cartons against the side walls of cars or trucks. Another is the shocks that occur when vehicles travel over an uneven roadway.

In planning a package for distribution, the designer must consider the value of the product to be packaged and the length of time the product will be in the package. "Overpackaging" is the popular term for a carton stronger than it needs to be. Overpackaging may be an economical investment for a product of critical value. Overpackaging also is desirable for an item likely to be in the distribution channels for a long period of time, subjected to unusual environment, or to extensive transportation or rehandling.

Efficient Packaging—A Joint Responsibility

Most major U.S. firms have packaging design, development and test organizations. Organizationally, the packaging group may be placed within research and development, marketing, manufacturing, or distribution departments. All too often, container design and sizing are a result of marketing decisions without regard to the cost implications to the distribution system. By the time these implications are recognized, the design, development and manufacturing process is too far along to change.

Many companies have established packaging committees within their organizations to supervise all aspects and implications of container design. This group reviews designs for new products and modifies design of older products to reduce material, manufacturing and distribution costs. This committee should be made up of representatives from distribution, packaging, manufacturing, marketing, quality control and purchasing. When the products will routinely move through public warehouse facilities or distribution centers operated by the customer, the committee should consult them, too.

A Series of Compromises

The design of packaging and packaging identification always represents compromises between the desires of the manufacturer, packaging engineer, warehousemen and customer. The package must be designed for various available handling methods in the distribution cycle. Warehouse operators naturally hope that the packages they handle will be as strong as possible. Users of those warehouses must suffer with the economic limitations of packaging. The fire underwriter wants all packaging to be perfectly safe. The final result is always a compromise between the desires of the various people handling the package.

7

TRANSPORTATION

Warehousing is so closely allied with transportation that it is almost part of it. In fact, one old definition of a warehouse is "a box car without wheels." Transportation terms like terminal or depot have traditionally been used to describe certain warehouses. And in some countries the majority of public warehousing companies are subsidiaries of transport firms.[1]

For these reasons, it is not possible to consider warehousing without also examining its relationship to transportation.

There are four primary modes of transportation: rail, highway, water and air. The location of the distribution center, both its general geographical location as well as its specific position within a community, is affected by the transport mode selected. The selection of a transportation mode also will affect operating methods, labor, equipment and plant layout. The warehouse layout must allow for efficient materials handling by the primary transport mode, and also must be flexible enough to adjust to future transportation changes. Special conditions may influence transportation selection. These include changing costs, local service conditions, or transit time. Traffic management includes the selection of transport modes to be used and choosing specific carriers within the mode.

Location

Transportation has a major influence on distribution center lo-

---

[1] This chapter was revised with substantial assistance from Daniel Bolger, The Bolger Group, Lancaster, Ohio.

cation. In some cases, computer modeling techniques may be used to determine the optimum location or locations to serve a given market. The availability, speed and reliability of transportation are considered in determining how many distribution centers are needed. The type and quality of transportation must be considered in choosing cities to serve as distribution points.

Once a city has been selected, a detailed study must be made to locate a piece of land or an existing building. Again, transportation is the primary limiting factor in selecting the site. For example, if air freight has first priority, the site selector will give priority to airport accessibility. Waterfront sites will be sought if the user has determined that water transport must be available in his distribution program. A wider selection of industrial property will be available if rail is used. In some cases there will be economic or service advantages by seeking service from one particular railroad.

The ultimate flexibility for a shipper is to have access to two or more railroads. This may be achieved through reciprocal switching, a privilege which allows free switching to any carrier in the community. With deregulation, this privilege is negotiated less easily than in the past. Another alternative is service by a short line railroad. Many communities have terminal railroads which exist for the purpose of connecting and switching for several carriers serving the city.

Proximity to a switching yard and availability of frequent switching service is important if rail volume is high.

The widest selection of sites is available to a truck-oriented center, since virtually all industrial property has road access. The warehouse may be deliberately located close to certain truck terminals. Ease of access is a location factor sometimes overlooked. The fact that the distribution center is located along the border of an expressway means little if trucks must make a long detour to the nearest access ramp.

The major location criterion is transportation convenience. Any distribution center can be reached some way, but the question of how easily it can be reached is the most important issue.

Effects of Transport Modes

The type of transportation used will affect the nature of the distribution center. But the user should have a choice of transportation when circumstances warrant. The last few decades have shown remarkable changes in the percentages of freight moving on each transport mode. Recognizing that further changes in relative importance of rail, truck, air and water will occur, the ideal distribution center should be designed for flexibility. This could include provision to expand or eliminate docks dedicated to rail or truck.

Marine containerization has substantially changed warehousing of cargoes transported by water. In some cases containerization has eliminated the need for a waterfront distribution center. In other cases it has created a new demand for marine container consolidation services.

More labor may be needed to operate a warehouse that receives rail cars than one primarily receiving trucks. Railroads never provide unloading labor. Motor carriers will provide this service, except when they are operating on a shipper-load/consignee-unload basis.

An increasing amount of merchandise traffic on the railroads is moving via trailer-on-flat-car service—popularly known as TOFC or piggyback. Using rail service will eliminate the need for a distribution center served by a railroad siding.

As more industries are involved in overseas commerce, the use of land/sea container systems is increasing. The principle is quite similar to piggyback. The container is removed from wheels and frame when loaded onto a ship.

Selecting a truck-oriented distribution center makes use of the extensive highway system available in most communities. However, traffic congestion, access routes with load or height limits, and availability of ample dock spaces and turnaround areas can be critical. For example, the disadvantage of using a multi-story distribution center is not only the restrictions of elevators on materials handling, but also the ratio of truck doors to volume of freight handled. Materials handling equipment to be used is governed by weight and inside height limitations of trailers. Since the final shipment to consumers in any distribution program is almost always by truck, any distribution center must be adaptable to efficient truck utilization. It is rare to find a distribution center that does not make use of motor trucks.

Compromises can be made for modes of transportation not directly serving the distribution center. An off-rail warehouse can handle rail traffic either through piggyback, or by using a team track or rail terminal leased from others. Water transportation can serve an inland distribution center through containerization or a freight forwarder. Proximity to the airport is desirable, but not necessary for use of air freight.

Freight Rates

The choice among modes of transport has always been a cost versus time consideration. For nearly 100 years—until 1980—freight rates charged by common carriers were regulated by government. Costs varied in relation to time, distance and weight. They also varied in relation to density. For example, since a ton of lawn furniture would occupy more space in a vehicle than a ton of steel, the furniture carried a proportionately higher rate.

Since 1980, however, with the passage of deregulation laws, freight rates have generally been subject to negotiation between shipper and carrier. As a result, rates now may reflect the volume and buying power of the shipper. The warehouse user must recognize that any rate created by one person can be changed by another. Committing a large capital investment based on a changeable rate should be considered very carefully.

Transit Time

Typical transit times must be anticipated in evaluating both the number and placement of warehouse locations. Terminal handling, yard handling, highway systems, waterways and traffic congestion are all factors that must be considered in calculating overall transit time.

The warehouse user should emphasize reliability rather than speed, since predictable delivery is critical in most logistics systems.

For this reason, a transportation service offering a reliable two-day delivery may be more desirable than another service that delivers overnight some of the time and takes three days at other times.

Private Fleets

During the last few decades, the fastest growing mode of transportation has been private trucking fleets. The energy crisis might have been expected to curtail such fleets. Instead, private fleet operators only intensified efforts to increase backhaul tonnage. Deregulation of truck transportation removed the few remaining constraints on the private operator's efforts to secure revenue for the return trip. Some large users of private fleets have organized the trucking operation as a separate cost and profit center, operating the fleet as if it were a common carrier. Some have availed themselves of piggyback services as a means of moving private trailers over long distances.

Some private fleet operators have used quantitative techniques to restructure routes and increase backhauls using commercial carriers on uneconomic routes. This enables them to control or reduce transportation costs despite changes in costs for fuel, labor, equipment, insurance and highway taxes.

Acquisition of larger "high cube" trucking equipment and better fleet maintenance programs have helped control costs in the face of escalating prices of new trucks and fuel.

From the warehouse operator's viewpoint, one advantage of the rise of the private fleet has been the increased awareness of the cost of detaining a vehicle. In an effort to cut such detention, private fleet users have readily accepted unitized loads to speed up the loading and unloading of vehicles.

Warehouse Freight Consolidation

In the decade preceding deregulation of motor freight, few developments in warehousing caused more publicity and misinformation than warehouse freight consolidation.

Consolidation is the use of the warehouse as a gathering point to create a larger shipment and, therefore, reduce the per-unit shipping cost to the ultimate consignee. Three conditions must be present to allow a warehouse freight consolidation program to be economical:

1. Order patterns must permit the combination of at least two shipments for delivery to the same destination at the same time.

2. Rates charged by the transportation company must create sufficient incentive to justify the use of freight consolidation. In today's free market this may require some intensive rate negotiation with the carrier.

3. The warehouse user must be willing to compromise service schedules in order to allow time to gather, hold, ship and distribute orders moving to the same destination.

Reducing Transport Costs in the Deregulation Era

Transportation is the largest cost element in the delivery system; your task as warehouse operator is to purchase it for your customer, the shipper, more economically than he or she can purchase it. If your customer perceives only marginal service advantages from warehousing, whether public or private, the spread in cost must be great enough to offset the incremental storage costs involved.

Public warehouse managers have three freight purchasing advantages over individual shippers. First, they can accumulate larger loads by combining the freight of several customers. Second, they can become attractive sources of freight business to motor carriers, railroads and intermediaries because of their total volume potential. Third, they can become experts in selecting the best cost/service shipping alternatives. The fact that the transportation services market is moving and changing so rapidly makes this skill even more important and more marketable than in the past.

Private warehouse managers have an additional advantage over their public counterparts—they can directly or indirectly control inbound freight carriers. Thus it may be possible to specify the same carrier inbound as outbound. This technique allows you to achieve greater leverage over a carrier by awarding greater volume of traffic and perhaps allowing for uniquely low "turnaround rates." Turnaround rates are those which apply when a second or additional movement is awarded to an inbound carrier, minimizing his empty mileage.

In addition, the private warehouse manager is not necessarily limited to the freight volume of his or her own company in the effort to wield more clout with carriers. Through "co-loading"—the technique of sharing freight movements with other shippers in the same car, container or trailer—frequency of movements can be increased for better service, and the better utilization of cube or weight capacity coupled with higher overall volumes can often achieve dramatic rate reductions.

Reducing Freight Costs

Whether your operation is public or private, you need to develop a clear and comprehensive definition of what the delivery system needs to do. This involves collecting and assembling data to describe shipping patterns and service requirements. The next step is to identify through research, analysis and negotiation those shipping methods that meet on-time delivery requirements at low cost. Finally, you need to route the freight so that, in combination, these low-cost methods and your warehousing program provide the best cost/service delivery system.

The basic format for constructing a shipper's distribution database (whether a public or private warehouse operation) is shown in Figure 7-1. It lists all actual shipments from the manufacturing plant (or central warehouse) to—where appropriate—the interim (outside) warehouse and the final destination in terms of weight, cost, frequency, freight classification, special requirements (e.g., protective service) and intransit service requirements. Statistics from a recent three-to-six month period usually provide the base information for such a table. Be sure to incorporate in the raw data any significant anticipated variations in the system such as the expected opening of new markets (thus new destination points) and weight variations (caused, perhaps, by smaller order sizes due to tighter inventory control).

The current cost of each movement gives you the economic start-

PRACTICAL HANDBOOK OF WAREHOUSING

Data Format[1]

Shipments from Franklin Park, IL (Chicago) to Massachusetts

Destin[2] Zip Code	Origin[2] Zip Code	Ship Date	Weight	Total Charge	$/CWT	# PCS	Carrier	Customer[3]	In Transit Time[4]	Freight Classification
02134	60131	11/1	192	$ 26.88	14.00	4	Acme	A	3	77.5
02134	60131	11/1	118	16.52	14.00	3	Acme	A	3	77.5
02134	60131	11/1	208	29.12	14.00	4	Acme	A	3	77.5
02134	60131	11/2	2,248	244.39	10.87	38	CF	A	4	77.5
02143	60131	11/2	196	34.12	17.40	4	Yell	B	4	77.5
02155	60131	11/3	434	81.58	18.79	9	Rdwy	C	3	77.5
02155	60131	11/3	224	42.11	18.79	5	Rdwy	C	3	77.5
02169	60131	11/3	310	53.96	17.40	6	Rdwy	C	3	77.5
02169	60131	11/3	204	38.34	18.79	4	Rdwy	C	4	77.5
02172	60131	11/3	540	88.34	16.35	11	CF	D	3	77.5
02194	60131	11/3	732	102.48	14.00	14	Acme	E	4	77.5
02571	60131	11/1	206	39.70	19.27	4	Rdwy	F	3	77.5
02602	60131	11/2	210	43.26	20.60	4	Rdwy	G	3	77.5
02700	60131	11/1	154	32.45	21.07	6	Rdwy	H	5	77.5
02771	60131	11/1	488	64.94	13.30	1	PIE	I	2	60
		Total	7,986	$1,152.67	14.43					

Service Requirement: 5 days in-transit maximum
Special requirement: None

1. Should reflect all point-to-point movements, i.e., from manufacturing plant to outside warehouse, wherein use, or directly to a customer. Movements from outside warehouses to the final destination should be shown as well.
2. Franklin Park is origin city, various points in Massachusetts and Rhode Island, the destination cities.
3. Customer identity, like other data not essential to understanding this actual freight pattern, is disguised.
4. Days

Figure 7-1. Data Format

TRANSPORTATION

ing point from which possible charges can be measured. Standards, qualitative limitations and management policies (e.g., 95% service level, no back orders) round out your definition of what the distribution system must accomplish in physical terms.

Your next step is to divide the shipper's distribution database into two categories: movements inbound to the warehouse, real and prospective, and movements outbound to the final delivery point. Your objective is to minimize the cost of each.

Inbound freight. Many warehouse managers do not try to control or route inbound freight, incorrectly feeling that this is the responsibility of the shipper, who can better manage it. However, comprehensive study of real and prospective movements to warehouses suggests two ways you can lower this cost significantly without compromising service: (1) by engineering and positioning loads up to the maximum practical and legally allowable weight and cube limitations; and (2) by selecting the lowest-cost carrier available for each movement.

1. Engineering larger loads. The key step in load engineering is to set weight and cube objectives at legal maximums for a given mode of service. Here you can make a major contribution by identifying underweight and undercubed shipments and recommending rearrangement of load patterns and pallet heights to maximize loads.

Two standard sizes of trailers are in common use, both 102 inches wide. These are single-trailer units 48 feet in length, or 28-foot trailers which are hauled as a pair. For the shipper with a commodity which will reach cube limitations before exceeding legally-allowable weight limits, the larger trailers have economic advantages. A 48-foot by 102-inch trailer has 13% more capacity than the 45-foot by 96-inch wide units which were in popular use before regulations changed. A pair of 28-foot by 102-inch wide trailers has 32% more capacity than the older 45-foot units.

For rail shipment, weights may be increased in box cars by top loading (often with strapping) and using larger high-cube cars. By repalletizing product into standard-size units, you can frequently increase weight and so reduce freight cost.

The warehouse (whether public or private) can help customers raise payload weights by eliminating wooden pallets. The pallets can make up five percent of the weight of the shipment and are sometimes lost or broken. A major reason for continuing use of wood pallets is that receivers do not have the equipment to handle unitized or slip-sheet loads. Modern, versatile lift trucks with clamps and slip-sheet handling can cut both pallet and freight expenses.

2. Selecting least-cost carriers. You have several practical alternatives. The first is to make use of commercial carriers with empty runs between the customer and warehouse. Empty mileage typically represents between 5% and 12% of a commercial carrier's normal pattern. By surveying all carriers operating in a given geographical area, you can identify those carriers needing loads and negotiate favorable rates and possibly long-term contracts. Figure 7-2 shows a sample letter and Figure 7-3 shows the response form.

Manufacturers with private fleets having authority to haul other companies' products to balance their regular outbound patterns often have backhaul capacity.

For-hire carriers often use owner-operators rather than salaried drivers. This option is available to warehouses through single source leasing laws, or by obtaining operating authority independently. Direct use of owner-operators can provide a 15-25% cost saving.

Another source of low-cost carriage is piggyback service. Piggyback consolidators enter into volume contracts with railroads and mark up the cost of individual shipments to their customers. They can be found through the yellow pages under "Freight Consolidation" or through a call to your local railroad freight agent.[2]

---

[2] From an article by James H. Stone and David F. Burns, published in Vol 19. No. 3 of *Warehousing and Physical Distribution Productivity Report,* Marketing Publications, Inc., Silver Spring, MD. Revised by David Burns, 1988.

Transportation Brokerage—A Useful Tool for the Warehouse Operator

A transportation broker is an individual with the[3] power and authority to deal with and arrange freight movements for and between all transportation suppliers and purchasers. Brokers do not issue tariffs or bills of lading in their own names, but the carrier issues them and is responsible for loss and damages. However, brokers can provide assembly, consolidation, distribution, and—unlike agents, who can act only in the name of carriers—can also represent shippers.

A distinction also needs to be made between transportation brokers and freight forwarders. The freight forwarder—in essence, a carrier that uses other carriers—has the same responsibilities as the broker except that he is still required by law (even though otherwise deregulated in 1986) to assume responsibility for loss or damage and to carry cargo insurance. A broker carries a $10,000 surety bond.

Public warehouses were early entrants into the field of transportation brokerage starting as far back as 1973. The flood gate was opened for public warehouses when the Motor Carrier Act of 1980 reduced the requirements for brokerage authority to a "fit, willing and able" basis, and when the new 1980 revisions to 49CFR1045 lifted most operating restrictions.

Since their entrance into the field of transportation brokerage, public warehousemen have used their licenses in many different ways—some concentrating on distribution, others expanding to haul line traffic, and still others offering fully-licensed operations. All of these uses are based upon the public warehouseman's previously-established reputation with his customers, and on the fact that many public warehousemen already serve as an extension of their customers' traffic departments—handling distribution of customers' products.

The private warehouse operator has no need to attract new busi-

---

3 From an article by John P. Martell, Transmart Company, published in Vol. 20 No. 5 of *Warehousing and Physical Distribution Productivity Report* Marketing Publications, Inc., Silver Spring, MD. Revised by the author, 1988.

ness. However, if the private warehousing company obtains a broker's license, the warehouse operator can use it to consolidate outside freight with his own company's traffic.

While private warehouse operators cannot accept commissions on their own freight, the addition of other freight moving over the same routes can provide the volume needed to achieve lower rates as well as commissions on other shippers' freight. In this way, a private transportation and distribution operation can be converted into a profit-making corporate activity.

Private warehouses can use the brokerage license to solicit outside freight. They can then consolidate this freight from other companies with their own freight and thus reduce their overall shipping costs. For example, private warehouse X furnishes a full product-line distribution in a major market area. Though it uses common carriers, warehouse X has sufficient volume to achieve charges which are 20 percent below class rates. Using a broker's license, company warehouse X obtains additional volume by consolidating freight from nearby public and private warehouses which is destined for the same customers. To avoid antitrust and confidentiality problems, only non-competing products are included in this consolidation. Through the increased volume obtained by consolidation, savings move from 20 percent up to nearly 40 or 50 percent. Even after brokerage fees are charged, the savings are quite attractive.

The private warehouse operator should look further at the advantages of improving private fleet economics. The broker can be used as an advocate to support contract authority for the private fleet, enabling it to fill backhauls or to carry third-party freight as well as owned goods on company vehicles. Contract carriers, unlike common carriers, are not required to publish tariffs or file reports.

Any individual in warehouse management can become a broker by filing an application with the ICC, posting a $10,000 service bond, and establishing service agents. The total cost of becoming a broker is usually $1,000 or less.

Obtaining a license is usually easy: the ICC fee is $150 and the one-time fee for service agents is less than $100. The required surety

TRANSPORTATION

Dear Carrier:

    We are a major manufacturer of household and industrial chemical products located in Providence, Rhode Island. We ship volume loads of finished products to customers in the eastern portion of the United States.

    The purpose of this letter is to determine your interest in providing service on some or all of our traffic movements. Representative original and destination points and estimated monthly volume are shown in the attached proposal forms. These are volume movements, generally occupying full capacity of the trailer or significant portions thereof.

    The footnote page provides information on product loading, delivery appointment requirements, temperature requirements, and so on. If you wish to participate in any of this traffic, please return the enclosed forms, indicating the proposed rates for those origins and destinations which you feel your company can service and your anticipated weekly equipment availability for each lane.

    You will note that many outbound movements require temperature control or protection from freezing during the winter months. Should you have a separate rate structure for this type of service, please submit rates for both protected and regular service.

    Your reply by _____, 19__ would be appreciated. Please feel free to contact me if you have any questions.

                                                   Sincerely,

Figure 7-2. Sample Survey Cover Letter

Carrier Name _____
Page 1

Outbound Shipments from Providence Rhode Island

Destination	Approx. Miles	Average Weight	No. Loads Per Month	Commodity (1)	Apt. (2)	Temp. (3)	Unit (4)	Proposed Rate	Per (5)	Min. Wt. (6)	Weekly TL Available (7)	Equipment Required
New Brunswick NJ	215	30-40000	6-10	4,12	3		2,4					45'
Bronx NY	178	25-40000	2-6	4,6	3	1	4,5					Conform to NY FD code
Syracuse NY	283	40-42000	5-20	1	3	1	4					
Mechanicsburg PA	347	30-43000	7-18	8	3	1	2;4					45'

(1) COMMODITY CODES
  1 Cleaning or washing compounds or soap, liquid
  3 Buffing or polishing compounds, liquid
  4 Toilet preparations
  6 Shaving cream
  8 Starch, liquid
  12 Insecticides or insect repellant
  22 Insecticides—flammable
  28 Corrugated boxes
  30 Pallets
  31 Dividers, pads
  32 Boxes
  33 Misc. items

(2) APPT.—indicates whether customer requires a delivery appointment
  1 No appointment (or rarely) required
  2 Appointment usually required
  3 Appointment always required

(3) TEMP.—Indicates whether temperature control is required.
  1 Freezable; must be kept from freezing during winter months. In general, product delivered directly with no stops in transit, will not freeze if kept in motion.
  2 Maintain at 70-75°
  3 Maintain at 55-65°

(4) UNIT.—Indicates whether product is unitized, and method of unitization.
  1 Floor loaded
  2 Palletized
  3 Slip sheets
  4 Stretched pallets
  5 Slip sheets on pallets
  6 Skee pallets

(5) PER—Please indicate the basis upon which rate is quoted (e.g., cwt., mile, flat rate, etc.)

(6) MIN. WT.—If rate is based on cwt., please indicate the minimum weight applicable.

(7) WEEKLY TL AVAILABLE—Please indicate the number of loads per week you can handle for this lane.

Figure 7-3. Sample Survey response form to be filled out by carriers and returned to warehouse/distribution manager.

bond usually costs less than $200 per year, and applications are normally approved forty-five days after they are published in the ICC register.

The Second Industrial Revolution and Its Effect on Transportation

Today five factors are causing major changes in the relationships between transportation and warehousing:
- The increasing degree of deregulation of all kinds of transportation services.
- Scarcity and/or high cost of fuel.
- Scarcity and cost of money.
- A pervasive and worldwide economic uncertainty.
- A second industrial revolution.

These factors are causing warehouse operators to ponder whether they should have more warehouse locations or less. They also are changing the configuration or design of many warehouses. High-priced money means continued pressures from corporate financial departments to reduce inventory. As long as the price of money is high relative to other commodities, the emphasis in physical distribution will be on inventory management. High-priced money probably will be accompanied by pressure to reduce the number of warehouses, and to use premium transportation over long distances. As private trucks replace railroad box cars, the area that can dependably be served from a given distribution center is widened.

There are other pressures that could cause an increase in the number of stocking locations or changes in present locations. One of these is the popularity of "just in time" inventory deployment that requires schedules as precise as the quarter-hour in which a shipment will be received.

In the 1970s, a key marketing strategy of the public warehouse industry was freight consolidation—the ability of the warehouse operator to merge the shipments of several manufacturers shipping to the same destination with a savings for all of them. Today, the dereg-

PRACTICAL HANDBOOK OF WAREHOUSING

ulation of transportation has made the whole concept of freight consolidation irrelevant. Since freight consolidation was based on the assumption that common carrier prices were based on fixed tariffs, the only way to achieve a lower price was to move a greater volume on the same shipment. Most of the advantages of consolidation can be achieved today by individual shippers with the power to negotiate for them. Larger companies no longer need freight consolidation to achieve the same advantages that seemed so important five years ago.

Economic changes also have affected the design of distribution facilities. At least one feature in most high-volume warehouses today is an obsolescent rail dock layout. Only a few years ago, many high-volume distribution centers were constructed primarily for rail and storage-in-transit movements. Those buildings have rail docks, many of them indoors. The outdoor idle rail dock does not represent any waste, but the indoor one does. Transportation futurists are predicting the demise of the box car. If merchandise traffic moving by rail continues to change from box car to piggyback, what will be done with those idle inside rail docks? The most obvious solution is to fill in the dock and cover it with a new warehouse floor. There are several ways to put a railroad siding in mothballs so that it can be restored at a later date. One warehouse operator stores goods on flat cars which were purchased and then parked in the idle rail pit. Another way is to fill the rail pit with sand, and to lay over that a removable slab. A third option is to use the rail pit area for storage of seasonal or slow-moving items—so-called dead storage. There are two problems in doing this. First, the rail pit must be paved both for reasons of sanitation and the maneuvering of materials handling equipment. Secondly, a ramp is needed to allow materials handling equipment to move easily in and out of the rail pit.

A contrary problem is the truck dock. As truck movement grows to replace tonnage removed from the rails, the warehouse may not have enough dock capacity. Fortunately, this is usually easier to correct. In any new construction, including additions to existing facilities, it is important to have a dropped foundation footer along the entire wall of any area planned for trucking. Truck dock doors that may

TRANSPORTATION

not have been planned in the initial construction may then be easily added.

Federal legislation has changed the size of truck bodies, and this also could change the design of the docks those trucks use. While relatively few warehouses have truck doors so small that they will not accommodate the 102 inch-wide bodies, at least the dimension change must be considered in planning for new doors.

Both federal legislation and major economic shifts have caused more changes in the transportation business in the last few years than were seen in several decades. These changes will continue to affect warehousing in the future.

8

ACCOUNTABILITY

Under common business law and practice, someone is accountable for goods at all times. In distribution, goods change hands from manufacturer to carrier, to distribution center, then on to another carrier, and finally to the customer. In these exchanges, the responsibility limits of each handler of the goods must be clearly defined.

Accountability in distribution covers the following: title to goods, inventory responsibility—overages, shortages and damage; product liability, and insurance and claims.

An "identifiable party," in the legal term, must be held accountable for the care of merchandise at each point in the distribution channel. Accountability does not become a problem until there are claims or losses. However, since claims are inevitable in distribution, accountability must be carefully defined.

The definition of accountability can be less rigid for a private warehouse, since the warehouse inventory is the property of the operator. In contrast, the public warehouseman is handling merchandise that does not belong to him, and must deal with freight carriers that also handle goods belonging to others.

Title

As the goods move through the distribution system, transfers of title occur. The precise point at which a transfer occurs is very important. Traditional selling terms for freight shipments define the owner of in-transit goods at different points in its movement. Various selling terms are used. When sales terms are "F.O.B.," the buyer assumes title to the goods from the time they leave the seller's premises. The seller retains title to the goods while in transit if the transaction is "F.O.B. Destination" (see Figure 8-1).

The responsibility of the public warehouse is based on the Uniform Commercial Code (UCC) which is law in every state except Louisiana. Many warehousemen use standard contracts developed by the AWA based on that code. Under the UCC, carriers are legally liable for nearly all losses to their cargo, but warehousemen are not. Therefore, a loss that takes place at the warehouse is frequently handled differently from a loss of cargo held by a common carrier. The warehousemen's liability for goods is limited to "reasonable care," defined in section 7-204-1 of the UCC as follows:

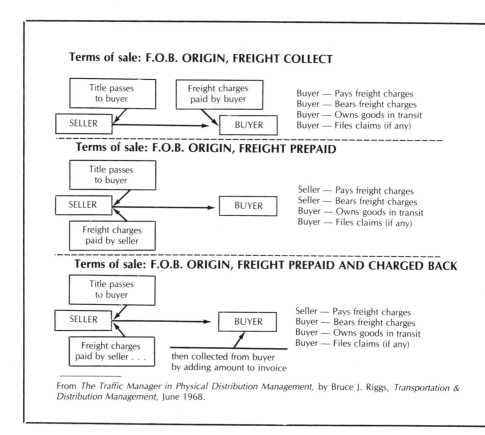

From *The Traffic Manager in Physical Distribution Management,* by Bruce J. Riggs, *Transportation & Distribution Management,* June 1968.

ACCOUNTABILITY

A warehouseman is liable for damages or loss or injury to the goods caused by his failure to exercise such care in regard to them as a reasonably careful man would exercise under like circumstances, but unless otherwise agreed he is not liable for damages that could not have been avoided by the exercise of such care.

When the public warehouseman's responsibility for goods in his care begins and ends is not always clear. The point of transfer of responsibility may vary with different operations. For example, if the warehouseman operates delivery trucks, he retains custody of the cargo until it is delivered. If goods are received by rail, responsibility may begin when the rail car enters the warehouseman's property,

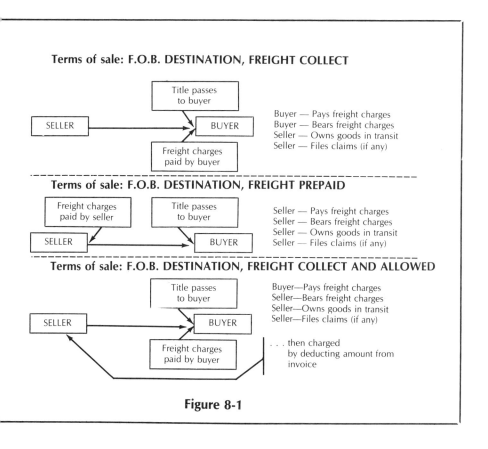

Figure 8-1

even though he may not have broken the seal to open the car. A question may arise concerning freight that is still on common carrier trucks when those trucks are parked on the warehouse company's property. The creation of a public warehouse receipt is a legal acknowledgment of responsibility for care of those goods. The signing of a bill of lading transfers responsibility from the public warehouse to the carrier.

Goods consigned to a public warehouse should always be addressed to the owner of the goods in care of the public warehouse company. This is the legal way to show that title to the goods remains with the owner, and that the public warehouse is acting as his agent.

The "reasonably careful man" theory is as old as English common law, though standards of what is "reasonably careful" are changing.

For example, standards of fire protection have been upgraded in recent years. Advertising a "fireproof warehouse" is almost never seen today, though it was once common practice. Because contents are nearly always flammable, there is no such thing as a fireproof warehouse. Claims to the contrary are misleading and could be considered an assumption of additional liability.

Protection standards against theft and vandalism also have become more stringent. Increases in national crime rates have affected warehousing practices as much as they have other aspects of life. Burglary protection systems believed adequate in the past must be constantly reevaluated by warehouse management.

Effective security against fire and theft is partially a matter of attitude. The finest systems in the world are worthless if a conscientious interest in loss prevention has not been instilled in all distribution center personnel.

Inventory Responsibility

Accurate maintenance of records is the keystone of any inventory responsibility system. The correctness of the book inventory must be checked by physical counts. A complete physical inventory should be made at least once a year. Cyclical inventory checks, the

ACCOUNTABILITY

counting of a few line items each period until the entire stock has been counted should be used as a back-up.

Since book inventory records often have clerical errors, one way to ensure accuracy of warehouse inventories is to maintain two book inventory systems—one posted by the warehouse user and one by the warehouse operator. When the two book inventories are not in balance, they can be compared for omitted, duplicate or incorrect entries.

Inventory discrepancies also result from errors in shipping and receiving. If the wrong item is shipped, but the correct item is posted to the book inventory, a compensating overage and shortage (a crossover) will show up when a physical count is taken.

The warehouse user should develop a policy for handling overages and shortages or the warehouse operator will be constantly penalized for declaring overages. To assume the warehouseman is responsible for all shortages, but does not receive credit for overages, would be a violation of the "reasonable care" standard outlined in the Uniform Commercial Code.

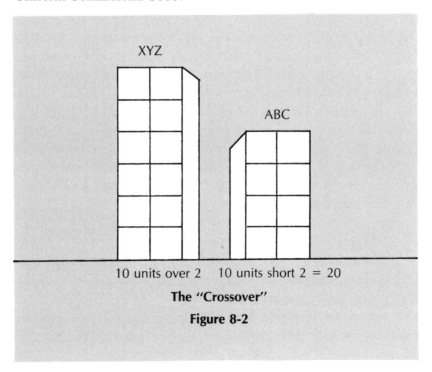

10 units over 2   10 units short 2 = 20

The "Crossover"

Figure 8-2

An occasional shipping error is normal in any warehousing operation. If the warehouseman must pay for the shortage without receiving credit for the overage, the owner of the merchandise "realizes a monetary benefit that ordinarily would not be gained in the conduct of business," as the lawyers would say. Most public warehouse users today recognize that inventory responsibility is best maintained through policies that deal reasonably with a predictable percentage of errors.

Product Liability

Some kinds of losses may result from the nature of the product itself or damage which occurs through use of the wrong product, resulting from an identification error in shipment. An extreme example would be two similarly marked products in an inventory—one a food ingredient, the other a highly poisonous compound. If the items are confused in shipment, the results could be disastrous.

The manufacturer must specify the conditions necessary for protecting such products. He should determine whether his product would be damaged by temperature and humidity variations, or cross-contamination through odor transmission or leakage. Infestation may also be a product liability risk, because rodents and insects are attracted to some products, but not to others.

Regardless of efforts to foresee them, unexpected product liability situations will occur. For example, an appliance manufacturer discovered by accident that its product's white enamel finish turned pink when stored near tires. Any manufacturer, carrier or warehouseman who discovers a risk due to the nature of a product should share this information with others handling the product.

Insurance

The major risks in accountability in distribution are usually covered through insurance. In private warehouses, broad insurance policies cover factory and warehouse alike.

When contract or public warehousing is used, the insurance contract must be more precisely defined. Fire and extended coverage insurance for goods in storage normally is carried by the warehouse user, since he owns the goods.

The public warehouse operator insures only his building, not the goods stored within it. However, the rate the user pays for his coverage usually depends on the insurance rate on the building. The building rate represents an assessment of risks made by the insurer of that building. Obviously, a building with a relatively high insurance rate is one considered unsafe for one reason or another.

Although the public warehouse operator is not legally liable for cargo casualty losses, he is liable for losses resulting from failure to exercise ordinary care.

The Uniform Commercial Code allows the public warehouseman to limit his liability for loss or damage. For example, a public warehouseman may limit claims to an amount not to exceed 200 times the monthly storage rate for each item. Thus, if the storage rate is 50 cents per-case, claims are limited to $100 per-case.

Most public warehousemen carry insurance to protect themselves against legal liability claims. However, such insurance policies include a clause that may cancel the policy if the warehouse operator signs any contract extending his responsibility beyond the standard set forth in the UCC. For this reason, most public warehousemen will not sign any agreement drafted by their customers without first securing approval from their insurance company. A customer's contract that cancels the warehouseman's insurance coverage would obviously remove protection for the warehouse user in the event of a major loss.

Insurance protection through fidelity bonds covers the risk of dishonesty of warehouse employees.

Summary of Claims Procedures

The distribution center user should set up procedures for handling claims. Claims will inevitably occur, and without procedures, the necessary supporting documents may be lost.

Procedures for handling overages and shortages and warehouse damage are good business practice and the distribution center user must determine whether he will provide an allowance for this damage.

A procedure to appraise damage claims must be developed. If an insurance company is required to pay the claim, there must be proof that the claim is fair and reasonable.

Prompt handling of carrier damage claims is important. Some carriers are notoriously poor in handling damage claims, and delay in submitting the claim will only aggravate the problem of prompt settlement. In today's deregulated environment, negotiation may be as important as documentation in settling carrier damage claims.

Repackaging is often possible when the product is not totally destroyed. A high percentage of some types of damaged products can be salvaged by repackaging. But this action, known as recooperage, could result in contamination. If recooperage is not handled carefully, a damaged product may reach a customer in a new carton. Training and inspection will help prevent improper recooperage at the warehouse.

Other methods of handling carrier damage include scrapping all damaged merchandise, with costs of the product and its scrapping paid by the responsible party. Damaged merchandise may be disposed of through a sale handled by the owner, the warehouseman, or the carrier.

Accountability is essentially the assigning of responsibility. Both the user and the operator of a distribution center must clearly understand the assignment of responsibility for property stored. The points at which responsibility passes from one party to another must be defined. Accountability covers title to goods, inventory responsibility, product liability and claims. Insurance is a means of controlling the risks. Since someone *must be* responsible for goods in distribution at all times, these responsibilities should be clearly defined.

9

STARTING-UP OR MOVING A WAREHOUSE OPERATION

Private, public or contract warehouse?(*see Chapter 4*). Before deciding, the user should develop a general job outline for the distribution center. This job outline should include a review of existing customer service needs and how these customer service levels are to be achieved.

A key step is a simulation of the costs of operating the center, regardless of whether the proposed center is public, private or contract. Many users overlook the fact that a private warehouse has its own internal costs, which can be expressed as imputed rates. If the user employs a public or contract warehouse operator, that contractor will calculate costs plus profit to quote a rate. However, if the user has a private distribution center, imputed rates allow him to evaluate the alternative of using an outside contractor. Therefore, a basic understanding of the costs of operating a warehouse helps control the new operation.

Finally, the user should develop a detailed plan for the opening to allow proper lead times and to minimize confusion during the start-up phase.

Initial Planning

Early in the planning stages for the distribution center, the user must consider communications, packaging, transportation, security and perishability of stored products. "Communications" includes the method of data transmission and its frequency. Examine package sizes, their handling characteristics, opportunities for unitization, and bulk storage *versus* warehousing of packaged goods. Consider the best mode for both inbound and outbound transportation. Problems of security include both fire and theft protection. Nearly everything is perishable to some degree, and a means of controlling perishability is one of the first considerations.

Cost Considerations

Warehousing costs divide into two categories—external and internal. External costs are those that originate outside the warehouse, but would not occur if it were not there. Internal costs are those generated within the facility and are directly under the control of warehouse management.

External costs cover:

- Transportation charges to and from the warehouse.
- Inventory taxes on goods stored at the warehouse.
- Insurance on warehoused inventories.
- User's costs of controlling the warehouse.

Internal costs include:

- Storage.
- Handling.
- Clerical services.
- Administration.

While some cost accountants would assign other headings or cost categories, the eight cost centers listed above are those usually found in distribution center operations.

Transportation

The largest external cost in most warehousing projects is transportation. Since deregulation, some warehouse locations are no longer economical, because they were based on artificially regulated transportation rates. Freight rates for goods moving to and from a distribution center may now be affected more by negotiation than the exact location of the distribution facility. On the other hand, selecting a location to maximize backhaul opportunities is always valuable to a private fleet operation.

Design of the facility also influences transportation costs. If dock doors are inadequate, or if dock ramps are congested, the building may be costly for either the private or common carrier to serve. Design of the building and the ability to turn around vehicles quickly can directly influence overall distribution costs. Locating close to a freeway system will reduce driving time for private trucks. When air or marine freight is used, proximity to airport or wharf is obviously a cost consideration.

Taxation

Second in importance as an external cost is the local tax on inventories, which can vary significantly from one location to another. A few states or other taxing authorities have used inventory as the basis for levying franchise, income or other taxes on the owner of property in storage. A user may shift inventories from a state that has created a significant tax burden by radical changes in tax policy. As transportation has improved, manufacturers can provide reasonable delivery service to their customers and still avoid those states and municipalities that have created what amounts to tax harassment.

The various states of the nation differ widely in their approach to inventory taxes. At the same time, rates of taxation within states or even within counties or towns also can show significant differences.

Taxation rates and attitudes toward taxation are in a constant state of change. The trend seems to be toward eliminating inventory taxes, as state governments have recognized they must take this step to remain competitive with their neighbors. Because state tax situations change frequently, expert advice should be sought when making comparisons about tax policies of individual states or communities.

Insurance on Goods in Storage

In both private and public warehouses, casualty insurance for goods in storage is the responsibility of the owner of the inventory.

A building considered to be a firetrap will have high insurance rates for both the building and its contents. Types of goods stored in the warehouse affect the insurance rates. The quality of the fire department serving the area, the underwriter's assessment of housekeeping, and management interest in property conservation also will affect rates.

While insurance costs are normally a minor part of total physical distribution cost, insurance rates are an excellent yardstick for measuring the quality of a warehouse operation. Because the underwriters have a good reason for giving a warehouse a high risk rating, a public warehouse with unusually high rates is likely to be badly managed.

To Move or Not to Move[1]

The question of whether or not to move a warehousing operation may arise because a more desirable building or location becomes available. Or it may be a result of transportation or other customer service considerations which have changed, thus making the existing location or building less desirable. Whatever the reason, consider these important factors while still in the "should we move?" stage.

—Total storage space cost comparison. Include all the rent or ownership costs, taxes, utilities, operating costs, etc., of the facility you're in now and the facility you're considering.

—Comparative storage space ratio of gross to net space. This important factor is too often overlooked. Net storage space is that which will actually have inventory stored in it. Normally, 55% to 65% of the total or gross cubic space available is all that can be actually used; the rest is in use for aisles, docks, rail siding or offices. When comparing the existing and proposed facilities, consider only the net storage space. (To do this properly, a tentative layout of the proposed facility will be required.)

—Total labor and clerical costs comparison, including both pay

---

[1] This section is taken from "A Step by Step Approach to Moving a Warehouse," by Lee P. Thomas, published as Volume 19 Number 2 of Warehousing and Physical Distribution Productivity Report, Marketing Publications, Inc., Silver Spring, MD.

and benefits. Consider relative employee efficiency if you anticipate a difference. Often there is a difference resulting from a more efficient layout or the opportunity to eliminate wasteful practices that have become ingrained at the existing location.

—Availability of a good labor supply. Will key people be retained at the new location? Will a good supply of high quality replacements be available?

—Labor union consideration. Will the new location make a successful union organization attempt likely?

—Accessibility to major highways and carrier terminals for motor transportation. In real estate decisions, the most important factors, in order of importance, are location, location and location!

—Rail service considerations.

—Customer pick-up convenience. (This may or may not be a factor in your business.)

—The prestige of the location. Be careful not to weigh this intangible factor too heavily so that it overshadows other, more important factors. A warehouse's customers or users should enjoy better service, lower costs or additional space for future needed growth as a result of the relocation, otherwise the move should perhaps not be made. Prestige factors should only provide a tie breaker in the "go" or "no-go" decision.

Timing Can Be Critical

When considering making a warehouse location change, management must estimate the cost of the move and include that cost in the primary decision of whether or not to move.

To calculate this cost, first start with a target date for the move. This may subsequently be changed several times, but for planning purposes a target date is essential. Weather factors and seasonal inventory level variations can be taken into consideration when the cost estimate is based on a specific future date.

Be extremely wary of basing a move target date on a construction contractor's promise . . . contractors are eternally optimistic. In many instances, managers have had to move an operation into an incompleted warehouse because relocation plans were based on a contractor's completion date estimate and it later became impossible to postpone the move. To avoid this, be sure to add a comfortable cushion to the contractor's promise date.

Estimating the Cost of Moving

After setting the target date, make an assessment of the inventory to be on hand at that date. The simplest approach to this is to forecast the future date's inventory levels as a percentage of today's inventory levels.

Remember that it's always possible to plan the moving date to coincide with a lower inventory level period or a period of lesser shipping activity—as long as the new building is ready!

After estimating the level of on-hand inventory, determine the number of transfer loads currently on hand. You can do this by counting the full and partial pallet loads and pallet equivalents on hand, and then dividing the number of anticipated pallets per transfer load into the total currently on hand.

For example, assume the current inventory level consists of 3500 full and partially-loaded pallets and pallet equivalents. (Pallet equivalents are items, not palletized, which equal a pallet in space and cube requirements.) Anticipate 26 pallets per transfer load. The number of loads of inventory to be transferred may be calculated as:

3500 pallets/26 pallets per load = 134.6 loads

134.6 current inventory loads adjusted by a 92% factor to compensate for reduced inventory levels at the move date yields:

134.6 x 92% = 123.8 loads of inventory to be transferred on the targeted move date.

How to Load the Trucks

Experience has shown that, for a short-distance move, floor loads (goods stacked just one pallet high) allow for quicker turnaround of transfer vehicles and are more efficient; for longer moves, fully-loaded vehicles are more economical. When the trucks used to

STARTING-UP AND MOVING A WAREHOUSE OPERATION

move the inventory can be turned around quickly by using a floor load, you'll be able to reduce driver waiting time.

Cost of Transferring the Load

The cost of each load to be transferred can be calculated by totalling the following costs:

—Transfer vehicle operating cost, or vehicle rental cost plus driver labor and fuel, per round-trip.

—Outloading manpower and machine cost per load.

—Unloading and "put away" manpower and machine cost per load.

—A small additional allowance for damage, extra clerical labor and contingency costs.

The sum of these costs is an indicator of the cost per load and, when applied to the number of loads to be transferred, will yield the total cost of transferring the inventory.

Example:
Truck cost: 20 miles/round trip @ $1.80/mile . . . 36.00
Outloading cost: 0.3 hours/load @ $22.00/hr 6.60
Unloading/putaway cost: 0.25 hours @ $22.00/hr . 5.50

Subtotal
Overtime allowance @ 25% x labor cost 3.03
Damage/clerical/contingency allowance @ 1.44

Total cost per transfer load $52.57

In addition to relocating the inventory, the warehouse "business" must itself be moved. So the following costs should be added to our example:

—Cost of relocating the office operation.

—Cost of relocating the warehouse maintenance shop and recouping operation.

—Cost of transferring the materials handling equipment.

—Cost of disassembling, transferring, reassembling and lagging down storage racks and other equipment.

These can be costed by estimating the number of transfer loads involved, multiplied by the cost per load previously calculated, and then adding additional estimated labor costs such as rack relocation and lag-down labor.

How Long Will It Take?

The following is an example of how to calculate the number of days that will be required to make the move:

Assumptions:
Inventory loads to be transferred .... 124 loads
Additional loads to move the office
equipment and files .... 5 loads
Total loads to be transferred .... 129 loads
Further assumptions:
Truck travel time, one way loaded .... 25 minutes
Return trip time, one way not loaded ..... 25 minutes
Outloading time, one lift operator .... 18 minutes
Unloading and putaway time, one lift operator 15 minutes
Total round trip cycle time .... 83 minutes

Time available, one shift operation
Available truck time, 7.5 hours .... 450 minutes
Minus allowance for delays/
interruptions, 10% .... 45 minutes
Minutes available per truck per shift ...... 405 minutes

405 minutes available per truck per shift / 83 minutes per round trip cycle = 4.9 loads/truck/shift.

With moderate overtime or extra efficiency, five loads may be achieved per shift per truckload with one lift operator loading and another unloading.

The 129 total loads to be moved will require:

129 loads / 5 per day = 25.8 days with one truck, two lift operators on one shift

129 / 10 = 12.9 days with two trucks

129 / 15 = 8.6 days with three trucks.

Deciding to Continue Services or Suspend Operations for the Move

Second in importance to the question of whether or not to relocate is deciding whether to continue services or suspend operations during the move. Customer service considerations should take priority.

The warehouse operating manager's preference will probably be to shut down because moving costs can be minimized and coordination of the move is much simpler. When operations are suspended, all that is required is thorough planning, having everything ready in advance, and a few hectic days of concentrated activity. Overtime costs may be significant but the period of disruption can be kept to a minimum. Long days and weekends are often thought best for this type of move.

However, the customers who are ultimately paying the bills probably should be the deciding factor in whether or not to close during the move. Moving a warehouse without a break in normal customer service requires special coordination.

First, and of critical importance, is communication. Each customer must be contacted on a management-to-management level during the early planning stages to discuss the move and to outline the tentative plan and timetable. As the move date approaches, there must be increasingly frequent contacts on the administrative and clerical levels between the warehouse and its users to assure that loads dispatched to the warehouse are routed to the new location on the proper dates, and that carriers are sent to the proper location for pick-up of outbound shipments. There may be a short-term requirement for carriers to pick up outbound shipments from both the old and new locations.

What to Move

The move of each distinct class or grouping of inventory should be made in the shortest time span possible in order to minimize the time that the customer's inventory is in two locations. In a public warehouse, this is the inventory of each individual user or customer. For a private warehouse, it is each class of stock.

Moving by inventory grouping is better than starting in one corner of the warehouse and relocating a day at a time. A move by groupings provides an opportunity to correct inventory count variances, identify, isolate and report possible damaged goods, and better consolidate like items at the new location.

This method also makes it easier to coordinate the warehouse activity with the sales department, order department or inventory control department, whichever is the prime contact and coordinator with the warehouse. This is a time when close communications between warehouse and user are most important. The warehouse manager should, with reasonable lead-time, tell the user when each class of inventory will be moved and also report when the move of that class has been completed. The period of time during which inventory is actually being moved is critical to both parties.

Customer Communications

Customer communications are the priority in a warehouse move. Customers need to be sold—tactfully—on the benefits of the proposed move. All customers must be informed well in advance of the move, and then provided with frequent progress reports. An over-informed customer seldom complains.

Points that must be covered in communications with customers include:

—Out of service dates, if the warehouse is to suspend operations during the move. Also provide for emergency service requirements, which will surely arise.

—Date and time of transfer of each class of inventory. This is needed to coordinate delivery and pick-up carriers. It requires daily, even hourly, contact.

STARTING-UP AND MOVING A WAREHOUSE OPERATION

—Phone number and address changes and their effective dates.

—Dates and times when data communications will be out of service. These breaks in service should be held to an absolute minimum.

—Directions to the new location for users to provide to their carriers and customers for pick-up and delivery.

—New hours of service, if these are to be changed temporarily or permanently.

—Changes in service charges or standards of service should be announced and explained.

Other Communications

Customer communications have top priority, of course, but carriers, employees, suppliers, the phone company and letter carriers need to have timely notice of the move.

Employees should hear about the relocation well in advance; if you don't inform them, the grapevine will. Employees should be informed—in the early stages of planning—what is happening, why, and what to expect at the new site. Employees should be given regular information updates. It is especially important that they tour the future facility while it is in the later stages of preparation. A complete walk-around, explanation of the location system, and any needed training in the use of new or different equipment will pay off in higher employee morale and a more favorable employee learning curve at the new site.

Bear in mind that union organizers enjoy taking advantage of any uncertainty—such as a move—to play on employee concerns. Good communications can help you avoid this.

Carriers also need to be informed, even though they may already know about the move from industry contacts or your warehouse employees. This is a good time to look at any special arrangements with carriers. For example, for night-time rail switching in or out of a locked warehouse, the rail crew may need to be issued a key to the rail door and the security system. The rail crew may need instructions for correctly blocking out and resetting the building security system.

Opening the Relocated Warehouse

Relocating a warehouse inventory or opening a new warehouse is a major undertaking. Before making the "go" decision, anticipate all the costs and be sure there are ample benefits to offset these costs. Moving or opening a new center involves planning, communications, planning and more communications. You cannot possibly have too much of either. Checklists are always an important part of planning. Shown as exhibit 9-1 is a checklist which covers warehouse administration, receiving, shipping, operations, space utilization and safety.

Figure 9-1
A Warehouse 'Start-up' Checklist

Using the Checklist: This checklist is divided into six sections. It is designed to provide reminders to help insure that nothing is overlooked during the busy time before you open a new facility. This checklist is a starter. The user should add additional checkpoints to cover specific features of his own warehouse operation.

Personnel Administration   Yes   No
1. Is the application form used for hiring thorough and in compliance with Federal and local law? ___ ___
2. Are adequate personnel available to carefully and thoroughly interview each job applicant? ___ ___
3. Has a company doctor been selected to perform physical examinations for each new person hired? ___ ___
4. Has a personnel manual been prepared to explain responsibilities, benefits, and personnel procedures? ___ ___
5. Has a detailed training program been prepared for both warehousing and clerical personnel? ___ ___
6. Has the training supervisor been appointed? ___ ___
7. Is an adequate cadre of experienced people available to train the new people in the new operation? ___ ___
8. Are all supervisory personnel hired and already in place before hiring hourly people? ___ ___
9. Are checking procedures available to detect problems with substance abuse, dishonesty, or credit problems of job applicants? ___ ___

STARTING-UP AND MOVING A WAREHOUSE OPERATION

(Start-up Checklist continued from previous page) Yes No

Receiving
1. Has a detailed procedure been prepared for receiving of freight? ___ ___
2. Is a manifest or other form designed to cover receiving? ___ ___
3. Will bar coding be used to identify received materials correctly? ___ ___
4. Is there a procedure to be used for shipments which arrive without manifests or without any advance notification of what is in the load? ___ ___
5. Is a procedure established to handle overages, shortage, and damage (OS&D)? ___ ___
6. Has someone been designated to check OS&D reports? ___ ___
7. Will someone be responsible for checking the accuracy of each receipt? ___ ___
8. Is there a procedure to document the time which each vehicle is held at the dock in order to approve or dispute carrier detention charges? ___ ___
9. Have procedure variations been established for receipt of merchandise returned by customers? ___ ___
10. Has procedure been established for routing receiving reports and other reports for the receiving dock through the warehouse office? ___ ___
11. If lot numbers are used, has it been decided how they will be assigned at the receiving dock? ___ ___
12. When produce is palletized at the receiving dock, has pallet pattern been determined? ___ ___
13. Will control be exercised to be sure that all product is palletized according to the prescribed pattern? ___ ___
14. Has it been decided how storage location will be determined when goods are received at the dock? ___ ___
15. As merchandise is staged to be moved to storage, is there a plan for communicating stacking limitations, stock rotation and other storage specifications? ___ ___
16. If a locator system is used, will checks be made to be sure that the product is actually stored where ordered? ___ ___
17. Has an appointment procedure for inbound carriers been established? ___ ___
18. Is there an enforcement plan for the appointment procedure? ___ ___
19. Do all inventory procedures go as far as they could to prevent fraud or dishonesty in receiving? ___ ___

PRACTICAL HANDBOOK OF WAREHOUSING

(Start-up Checklist continued from previous page)   Yes   No

Shipping
1. Has a detailed shipping procedure been prepared? ___ ___
2. Will there be a priority system for handling of outbound orders? ___ ___
3. Has a checking procedure been established to insure accuracy in shipping? ___ ___
4. Has an appointment procedure for outbound carriers been established? ___ ___
5. Is there an enforcement plan for the appointment procedure? ___ ___
6. Is there a procedure to follow for including a manifest or load plan with the outbound shipment? ___ ___
7. Has it been determined if these documents will be sent with the freight or mailed separately? ___ ___
8. Will special procedures be developed for shipping of hazardous products, freezable merchandise, or other goods requiring special treatment in transit? ___ ___
9. Do all inventory procedures go as far as they could to prevent fraud or dishonesty in shipping? ___ ___

Materials Handling Operations
1. Will fork lift trucks or other mobile equipment be owned? ___ ___
2. Will they be leased? ___ ___
3. Are the specifications for this equipment appropriate for the new warehouse operation? ___ ___
4. Will aisle turning radius be adequate? ___ ___
5. Are lift heights adequate? ___ ___
6. Is each piece of mobile equipment properly identified, equipped with an hour meter, and covered by a thorough preventive maintenance procedure? ___ ___
7. If used equipment will be acquired, is all of it in perfect operating condition? ___ ___
8. Is there a training procedure to orient equipment operators on safety, preventive maintenance, and productive use of the equipment? ___ ___
9. Is storage equipment adequate to save space, allow proper control, and to maximize cube utilization in the new building? ___ ___
10. In storage equipment, has it been decided what mix of pallet rack, drive-in rack, drive-through rack, flow rack, self supporting pallet rack, or other materials will be used? ___ ___

STARTING-UP AND MOVING A WAREHOUSE OPERATION

(Start-up Checklist continued from previous page)   Yes   No

11. Has all rack been carefully installed and lag bolted to the warehouse floor?
12. Is there a training procedure against abuse of storage rack through careless handling?
13. In materials handling equipment, has it been decided what mix of pallet rack, drive-through rack, flow rack, self supporting pallet rack, or other materials will be used?
14. Have the safety hazards for each piece of mobile and storage equipment been determined and properly communicated in personnel training procedures?
15. Are records adequate to record the age, maintenance cost and maintenance cost per year for each piece of equipment used in the warehouse?

Use of Space
1. Will one person be designated as a space planner?
2. Has it been decided what space planning procedures will be followed?
3. Is there merchandise which must be controlled on a first-in/first-out basis?
4. Has a detailed layout for the warehouse been prepared and checked?
5. Has the layout been reconciled with existing fire regulations, floor load limits, and safety procedures?
6. Are there procedures to follow to enforce proper storage procedures, including maintenance of integrity of aisles, housekeeping, and stacking limitations?
7. Will procedures be used to minimize space losses through honeycombing?
8. Is the location, width and number of aisles adequate to allow effective storage and movement of all materials?
9. Is adequate space provided for staging of inbound and outbound freight?
10. Are storage pallets of uniform specification, in good repair, and in sufficient quantity to hold the planned inventory?
11. Is there a place to store surplus pallets?
12. Has it been determined if storage locations will be random, fixed, or a combination of the two?
13. Will a pick line be used for all or a portion of the inventory?

PRACTICAL HANDBOOK OF WAREHOUSING

(Start-up Checklist continued from previous page)   Yes   No

14. Do you know how many units of product can be stored in your new warehouse? ___ ___
15. Will you have a means of determining the percent of capacity occupied at any given time? ___ ___

Sanitation, Security and Safety

1. Has a detailed housekeeping procedure been prepared? ___ ___
2. Do you know which stockkeeping units, if any, are subject to inspection by the Food and Drug Administration (FDA)? ___ ___
3. If a portion of the product is FDA controlled, do you know FDA's requirements for safe storage of this product? ___ ___
4. Will one member of the warehouse staff be specifically assiged to sanitation maintenance? ___ ___
5. Will that individual also be responsible for safety and security? ___ ___
6. Has a professional sanitation and pest control service been retained? ___ ___
7. Have specific training procedures for sanitation maintenance been established? ___ ___
8. Will an independent sanitation inspection service be retained? ___ ___
9. Is there a specific safety and accident prevention training program? ___ ___
10. Is there a procedure for checking the percentage of lost time accidents to total hours worked and comparing the safety record with other operations? ___ ___
11. Have the safety hazards of all equipment been identified and made part of the training program? ___ ___
12. Have all appropriate safety devices been acquired and put into place with proper training for their use? ___ ___
13. Have any safety guards or other safety devices been deactivated for any reason? ___ ___
14. Have you anticipated all the ways in which equipment or tools might be used in an unsafe manner? ___ ___
15. Have refueling procedures been reviewed from the standpoint of safety? ___ ___
16. Have all warehouse operations procedures been reviewed from the standpoint of safety? ___ ___
17. Have all warehouse operations procedures been reviewed by insurance underwriting inspectors to check on fire safety? ___ ___

(Start-up Checklist continued from previous page)	Yes	No
18. Is a no smoking policy strictly enforced? If not, where are exceptions made?	___	___
19. Are hazardous materials segregated from other materials?	___	___
20. Have you determined who will be responsible for maintenance of loss prevention inspection reports and compliance with inspecting authorities?	___	___
21. Has it been decided what specific procedures will be established to maintain or improve plant security?	___	___
22. Will random unloading and reloading of outbound shipments be performed?	___	___
23. How frequently will random checks be made?	___	___
24. Will check weighing procedures be established to prevent deliberate overloading of outbound vehicles or under-receiving of inbounds?	___	___
25. Do you know how and when physical counting procedures will be used to improve security and control?	___	___
26. Are procedures planned to control pilferage?	___	___
27. Do training procedures adequately warn every employee of the consequences of pilferage or theft?	___	___
28. Will undercover procedures be used to detect dishonesty?	___	___
29. Are procedures planned to prevent unauthorized people from entering into parking lots, grounds, or other outdoor property adjacent to the warehouse facility?	___	___
30. Will lighting be used to discourage unauthorized entry?	___	___
31. Will alarms, watch services, or other procedures be used to detect unauthorized entry?	___	___

Relocating a warehouse operation is a significant challenge, but with sound planning and thorough communications, you can do it at minimum cost and disruption to customer service.

Planning the Opening of the Distribution Center

After establishing goals and considering costs, the warehouse user then should develop a detailed plan to open the center. It is almost impossible to carry out a detailed start-up plan without checklists or charts.

Table of Contents

Introduction
1. Warehousing Objective and Design Criteria
2. Layout and Product Location
3. Manning and Labor
    a. Picking and Replenishment Systems Overview
    b. Standards
    c. Single versus Double Shift Operation
4. Integrity Management
    a. Receipt and Put-away
    b. Finished Goods Cycle Counting Procedures
    c. Order Releasing
5. Job Descriptions
    a. Warehouse Operations Manager (Supv.)
    b. Shipping Clerk
    c. Shipper Clerk
    d. Shipper Group Leader
    e. Shipper (Warehouseman)
    f. Truck Driver
    g. Warehouse Temporary Help
    h. Mill Shipper
6. Glossary of Terms and Definitions

Figure 9-2

STARTING-UP AND MOVING A WAREHOUSE OPERATION

Writing a Warehouse Operating Manual

The best time to revise your operations manual is when the new warehouse is about to open.[3]

What should go into an operating manual? Exhibit 9-2 is the table of contents for an actual warehouse management operating man-

---

[3] From Volume 2, Number 9, Warehousing Forum, Ackerman Company, Columbus, Ohio.

Job Description

Shipper Group Leader

Reports to: Warehouse Operations Manager (Supervisor)

Subordinates Reporting: Shippers and warehouse personnel in the absence of the Supervisor

Overview: Perform all shipping and receiving activities at the dock with safe loading practices and proper packing, marking and documentation. May in the absence of the Supervisor, direct the other employees in their routine work. May be responsible for opening up and starting a daily operations or closing and securing the warehouse at the end of a shift.

Elements of the Job

Receiving all inbound materials including returned goods coming in over the shipping dock with proper documentation and disposition of the merchandise.

Shipping all outbound merchandise:
- ensure proper package markings
- ensure correct items and counts
- load for safe transit
- complete all outbound paperwork legibly
- keep area clean and orderly
- May pick orders or perform other warehouse activity
- daily inspection of lift trucks

Follow safety rules

Figure 9-3

ual. It of course was designed for one specific operation. Note that it starts with a statement of objectives.

"Integrity management" in this manual means accuracy—of locations, item quantities, and package markings. It depends upon careful checking of three critical procedures: receipt and put-away of inbound merchandise, cycle counting, and accurate order selection and shipping. Accuracy and control in all these areas is essential.

The job description section covers each specific work classification in the warehouse. Such descriptions must be handled carefully. Good warehouse operations typically demand great flexibility. The "not-my-job" attitude should never exist in a productive warehouse; job descriptions should refer to tasks which are typically done rather than to define the only job which an individual is allowed to do. Exhibit 9-3 shows a typical job description.

Your operating manual must consider seasonal factors. If you have dramatic swings in volume between peak shipping periods and slack periods, a prime function of your operating manual is to show how you can meet the seasonal peaks and avoid undue waste of space and labor during the slow season.

The manual must also anticipate probable product changes. Product families change as new items are added and others deleted: Will the systems controlling them need to be changed? How will these changes affect the way in which the orders are selected, or the storage layout? Which has priority—integrity of operations or labor efficiency?

Before you prepare an operating manual, think of all these questions, and then try to provide the answers in the operating manual. Here are a few questions which might be answered, and no doubt you can add to the list:

- How should we handle damaged merchandise discovered at the receiving dock?
- How should we handle damage discovered in the warehouse?
- How should we handle overages or shortages discovered in the warehouse?
- If there is a date coding system, how does the code work?

STARTING-UP AND MOVING A WAREHOUSE OPERATION

- If stock rotation is necessary, what should be done when older stock is not shipped through error?
- What procedure is followed if we cannot find an item which is shown on the book inventory?
- What do we do when we cannot find sufficient space to put away an inbound shipment?
- What are the ground rules for hiring new warehouse employees—and who is responsible?
- How will new employees be trained?

No outsider can define what an operations manual should be for *your* distribution center. Only *you* can set a priority for each of the questions above.

10

WAREHOUSE PERFORMANCE AUDITS

Many managers are asked to evaluate or compare the quality of service given by a group of branch warehouses, regardless of whether these are contracted-for or privately-operated. And occasionally branch warehouses should be audited to ensure that established procedures are being followed. Warehouse performance audits serve both purposes.

For example, one chain of warehouses divides its branch audit procedure into four sections: management, operations, administrative and financial.[1]

The management section includes a review to be certain that operations profit analysis is maintained for each product line carried in the warehouse. This ensures that costs are current and that handling-cost standards also are updated. Management appraisals are reviewed for accuracy and timeliness. Attendance and punctuality reports, as well as other personnel procedures are reviewed. Appraisals of office clerks, warehouse and maintenance workers are checked. Maintenance of records and graphs showing results in key areas are reviewed.

Under the operations section, responsibility for space analysis is checked. Proper marking of aisles and observation of aisle markings is inspected. Overall warehouse discipline is measured by checking for damage in product stacks, paperwork control of current damage,

---

[1] The following is taken from information provided by William G. Sheehan, President, Distribution Centers Inc., Westerville, Ohio.

maintenance of storage racks, neatness and proper stowage of empty pallet stacks, placement of trash receptacles, and segregation of hazardous materials or foods. Pallet exchange records, where maintained, are inspected. A general evaluation of warehouse housekeeping and cleanliness is made. Receiving tally logs, truck scheduling sheets, and rail track records are all inspected. Carrier damage records are checked.

Under equipment and building maintenance, an inspection is made to be certain that lift trucks and attachments are clean and in good order. The maintenance shop is reviewed, as well as the preventive maintenance inspections. The fire protection system is checked, particularly pump houses. Record keeping of fork lift repair costs is inspected.

The review of plant security includes a check for the last date that keys and lock combinations were changed. Maintenance and security of a master key box is checked. Security and distribution of bills of lading or other paperwork controlling stock movement is checked. Maintenance of the security access list for the electronic alarm is reviewed. Grounds lighting is checked to be sure that bulbs work.

Frequent sanitation inspections are important. If the service is performed by an outside pest control, inspection records are reviewed. Checks should be made for cleanliness and lack of infestation. Keep a log for rodent bait stations or mechanical traps. Lunch rooms, toilets and vending machine areas should be clean. Where needed, the sanitation stripe along the interior walls of the building is inspected. Control of weeds and bait stations outside the building, as well as trash pick-up should be reviewed.

Loss-prevention procedures should emphasize protection from fire. Is there a current and posted disaster plan? Are fire pallets and hose reels properly maintained? Fire drills should be held, with date of the last one noted.

Fire extinguishers must be checked periodically, with date of the last inspection noted. Safety and first aid are part of loss prevention, and periodic safety meetings should be held. First aid equipment should be checked and in the right place. Accident reporting forms should be reviewed, and maintenance of trailer chocks or other vehicle-securing devices inspected.

The last of the operations areas is training. New employees should have some formal training and further training should be available for employees who desire to upgrade their skills.

Duties of administrative personnel are reviewed separately. A bill of lading should have been prepared for every piece of merchandise that left the warehouse. Control of out-of-stock conditions or back orders is reviewed. Are timecards checked? Are monthly activity reports and other management reports audited? Are records for Worker's Compensation, insurance and unemployment claims verified? "Over, short and damage" reports should be reviewed.

The central point of audit for the financial function is billing. Where invoicing is done at the warehouse, the audit should determine that the invoicing procedure is accurate and timely. A review of accounts more than 90 days old is important. If purchase orders are made at a local level, a review of the whole purchasing procedure should ensure that property acquisition is handled correctly. Finally, payroll procedures should be thoroughly checked. Periodically, a physical inventory of company-owned property should be made and the local checking account reconciled.

Tracking Accuracy

A multi-city warehouse operator should try to compare the service performance of the different cities where warehouses are located. The most obvious way to track service performance is to count the number of reported errors. However, those errors should be compared to the volume of orders, since a warehouse which is doing very low volume should also have a low number of errors. A service rating is arrived at by dividing the number of errors by the total number of orders, but one should consider the order volume compared to a six-month rolling average. If order volume is above normal, the overload could be the source of difficulty. Tonnage throughput should also be considered. If the tonnage is above normal, this could create a different type of overload. An overcrowded warehouse frequently causes excess damage and shipping errors, and for this reason space utilization should be examined. By using a service performance report, the multi-city warehouse operator can track service rating and compare it with two measures of volume, throughput and order count. Space utilization must be considered not only in connection with service performance but as an indicator that the warehouse has become too full (or too empty).[2]

Account Profitability

For the third party warehouse operator, the most difficult cost tracking is that which measures profit contribution for each customer. This is done by establishing targets for earnings per manhour in the warehouse and earnings per-square-foot per-year. When total earnings for time and space exceed the targets, there is a contribution to profit which is recorded. In any warehouse, there is some time or space which cannot be allocated, and this is identified and tracked. Then that time and space which can be allocated to individual customers is separated, and the contribution is calculated for each

---

[2] This sub-chapter and the exhibits were developed by John T. Menzies for "Tracking Warehousing Costs and Performance," Warehousing Forum, Vol. 3, No. 8 copyright Ackerman Company, Columbus, Ohio.

Service Performance Report

Warehouse	Orders	% of 6 Mo. Avg.	Errors	Service Rating	Thruput	% of 5 Mo. Avg.	Space Utilz.	Space
Boston	1,870	105%	6	99.68%	67,521	101%	90%	78,000
Atlanta	1,746	85%	11	99.37%	105,879	87%	84%	105,000
Baltimore	1,571	98%	13	99.17%	95,072	102%	75%	98,000
Los Angeles	1,412	107%	21	98.51%	140,238	103%	85%	140,000
Chicago	1,268	97%	18	98.58%	117,816	99%	95%	125,000
Houston	900	95%	30	96.67%	130,965	89%	77%	127,000
TOTAL	8,767	96%	99	98.87%	657,491	100%	83%	673,000

Figure 10-1

customer. This allows the operator to know whether or not each customer is making a reasonable contribution to profit. If the unallocated portion is significant, then an improved method of assigning costs must be found. If a given customer is not producing a satisfactory contribution, an upward adjustment in warehousing charges should be negotiated.

The private warehouse operator could adapt this report to measure costs according to different products rather than different accounts. By measuring in this manner, the warehouse operator is able to measure not only overall results, but those which pertain to specific accounts. Note that there is always a small amount of space and labor which cannot be assigned to any particular account, and this is shown as unallocated.[3]

Qualitative Performance Ratings

Not all judgments on warehouse performance are quantitative. One organization does a quality evaluation of warehouse management. A warehouse operations evaluation survey asks the evaluator to rate management on a scale of 1 to 10, or any other mutually agreed-upon evaluation scale. The evaluation is based upon quantitative factors rather than management responses.

An unusual performance measurement is one in which the user asks the warehouse to rate *him*. Du Pont has tried to learn how it compares with other warehouse users. This is done through a questionnaire which asks the warehouse operator to compare Du Pont with other customers.[4]

Other Performance Guides

Another source of warehouse inspection guides is the Department of Defense, which has developed systems for evaluating warehouse contractors providing services to the government. One military inspection guide divides inspection features into two groups—struc-

---

3   Ibid.
4   Ibid.

PRACTICAL HANDBOOK OF WAREHOUSING

Account Performance and Contribution
$30.00 Hourly Target
$3.84 Square Foot Target per Year

Date	Square Feet	Hours	Storage	H$Earn	Total Revenue	$/Ft	$/Hr	Contribution	Contribution %	Contribution /Ft
Total*****	120,000	1,559	36,533	47,562	84,095	3.35	32.08	(1,085)	-1	-0.01
Total May-87	120,000	1,645	37,500	53,918	91,418	3.75	32.78	3,668	4	0.03
Total Apr-87	120,000	1,605	36,600	46,930	83,530	3.66	29.24	(3,020)	-4	-0.03
Total Mar-87	120,000	1,550	38,399	48,991	87,390	3.84	31.61	2,490	3	0.02
Unalloc*****	9,291	66	0	0	0	0.00	0.00	(4,943)	0	-0.53
Unalloc May-87	6,530	65	0	0	0	0.00	0.00	(4,040)	0	-0.62
Unalloc Apr-87	8,765	32	0	0	0	0.00	0.00	(3,765)	0	-0.43
Unalloc Mar-87	9,152	65	0	0	0	0.00	0.00	(4,879)	0	-0.53
Cust A*****	28,835	461	9,224	16,523	25,747	4.11	35.60	2,685	10	0.09
Cust A May-87	24,428	475	8,196	18,021	26,217	4.03	37.94	4,150	16	0.17
Cust A Apr-87	23,498	458	8,123	15,540	23,663	4.15	33.93	2,404	10	0.10
Cust A Mar-87	24,171	489	8,863	16,636	25,498	4.40	34.02	3,094	12	0.13
Cust B*****	20,390	122	5,422	3,415	8,836	3.19	28.43	(1,358)	-15	-0.07
Cust B May-87	22,466	128	5,355	3,812	9,167	2.86	29.78	(1,862)	-20	-0.08
Cust B Apr-87	21,160	117	5,601	3,470	9,071	3.18	29.66	(1,210)	-13	-0.06
Cust B Mar-87	19,941	108	5,302	3,558	8,860	3.19	32.94	(761)	-9	-0.04
Cust C*****	8,841	71	2,634	1,927	4,562	3.50	30.01	(382)	-8	-0.04
Cust C May-87	8,582	73	4,034	1,947	5,981	5.64	26.67	1,044	17	0.12
Cust C Apr-87	9,383	61	3,435	1,579	5,014	4.39	25.89	182	4	0.02
Cust C Mar-87	9,897	84	3,421	1,906	5,327	4.15	22.69	(360)	-7	-0.04

Figure 10-2

In the illustration shown above, Customer A has a healthy contribution to profit, Customer B is a loser, and Customer C has a mixed result. The lines with asterisks are a 6-month rolling average. This shows that Customer C performed considerably better in the last two months, which shows either an improvement in rates or productivity.

Questions Comparing Du Pont with Other Customers

1. Are Du Pont's service requirements higher, lower or about the same on:
   - Lead times
   - Administrative requirements
   - Information
   - Special Services
   - Package quality

2. How does Du Pont's communication of expectations compare to other customers?
   - Better
   - About the same
   - Not as good
   - If "Not as good," can you be specific as to functional areas?

3. How does our responsiveness to problem situations compare to other customers?
   - Better
   - About the same
   - Not as good
   - If "Not as good," can you be more specific?

4. How does our receptiveness to suggestions compare to other customers?
   - Better
   - About the same
   - Not as good

5. Do you have a more "partner-like" relationship with Du Pont or other customers?

6. How many other customers are striving for a "partnership" relationship with you?
   - None
   - Few
   - Many

7. Do you feel all of the Du Pont groups you interface with have a "real" interest in a partnership relationship with you? If no, would you care to be more specific?

Figure 10-3

tural and operational. The structural inspection includes the storage buildings, as well as their heating and protective systems and storage equipment. The operational inspection includes materials handling methods, handling of documentation, personnel, housekeeping and all other aspects relating to use of the warehouse facility.[5]

A major grocery products company uses a warehouse inspection and operational review highlighted by an annual award to that warehouse performing best under the review. The review has a section on clerical operations, another on warehousing operations, a third to check maintenance of code dates on inventories, and a separate checklist of sanitation requirements. Procedures currently followed are compared with those specified in the company's warehouse instruction manual. Quality and maintenance of pallets is observed, as well as prevention of contamination from strong odors. Results of recent physical inventories are noted and compared. This inspection review includes a detailed check of traffic operations, including compliance with current routings and maintenance of scheduled ship dates.[6]

The warehouse performance audit will be no better than the auditor who performs it. If the audit is not done frequently, its effectiveness will be diminished. Performance audits can do a great deal to ensure consistency of service in multiple locations. When used as a basis for recognition or reward, they can be a powerful motivator.

---

5   Taken from C22, DOD 4500.34-R.
6   Based on information provided by J. W. Snakard, Director of Distribution and Sales Service, Thomas J. Lipton Inc., Englewood Cliffs, N.J.

Part III

REAL ESTATE ASPECTS OF WAREHOUSING

11

FINDING THE RIGHT LOCATION

Site selection is an art as well as a science. Decisions usually involve weighing priorities, determining which features are critical and a process of elimination. Since every location has both advantages and disadvantages, the final selection of a site will likely involve some compromises.

Reasons for Site Seeking

First, consider why you are looking for a new warehouse site. There are four common reasons:

1. It is necessary to relocate an existing warehouse operation.

2. The business is expanding and must move inventory into a new market.

3. More warehouse space is needed to accommodate a growing inventory.

4. Contingency planning requires some decentralization of existing warehousing—in other words, there are too many eggs in one basket.

Depending on which of the above reasons is the motivator for seeking a new warehouse, the site search can take many different forms.

Location Theories

Location theory is a useful exercise in picking the spot for a new warehouse. Whether you accept or reject any of the theories that follow, at least consider them as helpful guidelines to the location decision.

First, move from a macro to a micro decision. We might first determine that the new warehouse must be somewhere within the continental U.S. Let's say that the facility should be in the southwestern region of the country, in the state of Texas, in the Dallas-Ft. Worth metropolitan area; and that it be in the southwest area of the city of Dallas, and finally on a specific site on Duncanville Road.

Over a century has passed since J. H. Von Thunen wrote about the advantages of growing the bulkiest and cheapest crops on the land closest to the city needing that commodity. The same principle can certainly be applied to all kinds of manufactured products—location of inventory is most critical for those that have the least value added. An inventory of gold or diamonds, for example, can be economically kept in one spot and moved by air freight all over the world, but the producer of bagged salt may find it necessary to place inventories close to the market.

The theory of postponement (*see Chapter 17*) also has an impact on site selection. Some products can be manufactured to a semi-finished state, or finished but not given the final brand or identification. Inventory of postponed goods can be held until specific customer orders are received, at which time the goods are finished or branded to the customer's specifications. The postponement warehouse then becomes an assembly center, as well as a distribution center. The need for these assembly capabilities could affect warehouse location.

Outside Advisers

It is difficult to pick a spot for a new warehouse without eventually turning to people outside your organization for advice. For this reason, we should consider some commonly-used sources of such advice.

Real estate brokers are often involved in any search for new property. Remember that the broker's compensation for this activity is based on successful completion of a transaction, so the broker is motivated to complete a commissionable sale. When the value of the broker's time invested in the search begins to approach the value of the commission, the broker would be foolish not to abandon it.

Not all brokers have multiple-listing services. Furthermore, commission arrangements that may not be known to the prospective buyer could make it difficult for the broker to maintain objectivity as a source of outside advice. Not all real estate brokers are biased, of course, but the user should be aware that the commission system for brokerage tends to create pressures that could affect the results of the site search.

If existing warehouse space is sought, a warehouse sales representative could be useful. The public warehouse industry has several national marketing chains, each with representation in the Chicago and New York metro areas. Typically, the chain sales representative is not paid a sales commission, though a few may work on a commission as well as a salary basis. Since the chain sales agent normally cannot represent competing warehouses, each chain has only one outlet in each major city.

Major railroads are a source of advice on sites for new buildings, as well as existing warehouses. Railroad people do not work on a commission basis, but they naturally will want to have the new warehouse located on their line rather than a competing one. If railroad competition is not a factor in the decision, the railroad representative can be a reliable source of advice. If only two railroads are under consideration, the use of two competing railroad development agents may be a valuable exercise.

The area development departments of electric and gas utilities also are an excellent source of outside advice, once the general area for the facility has been determined. With a few notable exceptions, there is usually just one electric or gas utility to service a city and its suburbs. Therefore, the representatives of that utility should be unbiased in recommending sites within the entire metro area. Utility people try to work with all local real estate brokers to become a clearinghouse of information on available buildings and land within their service area.

State and local government development agencies, as well as

local chambers of commerce, are also reliable information sources, once the choice of community has been made. Some large metro areas have several chambers or government development officers. Since these people are motivated to attract new industry into their own area, make sure their area is where you really want to be before relying on this source of advice.

Finally, the use of a management consultant may provide an unbiased source of advice. In selecting a consultant, one important consideration is to be sure that the individual is truly independent and objective in approaching the site search. Some individuals who use the term "consultant" also are involved in other activities, such as real estate brokerage, which may affect their objectivity in handling the site search.

There are three reasons to bring in an outside consultant: The most important of these is objectivity. Secondly, the consultant may have extensive past experience in site seeking, and this experience will save time. Finally, the consultant represents a source of executive time with no long-term commitments involved. A detailed site search takes many hours of time, and hiring new executives to handle it would be foolish if their job will end once the site has been selected.

Summing up, the most objective sources of outside advice are either a public utility or a consultant. Whatever source you use, be sure that the outside adviser is truly independent and objective.

Requirements Definition

Your search for a location should be modified and directed by your specific requirements as a warehouse user. For a successful site search, you need to define these requirements carefully.

Since transportation goes with warehousing like ham goes with eggs, the transportation requirements for the new warehouse are of utmost importance. Which modes of transportation are required? All warehouses need trucks and some depend heavily on rail, water, air, or a successful combination of all three. While delivery zones are less important in these deregulated times, the user also must consider zones and transportation costs to specific sites.

FINDING THE RIGHT LOCATION

To the extent that a warehouse is a marketing tool, it is important to consider the relationship of the proposed site to the locations of the planned users. Beyond just looking at locations, you should consider transportation costs and order cycle times to best serve various potential users. Remember that the warehouse user should not really care *where* the inventory is held, as long as product can be delivered in a timely fashion and at a reasonable price.

The labor market, probably less important for most warehouses than for the typical manufacturing plant, depends on the nature of the warehousing operation. If the proposed operation has a high degree of automation and relatively little touch-labor, the availability of well-motivated workers may be of minor importance. On the other hand, labor can be of critical importance in other warehousing operations.

There can be wide variations in taxes, particularly inventory taxes, within some metropolitan areas. Taxes can be a critical competitive factor when the warehouse inventory has high value.

Community attitudes toward the new warehouse should be carefully studied by every site seeker. In most communities today, a clean and quiet warehouse development is considered preferable to the many manufacturing operations that cause pollution, congestion or other conditions perceived as detrimental to the quality of life of a community. Yet, some communities are opposed to *any* new industrial development, even warehousing. If such opposition exists, it's best to recognize it and look elsewhere.

A requirement list should certainly consider whether a new building is necessary, or if an older one could be adapted for this use. Rehabilitating older buildings is described in the next chapter, but such sound, older buildings are not easily found, particularly in newer communities.

Financial questions are an important part of defining requirements. How important is the availability of mortgage financing at favorable rates? The question of whether to own or lease the warehouse facility also is covered in the next chapter.

If space needs are not fixed, the availability of overflow space either in public warehouses or short-term lease space is an important

Site Analysis Checklist

General Information
1. Site location (city, county, state):
2. Legal description of site:
3. Total Acreage: Approximate cost per acre:
   Approximate dimensions of site:
4. Owner(s) of site (give names and addresses):

Zoning
1. Current: Proposed: Master plan: Anticipated:
   Is proposed use allowed? __ yes __ no
2. Check which, if any, is required:
   __ rezoning __ variance __ special exception
   Indicate: Approximate cost   Time required:
   Probability of success:
   __ __ excellent __ good __ fair __ poor
3. Applicable zoning regulations (attach copy):
   Parking/loading regulations: Open space requirements:
   Office portion: Maximum building allowed:
   Whse./DC portion: Percent of lot occupancy allowed
   Other: Setbacks, if required:
   Height restrictions: Noise limits: Odor limits:
   Are neighboring uses compatible with proposed use?
   __ yes   __ no
4. Can a clear title be secured? __ yes __ no
   Describe easements, protective covenants, or mineral rights, if any:

Topography
1. Grade of slope: Lowest elevation: Highest elevation:
2. Is site: __ level __ mostly level __ uneven __ steep
3. Drainage __ excellent __ good __ fair __ poor
   Is regrading necessary? __ yes __ no
   cost of regrading:
4. Are there any __ streams __ brooks __ ditches __ lakes
   __ ponds __ marshlands __ on site __ bordering site
   __ adjacent to site?
   Are there seasonal variations? __ yes __ no
5. What is the 100-year flood plan?
6. Is any part of site subject to flooding? __ yes __ no
7. What is the ground water table?
8. Describe surface soil:
9. Does site have any fill? __ yes __ no

Figure 11-1

FINDING THE RIGHT LOCATION

10. Soil percolation rate __ excellent __ good __ fair __ poor
11. Load-bearing capacity of soil: __ PSF
12. How much of site is wooded: How much to be cleared?
    Restrictions on tree removal: cost of clearing site:

Existing Improvements
1. Describe existing improvements.
2. Indicate whether to be __ left as is __ remodeled
   __ renovated __ moved __ demolished
   Cost $ __

Landscaping Requirements
1. Describe landscaping requirements for building parking lots, access road, loading zones, and buffer if necessary.

Access to Site
1. Describe existing highways and access roads, including distance to site. (Include height and weight limits of bridges and tunnels, if any).
2. Is site visible from highway? __ yes __ no
3. Describe access including distance from site to a) interstate highways b) major local roads c) central business district d) rail e) water f) airport. Describe availability of public transport.
4. Will an access road be built? __ yes __ no
   If yes, who will build?   Who will maintain?   Cost?
   Indicate a) curb cuts   b) median cuts   c) traffic signals
   d) turn limitations
5. Is rail extended to site? __ yes __ no Name of railroad(s):
   If not, how far?   Cost of extension to site:
   Who will maintain?   Is abandonment anticipated? __ yes __ no

Storm Drainage
1. Location and size of existing storm sewers:
2. Is connection to them possible? __ yes __ no
   Tap charges:
3. Where can storm waters be discharged?
4. Where can roof drainage be discharged?
5. Describe anticipated or possible long-range plans for permanent disposal of storm waters, including projected cost to company.

Sanitary Sewage
1. Is public treatment available? __ yes __ no
   If not, what are the alternatives?
2. Is sanitary sewage to site? __ yes __ no
   Location of sewer mains?
3. Cost of materials (from building to main)—include surface restoration if necessary:

PRACTICAL HANDBOOK OF WAREHOUSING

4. Tap charges:
5. Special requirements (describe fully):
6. Describe possible or anticipated long-range plans for permanent disposal of sewage, including projected cost to company.

Water
1. Is there a water line to site? __ yes __ no
2. Location of main: Size of main:
3. Water pressure: Pressure variation:
4. Hardness of water:
5. Source of water supply: Is supply adequate? __ yes __ no
6. Capacity of water plant: Peak demand:
7. Who furnishes water meters?   Is master meter required?
   __ yes   __ no
   Preferred location of meters:
   __ outside   __ inside
8. Are fire hydrants metered? __ yes __ no
   If yes, who pays for meter installation?
9. Attach copy of water rates, including sample bill for anticipated demand if possible.

Sprinklers
1. What type of sprinkler system does code permit?
2. Is there sufficient water pressure for sprinkler system?
   __ yes __ no
3. Is water for sprinkler system metered? __ yes __ no
4. Is separate water supply required for sprinkler system?
   __ yes __ no
5. Where can sprinkler drainage be discharged?

Electric Power
1. Is adequate electric power available to site? __ yes __ no
   Capacity available at site:
2. Describe high voltage lines at site:
3. Type of service available:
4. Service is __ underground __ overhead
5. Reliability of system __ excellent __ good __ fair __ poor
6. Metering is __ indoor __ outdoor
7. Is submetering permitted? __ yes __ no
9. Indicate if reduced rates are available for a) heat pumps __ yes __ no b) electric heating __ yes __ no
10. Attach copy of rates, including sample bill for anticipated demand, if possible.

Fuel
Gas:
1. Type of gas available:
2. Capacity   Present: Planned:

FINDING THE RIGHT LOCATION

  3. Peak Demand   Present: Projected:
  4. Location of existing gas lines in relation to site:
  5. Pressure of gas:
  6. Metering is __ indoor __ outdoor
  7. Is submetering permitted? __ yes __ no
  8. Is meter recess required?
  9. Who furnishes gas meters?
10. Indicate limitations, if any, on new installation capacity requirements.:
11. Attach copy of rates, including sample bill for anticipated demand, if possible.

Coal
1. Source of supply: Reserves:
2. Quality of coal available: Cost (per million BTU) delivered:
3. Method of delivery:

Oil:
1. Source of supply: Volume available:
2. Quality of oil available: Cost (per million BTU) delivered:
3. Method of delivery:

Taxes
1. Date of most recent appraisal:
2. Real estate tax rate history, last five years:
3. History of tax assessments, last five years:
4. Proposed increases, assessments and tax rates:
5. Are any abatement programs in effect? __ yes __ no   If yes, describe.
6. Is site in an Enterprise Zone? __ yes __ no
    Duty free zone? __ yes __ no
7. Have any special taxes been assessed? __ yes __ no
    If yes, describe:
8. Indicate anticipated or possible major public improvements:
9. Services provided for taxes paid: Local: County:
    State:
10. What is state policy on inventory tax, floor tax, etc.?
    Is it a free port state? __ yes __ no   If no, describe assessment dates, procedures and tax rates.
11. Indicate rates for: Personal income tax:
    Corporate income tax: Payroll tax:
    Unemployment compensation: Personal property tax:
    Sales and use tax: Workmen's compensation:
    Franchise tax: Other taxes:
12. Indicate taxation trends:
13. Are industrial revenue bonds available? __ yes __ no

This figure is reprinted with permission from Warehousing and Physical Distribution Productivity Report, © 1982, Marketing Publications, Inc., Silver Spring, MD.

consideration. It is seldom possible to construct or purchase a building that is the right size for year-round space needs. Therefore, the warehouse user should acquire a relatively small permanent space, utilizing overflow space within the community for seasonal requirements.

Comparisons From Published Sources

Certain published statistics may help in obtaining regional comparisons of trucking costs, and quality and cost of labor. The Highway Users Foundation in Washington D.C. publishes reports which provide comparisons of state motor fuel tax-rates, and size, weight and access restrictions for motor vehicles. If trucking productivity is a critical factor in the site decision, these statistics for the 50 states should be carefully compared. Grant Thornton, an accounting and management consulting firm, publishes "General Manufacturing Climates," a study which rates the 50 states according to various economic factors. Three which could affect the location decision are average wages paid in each state, percentage and percentage change of union workforce, and two measures of labor productivity. The labor productivity measures consider the percent of available time lost because of work stoppages, and the average value added to manufacturing production for each dollar invested in payroll. The results show that differences among the 50 states are substantial.

The Final Selection Process

With all requirements defined, the user should now move through the selection process.

As sources of information are checked in the move from "macro" region to "micro" street location, be very skeptical. If one party advises that a specific site has never had a flood history, for example, get a second opinion from another party who cannot possibly be influenced by the earlier informant. If multi-source checking produces the same answers to critical questions, one can then have some confidence in the information.

As you move through the selection process, and particularly as you zero in on a specific site, it is essential to have a contingency plan.

FINDING THE RIGHT LOCATION

If site A is the preferred one, it is wise to have a site B that is almost as good and equally available. It is well to let the seller discover that there is a fallback site in the event that bargaining for site A should break down.

Because site selection is one of the most important decisions the average distribution executive ever makes, the process of making the decision cannot be discussed too often. Other distribution decisions are correctable, but a poor choice for a warehouse site is a decision that, while not irreversible, is very costly to correct.

The checklist on pages 128 through 131 can be a useful aid in analyzing a location decision.

12

BUILDING, BUYING, LEASING, OR REHABILITATING?

When establishing a distribution center, the choice is whether to build a new structure or to purchase an existing one. Secondly, should the warehouse operator own or lease the property? Third, how should the user finance the structure? Finally, if a new building is to be built, what size and shape should it have?

Acquisition

The question of whether to build or buy is easily decided if the operator determines that only one location will work for him, and there are not satisfactory buildings in existence at that location. Clearly he will build. On the other hand, the presence of a "bargain" building may determine the location for some operators. In considering acquisition, the operator must compare the design characteristics needed with those in the available building. Those design parameters in themselves may dictate new construction if no existing facility fits the specifications. When a distressed local real estate market exists, it may be possible to purchase a nearly new building for a price below the cost of reproducing it. On the other hand, in some communities, suitable construction sites are in short supply, and the market for real estate may force purchase of an existing facility because a new one cannot be built at a reasonable price.

Financial considerations frequently affect the decision to build or buy. Sometimes it is easier to obtain money for a new building than for an old one. However, there are more times when an older building can be purchased with the assumption of a mortgage at a very favorable interest rate. Indeed, attractive financing is often a prime reason for buying an existing property.

Finally, the obsolescence factor itself must be considered. One characteristic that all used buildings have in common is that their original owners no longer want them. The buyer should carefully examine an existing property to determine whether its location, design or construction represent potential problems for the future owner.

A Worksheet for the Buy/Lease Decision[1]

A company normally acquires a property by buying or leasing it. The purchase may be with cash or through mortgage financing. With a cash purchase, the company avoids financing charges. However, the heavy cash outlay of an outright purchase may have a serious negative impact on working capital. But mortgage financing reduces working capital because of the cash down payment. In addition, the mortgage debt will probably reduce the company's borrowing ability for future opportunities. Sometimes it is possible to acquire the property with 100 percent financing and no down payment. With either cash or mortgage financing, the buyer has ownership of the property, receives full benefit of the depreciation and, if applicable, the investment credit.

Leasing. Under the simplest form of lease, the company could direct the construction of the building and enter into an agreement with the owner of the property, who could be the contractor or a third party, to lease it. In this situation, the company would deduct its lease payments in lieu of depreciation and would not own the property at the end of the lease term. There are several alternatives under the leasing method. And with the Economic Recovery Tax Act of 1981 and Reagan Tax Plan of 1984, the alternatives have increased and deserve the attention of both parties to a lease agreement.

---

1 This entire subchapter was written and revised by David P. Lauer, partner-in-charge, Deloitte, Haskins & Sells, Columbus, OH.

BUYING, BUILDING, LEASING, OR REHABILITATING

The basic concept used in the worksheet below is the "present value" approach. In financial transactions involving money at different dates, every sum of money mentioned should be thought of with an attached date. That is, the mathematics of investment deals with dated values. On the time scale, A represents an initial date, called the "present," and Z is the date at the end of T years and called "future." We place $P (present dollars) at A and $F (future dollars) at Z. $P and $F are related through time and have equivalent values. The company would be in the same financial position with $P or $F—it would just be in that position at a different time. $P is the present value of the future amount of $F.

This computation assumes a certain rate for the time-value of the money. In this example, assume that the interest rate will be 10 percent, $P will be $1,000 and T years will be 1. Based on these assumptions, F would be $1,100. Assuming that F is $1,100 and all other assumptions remain the same, $P is computed to be $1,000. Therefore, the company should be equally satisfied with $P at time A as it would be with $F at time Z.

The present value concept in a buy/lease decision enables the warehouse user to assign a value in today's dollars to cash flows at a given point in time. The worksheet shown below is primarily for a decision on a warehouse building, but with a few changes it could be used for equipment or other property.

In addition to the financial consideration in the buy/lease decision, other factors must be considered, which include, but are not limited to, the following:

1. Present location *versus* future desired location.
2. Change in business operation that may require a different type of structure.
3. Innovations in heating that cannot be placed in an older building.
4. Obsolescence of product.
5. Effect on the asset value of the business.

Buy?

Positive cash flows

Present value of the interest
  deducted after tax $ _____

Present value of the depreciation
  deduction after tax   _____

Present value of the land at end
  of lease   _____

Present value of the building at
  end of lease   _____

Negative cash flows

Cash down payment   _____

Present value of the future
  mortgage payments   _____

                                      Total $ _____

Lease?

Positive cash flows

Present value of the lease
  payment deduction after tax   _____

Present value of the down
  payment   _____

Negative cash flows

Present value of the future
  lease payments   _____

                                      Total $ _____

Figure 12-1
The worksheet above provides a financial point-of-view in measuring the advantages of buying or leasing.

BUYING, BUILDING, LEASING, OR REHABILITATING

Assumptions

The Buyer
1. The price of a 50,000 square-foot building and land is $1,500,000: $1,300,000 for the building and $200,000 for land.
2. The buyer obtains a $1,500,000 fixed-rate mortgage for 30 years at 14 percent. Accordingly, 100 percent financing is obtained.
3. The buyer will depreciate the building cost of $1,300,000 on a straight-line basis over 18 years.

The Lessee
1. The building and land are leased at a cost of $5.50 per square-foot for a total of $275,000 per-year.

General
1. The maintenance, taxes, etc., are paid by the buyer and the lessee.
2. The company has an effective federal tax rate of 46 percent.
3. No consideration is given to state and local taxes.
4. Ten percent is used for discounting to compute the various present values.
5. The value of the land at the end of 30 years is $200,000.
6. The value of the building at the end of 30 years is $650,000, or half its original cost.
7. The computations are made on an annual basis.

The present value of the outflow is less under the "buy" alternative than the "lease" alternative for the given assumptions. Shown below is the way in which the buy/lease decision works out under the assumptions listed above:

BUY?
Positive flows

Present value of the interest
deduction after tax 639,000
Present value of the depreciation
value after tax 272,000
Present value of the land at end
of 30 years 11,500

Present value of the building at
end of 30 years 34,500

Negative flows
Cash down payment .... N/A
Present value of the future
mortgage payments .... -2,019,000

Total ....$-1,062,000

LEASE?
Positive flows
Present value of lease
payment deduction after tax 1,193,000
Present value of the down
payment .... N/A

Negative flows
Present value of the future
lease payments .... -2,592,000

Total ....$-1,399,000

The Rehabilitation Alternative

The preservation, conservation and adaptive re-use of older buildings has gained increasing attention during the past decade. Some of this interest was spurred by tax legislation which provided significant tax credits for rehabilitating both historic structures and buildings of some age.

Between 1978 and 1986, such tax credits provided a meaningful incentive for the warehouse user who would recondition an older building rather than construct a new one. By 1986, changes in the tax law had reduced the credit for non-historic structures to only 10%, and this applies only to buildings placed in service before 1936. In effect, the rehabilitation tax credit no longer makes it worth-while for the warehouse operator to rejuvenate an old building. This does not mean that rehabilitation may not sometimes have economic incentives regardless of taxation. Sometimes the best location for future growth is an available building in the center of the city, and it may be more economical to rehabilitate that structure than to build a new one. At times, the community in which an older building is located may offer tax abatement, favorable financing, or other financial incentives to persuade the user to purchase an older building. Obviously these incentives must be balanced against the cost of rejuvenating the building and the cost of operating that building *versus* operations costs for a new structure.

Rejuvenating Warehouse Roofs

The two most critical construction components in a warehouse[2] building are the floors and the roof. By comparison, walls are relatively easy to change or fix. A roof or a floor with a serious problem can be most expensive to correct.

The great majority of roofs in industrial and warehouse buildings in the United States are referred to as "built-up roofs" because they consist of a series of layers of felt material which are bonded and saturated with hot asphalt coating. On some roofs, a final coating of tar and gravel is applied, with the layer of gravel providing protection and insulation. Heat, moisture and ultra-violet rays cause the surface to deteriorate and wear out. The wearing process can be aggravated by the pressures of wind storms and occasional roof traffic. Ponding water and thermal shock can also cause built-up roofs to deteriorate.

One of the tools used for assessing the condition of built-up or composition roofs is a thermographic roof analysis, which may include field inspection, infrared scan, core cuts to determine the type and thickness of roofing materials, and an engineering report. This type of study can indicate if there are sections of the roof that should

---

2 This subchapter is based on material provided by the Ruscilli Construction Company, Columbus, OH. Most of this originally appeared in Volume 19, Number 7, *Warehousing and Physical Distribution Productivity Report*, Marketing Publications, Inc., Silver Spring, MD.

be completely removed prior to re-roofing, or whether you can safely bridge these areas and save money. The price for such an analysis varies from $.01 to $.05 per square foot, depending on the size of building and number of roof penetrations and levels.

Built-up roofs can be repaired in a variety of ways. A coating of either hot or cold asphalt material will seal spot leaks. New plies of roofing material can be installed over the top of the older roof, but if the surface is in very poor condition it must be removed before new material is applied.

A newer alternative for roofing problems is the single-ply roof. A single ply-system is similar to a giant rubber sheet, and in some situations it is the best way to stop a persistent and hidden leak. The single-ply system is lighter and simpler than installing additional plies of built-up roof. The single-ply systems can be installed either with or without ballast. It is advisable, especially with older buildings, to check with a professional engineer to see if your structure can support the additional weight of a ballasted system.

Some roofing problems are corrected by installing a foam roof. This is a plastic compound which foams as it is applied and creates a surface which provides additional insulation as well as protection from leaks. However, there are questions regarding the long-term life of a foam installation; in addition, foam material is susceptible to damage from roof traffic.

A more recent solution is to install a metal roof system over an existing roof. This may involve constructing a new ridge and eaves over the flat built-up roof, utilizing a sub-framing system to support the metal roof. Metal roof systems require some slope, and in fact this slope, which sheds water, contributes to the longer life of the metal roof. One manufacturer offers a panel roof system with a 20-year warranty against rupture, structural failure or perforation and has been supplying a metal roof system for new construction for over 50 years. This same panel system has now been adapted to re-roofing. The newer systems use flat metal panels with a vertical, standing seam rib. The standing seam between the panels is about two inches above the roof's weather surface which provides protection from leaks. The clips which hold panels of the roof in place allow for thermal expansion and contraction while keeping the roof tightly locked together.

Typically, new insulation can be installed as part of the roof system in order to help minimize heat loss and increase the overall value of the building. And the new metal roof goes right over the old roof, so there is minimal disruption of warehouse activity.

Checklist of Roof Maintenance

Metal roof
- Check all flashings and fasteners at least twice a year. Look for areas which are no longer watertight and may be rusting.
- Check the deck itself, looking for any rusting; color changes are a key.
- If rusting is found the area can be coated with an appropriate sealer or protective agent.

Built up roof
- Check flashings. Check for areas where they are no longer water-tight, and look for areas where they have split open.
- Recoat such areas with a polymer modified asphalt designed to recoat such surfaces. It is especially good for vertical surfaces because it will not slide off.
- Also check surface of roof itself. Look for puncture damage. Roof may need a top-coating from time to time to cover up punctures, especially if the felt on the roof is showing.
- Check also areas where drains are located to make sure they are air-tight and are not leaking. If they do leak, emulsion may solve the problem.

Single-ply roof
- Very little maintenance is needed for these roofs; they either work or not. But if coating is desired, do it when the roof is brand new.
- Also, as with all the other roofs, always check the flashings to see that they are not pulling or punctured.

Figure 12-2

Rejuvenating Warehouse Floors[3]

In some ways, a warehouse floor is like a highway, but forklift trucks have relatively small solid tires; the front tires of a counterbalanced forklift truck are carrying the weight of the load as well as a portion of the truck's weight. They thus load the floor more heavily than any highway and are heavier than most storage loads. The loaded legs which support storage rack also concentrate a heavy weight on a small area of the floor. For these reasons, floor composition of a warehouse is most important.

Many subsoils offer particular challenges. In some regions of the country, expansive soils are found, particularly clays and certain slags. An expansive soil swells when it gets wet and shrinks when it dries. The result is a shifting which can cause great damage to the floor above the soil. At times, uncontrolled expansive soils have virtually ruined a warehouse floor. Expansive soils are best stabilized by being treated with lime, which soaks up any water in the soil and creates a hard surface. Once the expansive soil is sealed, it cannot get wet again to expand and contract.

One way to correct floor problems is by putting a floor over a floor, creating a second concrete surface over the original floor, using a bonded overlay, a partially bonded overlay, or an unbonded overlay.

A bonded overlay is best used on a floor which has excellent subsoil but has been badly worn, for example in the aisle areas of a busy grocery warehouse where thousands of poundings by lift trucks have simply weathered and worn the floor. The floor is first cleaned and sandblasted to ensure that there is a perfect surface to allow bonding of the two layers of concrete, then covered with approximately two inches of concrete.

The partially-bonded floor is probably the least expensive means of getting a new surface. No effort is made to clean the old floor; a new surface is just put over it. Cracks in the old surface are likely to

---

[3] John Dixon of Ohio Ready Mixed Concrete Association provided substantial assistance in preparing this subchapter. Most of this originally appeared in Volume 19, Number 18, *Warehousing and Physical Distribution Productivity Report,* Marketing Publications, Inc., Silver Spring, MD.

migrate up to the new one. However, if there are no cracks and an inexpensive job is needed, this is a way to do it.

Probably the best method of treating a floor which is in very poor condition is the unbonded overlay. An unbonded overlay consists of an application of either asphalt slurry or two sheets of polyethylene as a buffer between the old concrete and the new. After the buffer is applied, between three-and-one-half and five inches of new concrete are put on top. The buffer keeps cracks from moving up to the new surface.

In cases of extreme deterioration, some preparation must be done before the unbonded overlay is applied. Primarily this would involve finding and filling voids in the floor. In a poorly-prepared subsoil, sections of concrete slab may have no soil directly beneath them. These voids can be discovered by tapping the floor with a metal rod or a small hammer, listening for hollow sounds. A mud pump should be used to fill any such voids.

The unbonded overlay can be successfully used for all but the most severe expansive soil deterioration as well as many other kinds of deterioration caused by poor subsoil. Before the overlay is applied, the subsoil condition should be corrected to the extent possible. In the case of expansive soil, no repair attempt should be made until the soil has been stabilized by time—until the floor over the soil is about two years old.

Less serious floor problems can be dealt with more simply. Concrete floors commonly develop joints or cracks which tend to widen or spall as they are repeatedly hit by lift truck wheels. Such conditions can be stabilized and corrected by filling these joints and cracks with a flexible sealant, a flexible epoxy or an elastomeric compound. Use a filler which is flexible rather than rigid, otherwise the movement between the two slabs on either side of the crack will quickly destroy the filler.

Spalling is the dislodging of pieces of floor at the edges of the joint, and it happens through the normal wear and tear of warehouse work. Where much material has been removed from the floor by spalling, the entire cavity needs to be patched. After the patching is complete, a saw cut is made to restore the joint. The joint in the patch must be at least as wide as the old opening and at least as deep, otherwise spalling will occur again.

Sometimes a carelessly poured-floor contains sections thicker than necessary or thinner than needed to support traffic and loading. In extreme conditions, the unbonded overlay may be the only way to save such a floor. In other cases, the floor can be fixed simply by tearing out those abnormally thin portions and replacing them after removing enough subsoil to create a slab of normal thickness. You can often replace small sections of a floor without great disruption by removing the defective areas with a chipping hammer.

13

CONSTRUCTION

Warehousing managers are often asked, or at least they should be asked, to offer advice on any project to construct a new warehouse. Both costs and techniques in construction change rapidly, but the opportunities that may be taken, or the pitfalls that can be avoided, are relatively constant. This chapter assumes that the user has already selected a construction site, but warehouse management has been asked to advise on construction of a building at a given location.[1]

Perhaps your first role in this project should be to question the development cost of the selected site. The developed cost of a site can vary significantly, depending on the condition of the piece of land. For example, if you own fully developed industrial land at a value of $60,000 per acre, and the land and zoning will support 45 percent building coverage, your land cost per square foot of attainable building is only a little over $3.00.

On the other hand, if you own undeveloped land at $39,000 per acre zoned for building coverage of only 30 percent, and you estimate the cost of completing the development as another $41,000 per acre, then the land cost per square-foot of attainable building size is more than $6.00. Therefore, in comparing your site with others that may be available, it is important to look at the total development cost and

---

[1] This chapter is based on an article by W. S. Ammons, University Mall Properties, Pensacola, FL. The article was published in Vol. 14, No. 7, of *Warehousing and Physical Distribution Productivity Report*, ©Marketing Publications, Inc., Silver Spring, MD. The article was revised with substantial help from L. Jack Ruscilli, Ruscilli Construction Co., Columbus, OH.

the attainable or desirable building coverage. The apparently inexpensive site may be more costly in the long run.

Test Borings

Soil tests to determine the load-bearing capacity of the site are absolutely essential. On some sites, the soil tests may determine the location of the building on the piece of land. Load-bearing capacities may easily vary from 1,000 pounds per-square foot up to 6,000 pounds per square-foot. No warehouse floor is any better than the quality of the land beneath it. Independent soil testing laboratories are found in all major cities.

Government

Municipal zoning and permits should be considered very early in the selection process. Environmental impact studies are needed in some states. In one community, for example, local government agencies require a minimum of three to six months to approve a site development plan. They will not permit the developer to apply for a building permit until after the site development plan has been approved. The building permit process then takes additional months, and the whole process takes up to a year. In looking at zoning and permits, perhaps your first question should be how much lead time is required to obtain necessary permits.

Storm Water Management

One site planning aspect for which local governments have been increasingly alert is how to deal with flash floods during sudden heavy rains. Many municipalities require that developers show considerable capability for handling storm water. There are two generally accepted ways of doing this. One is to excavate a retention pond to hold a temporary increase in water. The second is to put in a recessed parking lot that can be flooded to a depth of six inches to provide temporary water storage capacity. If properly designed, the recessed parking lot may inconvenience people parking in it, but will not flood to a depth that damages the parked cars. A typical storm water policy is that the

CONSTRUCTION

site shall not drain more water on neighboring land than was deposited before the new structure was built.

Utilities

Planning for all kinds of utilities must take place in the early stages of construction. The warehousing manager should be familiar with requirements for electricity and water for fire protection, electricity or gas for heat, and for telephones.

Layout Design

Every warehousing professional questions the advice of architects and engineers. Clearly, no building can be built without them, but the best source of advice on warehouse layout comes from warehousing people themselves.

Before the project is started, you should design your own layout, based on your knowledge of the necessary stack heights, aisle widths, staging areas, bulk storage, order-picking lines, etc. Then have an architect or engineer fit the building to your layout. Be sure you are apprised of the construction cost options, such as possible savings from using standard construction modules or varying your layout to permit more economical building practices.

By starting with your own preferred layout, and receiving feedback from the construction professional, you will find the tradeoffs between the most economical building from a construction standpoint and the most efficient building from a warehousing standpoint.

Since you are the one who must make the operation run efficiently after the building is built, remember the construction cost is a one-time expense, but the cost of operating the building is likely to increase each year.

As you develop your plans and specifications, it is important to understand the relationship of construction costs to layout. From a construction standpoint, the lowest-cost building is square, since a perfect square provides the maximum number of square feet per linear-foot of wall. Yet a square building might be inefficient from an operating standpoint, as it may have unreasonably long travel runs in the distance between rail dock and truck dock or dock areas and storage areas.

Bay size (space between upright columns) is a factor in reconciling warehouse layout with construction costs. For most structures, the most economical bay size is 25 x 50 feet or 30 x 30 feet. A 40 x 40-foot bay will require a relatively small premium. In one situation, a 40 x 50-foot bay proved to be more economical than a 40 by 40-foot bay. However, construction costs will escalate rapidly when bay sizes exceed 50 x 50 feet, particularly if the roof must bear heavy snowloads.

The Contract

Some construction firms now offer a contract where they take the responsibility for the design as well as the construction of the facility. These are known as "design-and-build" contractors. Such construction offers the advantage of expediting the total construction process, as the builder is working with a design used before.

The greatest disadvantage to the buyer is the difficulty in comparing competitive bids from different design-and-build contractors, since each proposes different details in construction.

One way to compare design-and-build plans is to provide a set of outline performance specifications to the bidding contractors, as well as an operating layout for the building. Ask the contractors to provide certain breakdowns.

For example, one might request a breakdown for the cost of the floor, and the cost of the electrical and mechanical system. This will aid you in evaluating the detailed proposals and provide information for use in final negotiation of the contract. The assistance of an architect, professional engineer or construction manager may be helpful here.

In drawing up construction contracts there are several different bases for an agreement. These include lump-sum contracts, cost-plus-fixed-dollar contracts, guaranteed maximum costs with cost-plus-fixed dollar fees, and guaranteed maximums with sharing of savings. Be aware that cost-plus percentage fees provide no incentive for the contractor. The builder should have a bonus for early completion or completion under budget, as well as a penalty for late completion. The maximum mark-up for changes should be specified.

Changing your mind is always expensive, particularly after construction is under way. A change order costs everyone additional money. For this reason, it is important to have all drawings and specifications checked while the project is still in the planning stage.

Risk Management

As the building is planned, it is essential to consult your insurance underwriter and to consider features that will reduce your insurance cost. Building insurance rates and contents rates are always related. Since most warehouses today contain an inventory worth many times the cost of the building itself, the contents rating is a much more important part of insurance costs than coverage of the building.

In addition to fire risks, the insurance underwriter will look at uplift and wind-load. A minor and inexpensive adjustment in building design can make a structure capable of withstanding 27 pounds per square-foot wind-load and 30 pounds per square-foot uplift. These are the numbers often required by the major underwriting organizations, In some areas of the country, particularly those subject to hurricanes, wind-loads may be increased. A review of construction plans by the underwriter can avoid errors which might be costly or impossible to correct.

Roofs

The type of roof selected will affect both construction and operating costs. A conventional built-up roof is constructed of layers of felt roofing and asphalt. It weighs eight to ten pounds per square-foot, whereas an insulated metal roof weighs one pound per square-foot or less. Both metal and built-up roofs are available in materials allowing the 20-year bond usually required by owners. The best metal roof currently available is constructed of aluminized steel and is built with a standing-seam process. The standing-seam metal roof provides superior protection against leakage, and permits the user to have a roof with a less-pronounced pitch than the older metal roofs. The earlier model of metal roof required a pitch of one inch in 12 inches, but standing-seam permits a pitch as gentle as one in 36. An excellent alternative to the built-up roof is the single-ply roof, now frequently used to replace worn out, built-up roofs.

Roof bonds should be examined with care, since they usually have many escape clauses. Inspect the roof during construction to be sure that the materials are correctly applied. Roofing is dirty, hard work not conducive to exceptional quality control by the person actually performing the job. Some contractors currently offer a 20-year weather tightness guarantee, a longer protection than is offered by many roof bonds.

Fire Protection Systems

The insurance underwriter should always be involved in the design of the fire protection system. Most protection standards specify space to allow fire-fighting equipment to be driven around the perimeter of each warehouse building, so lanes or roads should be designed with the proper width and turning radius to accommodate fire engines. Nearly all warehouses in the U.S. use sprinkler systems to fight fires. Many systems have booster pumps to provide additional water pressure in an emergency. If your project has such a pump, consider the benefits of installing a larger pump than the minimum required size. Sometimes a larger pump will increase the cost of the pump installation by only 10-to-15 percent while obtaining 55 percent more water capacity, which allows higher stacking and greater cube utilization in the warehouse.

The underwriter is likely to ask for hose stations, hose racks and, sometimes, intermediate sprinklers placed in storage racks. Don't be afraid to negotiate these items, testing to see if they could be eliminated by providing greater water capacity or other safeguards at lower cost. Underwriters are familiar with tradeoffs, and they will respond if asked to look for alternative ways to provide adequate protection.

CONSTRUCTION

Heating

Since most products in warehouses are subject to damage by freezing, there are very few facilities today that do not have some heat in the building. Additional insulation is one of the best buys in the construction market today, particularly in a time of accelerating energy costs.

In many northern cities now, energy codes dictate the steps the builder must take to prevent heat loss. Four-inch insulation is common, and some warehouse buildings have as much as 8-to-10 inches of insulation below the roof.

In examining heating paybacks, consider the lowest heat level possible in the building. Warehousing is a strenuous job, thus workers do not need very warm temperatures. By using low range temperature thermostats and reducing heat to 40 degrees, it is possible to cut energy consumption by nearly 40 percent.

Gas-fired unit heaters can be a source of maintenance problems. Some units have a heat exchanger that tends to wear out within five years. If a stainless-steel heat exchanger is specified, the additional cost will only be a fraction of the cost of replacing the heat exchanger. Therefore, it is a good investment to specify stainless steel at the initial purchase. The fans in these units should be designed to serve as ventilator fans in the summer.

Air rotation furnaces are a newer alternative to overhead gas-fired heaters. These furnaces are floor-mounted and move large amounts of air to equalize temperatures at the floor and ceiling. Each heater will handle over 100,000 square feet of warehouse space.

Roads

Roadway design and construction is a critical part of any warehouse development. The drainage of roadways is perhaps the most important part of their design. The asphalt or concrete on a roadway is merely a wearing surface, while the strength of the road is in the base below that wearing surface. If that base is not properly drained, it will lose its bearing strength and pavement failures will result. Inspection and testing of the base of roadways is as important as it is for warehouse floors.

Even if your company is not in the trucking business, the design of truck maneuvering areas to save labor costs should be important to you. Reducing costs for your suppliers will ultimately reduce your own costs. Many warehouses have truck areas apparently designed by people who knew nothing about driving a truck. Some architectural standards state that 85 feet from the edge of the truck dock to the outside edge of the maneuvering area is adequate. However, allowing a maneuvering area of at least 110 feet from edge of the dock to the outside edge of the pavement, and preferably 120 to 124 feet, will greatly improve truck movement. The truck berth should be a minimum of 12 feet wide.

The truck apron, like the roadway or the warehouse floor, is no stronger than the base material beneath it. However, with the extreme wear on the truck parking pad, it is best to install a concrete apron 7-to-8 inches thick, reinforced with wire mesh.

Many truck docks are constructed of asphalt. These must have a concrete dolly strip to provide support in the event trailers are dropped at the loading dock. If a dolly strip is planned, bear in mind that location of the trailer dollies will vary as much as 15-to-20 feet from one trailer to another. Therefore, the dolly strip must be wide enough to allow for maximum variance.

Concrete aprons are subject to severe damage from de-icing chemicals. Never use a de-icer unless you know its chemical composition. It is particularly important to minimize the use of chemical agents on any concrete less than a year old. Never use de-icers with ammonium sulfate or aluminum nitrate on any concrete.

Docks

An item once referred to simply as a "truck door" is now an important mechanism for the operation of the warehouse. It is likely to include a mechanical dock leveler and pit, a thermally-efficient overhead door, a dock shelter to provide a weather seal between the truck body and the warehouse, and a concrete pad for the trailer.

A few years ago, trucking industry standards called for a dock height of 50-to-51 inches above the pad. Truck tires have become smaller in recent years, and current industry recommendations are 48 inches.

The recommended door for warehouse construction is larger than it used to be. In 1984, the federal government encouraged adoption by all the states of a trailer width of 102 inches, or 8.5 feet. Trailers of this width were rare in most states. The typical door for a warehouse was an eight-foot-wide by nine-foot-high, vertical-lift insulated metal door. Today, a nine-foot by ten-foot door is preferred. Because of heat loss, uninsulated metals should never be used in door construction. Wood construction provides better thermal efficiency than plain metal, and is easily repaired when damaged. The nine-foot width brings the entire truck body within the door area even if the driver misses his mark by a few inches. A narrower door will create loading problems if the truck is off center by a few inches. The ten-foot height provides total clearance of 14 feet from door jamb to the truck pad. This ensures that any truck which is within the 13-foot, 6-inch highway clearance limitation will not damage the top of the door jamb or the building.

Wear and Tear

Protection of the building from the probable buffeting of highway trucks and fork-lift trucks is frequently overlooked. A few dollars spent on placing 6-inch diameter pipes at a height of about four feet in places where they will protect overhead door tracks, door jambs, electrical apparatus, building corners, or other vulnerable items is well worth the modest investment involved. These protector pipes are filled with sand, capped with concrete and painted yellow so they can be readily seen.

Such pipes also may be placed on critical points in the radius of a curve in a roadway or at sensitive points in the truck maneuvering area. Inspect the condition of roadways and maneuvering areas after the facility has been in operation for a month or so. You are likely to find other places where guard posts should be installed to further protect the pavement or the facility.

Floors

In any warehouse the two most important parts of the construction are the roof and the floor. Warehouse floors, like roadways, are merely a wearing surface with minimal tensile strength. They are no stronger than the material beneath them.

Close inspection and adherence to specified compaction standards should be made right up to the actual pouring of the concrete on the base. If inspection is not made constantly throughout the compaction process, early precautions can be wasted. Concrete cylinder tests should be made throughout the pouring of the concrete floor. If any such tests provide a warning signal, actual core samples should be taken from the floor itself.

Meticulous inspection by an independent source, such as a testing laboratory, is critically important in every phase of floor construction, beginning with soil compaction right up to completion of the floor. A floor constructed badly at the beginning may never be correctable at an economical cost.

Rails

Drainage is as important for railroad tracks as it is for roadways. This is true whether the rail lead is inside or outside the warehouse. Poor drainage is often indicated by a rail lead "pumping" when the locomotive rolls over it. Such action eventually makes it necessary to realign the rails, and will be a continuing problem until proper drainage is established.

The usual railroad standard side clearance is 8.5 feet from center line of the rail lead to the building wall. However, it is possible to obtain closer clearance for an inside rail dock, down to 5 feet 9 inches. The approval process usually involves the state public utilities commission and sometimes the railway union. Securing close clearance is well worth the trouble because it enables the warehouse operator to use a much shorter dock plate, which improves materials handling speed and safety. Since approval of a close clearance may take as long as three or four months, it is important to start this procedure early in the planning process.

Illumination

Lighting systems for warehouses offer a wide variety of choices. Systems have been designed using incandescent, standard fluorescent, high-output fluorescent, mercury vapor and sodium lighting systems. Perhaps your first determination should be the candlepower of light you really need in the warehouse. An acceptable design provides a range of 20-to-30 foot candles in the aisles, with minimal light in the stack areas away from the aisles. For this purpose, the high output 1500 milliamp, 277-volt, 8-foot fluorescent fixture with reflector will provide necessary light levels in the aisles. Such a system also provides maximum flexibility, since the 277-volt lighting saves on wire and conduit size.

The lighting system should be designed with sufficient extra wiring to permit movement of the fixture if aisle configurations change in the future. Since change is a probability in nearly every warehousing operation, it is best to select a lighting system designed to accommodate future realignment of aisles.

An important supplement to artificial lighting is the use of skylights. An insulated, fire-retardant translucent skylight is readily installed in a pre-engineered steel roof. Installing a similar skylight in a built-up roof is usually three or four times as expensive.

An ample number of skylights in the roof can save illumination on bright days. However, in a cold climate the heat loss through the skylights may be greater than the illumination saving. Furthermore, unless good warehouse discipline is maintained, the lights will be turned on even if they are not really needed. In mild climates, extensive use of skylights can be a good investment.

Walls

In wall construction, we have a variety of choices. Masonry is probably the most common component of warehouse walls. It may be either concrete block or block with a brick veneer. Concrete tilt-up is increasing in popularity for warehouse walls, as well as insulated metal panels or combinations of the two. The concrete tilt-up wall costs less than masonry, but the uninsulated tilt-up wall is an extremely poor energy conserver. If the building is to be located in a cold climate, any tilt-up walls should be well-insulated.

Insulated metal wall panels are less costly than either masonry or tilt-up, and provide maximum energy conservation. One compromise is masonry or tilt-up concrete to a height of 10 feet above the floor, with insulated metal panels above for maximum economy and energy conservation. The concrete wall provides a substantial barrier to truck and fork-lift traffic, as well as adding to the esthetics of the building exterior.

Landscape

When the entire structure is completed, don't forget landscaping. There is no building that cannot be enhanced with greenery, and there is no better value in improving the appearance of your building.

14

PLANNING FOR FUTURE USES

Whenever construction of a new building is planned, you must face the probability that the building will become inadequate before the structure is actually worn out. The building could become obsolescent for several reasons: First, the volume and storage pattern needed when it was planned has changed. Second, the design of the building may become outdated by the introduction of new warehousing technology, or by a change in the product line to be stored. Third, the function and design may still be good, but the location no longer effective for the use originally planned. In some cases, faulty construction or poor soil conditions may cause the building to be simply worn out. Finally, substantial expansion or contraction of warehousing volume may cause the structure to be either much too large for current purposes, or too small with no practical means of expansion.

Design Structure for Versatility

How does future obsolescence affect building design? Consider the fact that a building designed for your use alone may have very limited appeal if it must be put on the market for a quick sale. In general, there is a somewhat smaller market for very large warehouses than for those of a more common size. If your space needs are quite large, running from 500,000 to several million square feet, consider a cluster of separate buildings, rather than one huge structure. Separate buildings can be leased or sold to different users when no longer needed. Furthermore, a cluster of separate buildings reduces fire risk, since it is relatively easy to control the spread of fire from one building to another. While some warehousing operations require a single building for effective shipping and receiving, many others can be managed in a way that does not require all merchandise to be under the same roof. Don't assume that one building is essential just because one manager thinks so. Carefully simulate an operation that separates different product lines or inactive reserve stock into adjacent and separate buildings.

Planning the construction of a new building for easy expansion is a valuable means of extending the lifetime of the facility for its current use. Pre-engineered construction is frequently selected just because it is particularly easy to expand a pre-engineered building. On such buildings, the end walls can be removed and the structure readily extended. However, if the side walls (eaves) are to be expanded, either the roof must become lower than the present eave height, or the new roof must slope upwards, which leaves an interior drain in the warehouse.

Many warehouses are designed for easy conversion to a future manufacturing plant. When this is done, certain construction specifications for heavy-duty power, lighting or drainage may be added to the building for manufacturing that are not needed while the structure is still used for warehousing. The number of parking spaces needed for manufacturing workers is usually far greater than those needed for warehousing. At the same time, bear in mind that construction to manufacturing specifications could greatly escalate the cost per square foot.

In deciding between a low-cost square building and a higher-cost, long and narrow building, the planner should also consider future uses and resale, as well as the current use. The long building may cost more to build, but it is easier to subdivide into smaller spaces.

Changes in Basic Function

In essence, nearly every warehouse building is designed to accommodate storage, order picking, and a combination of receiving and shipping. Receiving and shipping can be considered together,

PLANNING FOR FUTURE USES

since they can be performed over the same docks. Storage and order picking may require specific building design features. As you plan a new building today, you are likely to have in mind certain specifications for these three basic functions. Consider what will happen to the building design if these specifications should change. In some industries, storage has been practically eliminated, as materials flow quickly through the building from receiving docks to shipping docks. In others, there have been radical changes in methods of order picking or methods of receiving and shipping. As mentioned in Chapter 3, the six most common functions of warehousing are stockpiling, product mixing, production logistics, consolidation, distribution, and customer service. The building you are designing presumably fills

Adequate room for expansion and additional parking are important considerations in site selection.

161

one or more of these functions and is designed to do so. How effective will that building be if the emphasis is on a different function ten years from now? Is it possible to design sufficient versatility into your building so that its function could change? At least, if the function does not change, the building should be designed for ready resale to another user.

Internal and External Changes

Every warehouse represents a combination of raw space, labor, and equipment. An inventory which turns 30 times per year represents a highly labor-intensive operation, and effective use of people may be more important than using nearly every foot of space. But under changing conditions, the space may become a great deal more valuable than the labor. These changes, because they effect the balance among the three prime elements of warehousing—space, labor and equipment—would be considered as internal changes.

A more common cause of building obsolescence is external changes, those which take place outside the facility and may well be beyond the control of the warehouse operator. These include a changing neighborhood, new developments in transportation, new taxes, legislative changes, or changes in fire regulations.

At times, the future of a neighborhood can be predicted with reasonable accuracy by local development people. Consider what the area is projected to be like 20 years from now, as well as its current character. For example, a sudden acceleration in crime rates in a neighborhood could make a facility undesirable for you and for most future buyers.

Transportation considerations must also be measured. You may choose a rail-served site for your warehouse even though no rail transportation is planned. A rail-served site has a wider resale market. Legislation which allowed wider highway trailers made some warehouse buildings obsolescent, because shipping and receiving doors were too close together to accommodate the wider trailers. While this defect is usually correctable, there could be situations where such changes greatly impair the utility of a building.

PLANNING FOR FUTURE USES

Taxation changes can make a building more or less attractive in the future. If the tax rates in your community were to be increased substantially compared to prevailing rates in nearby towns, your building might be less attractive to the user who could find a better tax cost in a comparable location. States and communities have changed their policy on inventory taxation, and when a community makes such a change, warehouse users often seek to take advantage of favorable inventory taxation. Inventory taxes are an example of changes caused by legislation. An increasing number of states and communities have recognized that offering "free port" status to warehouse users will create a substantial increase in business in the community.

Fire regulations are typically affected by severe losses. Within the past decade there have been some unusually severe fire losses in warehouses in the United States. As a result, some communities have become exceedingly strict in fire prevention requirements.

When buildings are planned for future use, the user frequently finds that the useful life of the facility will be greatly extended. However, if there is no way to extend its use, resale of the facility may be the best course. Proper planning can make your building more attractive for resale than a "white elephant" which once was a precise fit to your requirements but has little value for any other user.

PART IV

PLANNING WAREHOUSING OPERATIONS

15

SPACE PLANNING

In essence, the warehousing is nothing more than supplying adequate protection for stored goods in a minimal amount of space. Therefore, the goal of any warehouse operator should be to protect the inventory in his or her care and at the same time use as little space as possible. Achieving this goal depends primarily on careful planning.

Sizing the Warehouse

How much space is needed to hold a planned inventory?

Assume that a warehouse operator needs to develop a total space plan for toilet paper. A carton measures $20 \times 24 \times 10''$, and is received in box cars containing 1,500 cases, with just one line-item in each car. The product can be stacked 15' high. The operator determines that the product can be stacked in tiers measuring 2-by-2 units, five tiers high on a $48 \times 40''$ pallet. With a stacking limitation of 15 units, this converts to a height limitation of three pallets. A pallet load, allowing for normal overhang, will occupy 15 square feet.

Layout plans show that 40 percent of the building will be used for aisles, docks and staging areas, leaving 60 percent available for storage. From this 60 percent net storage, an additional 20 percent is lost by "honeycombing." ("Honeycombing" is the space lost in front of partial stacks.) After making the space calculation shown in the exhibit below, we find that 31.25 square feet are needed for each stack, or .52 square feet per-case of product. With this information and with inventory projections, you can plan your total space require-

PRACTICAL HANDBOOK OF WAREHOUSING

ment. Similar calculations should be made for each item planned in the inventory. The chart below shows how a storage analysis is done for this or any other item in inventory.

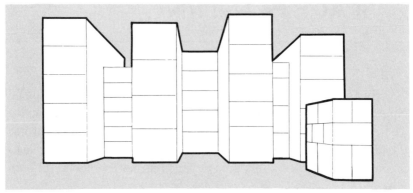

Lost storage space in front of partial stacks is called honeycombing. It exists in all warehouses to some degree.

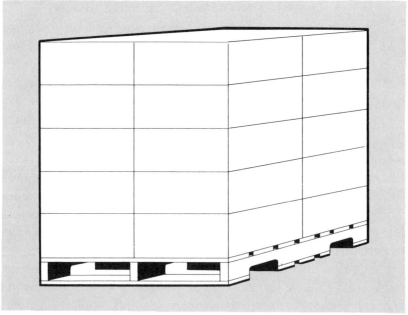

Twenty cases per pallet.

SPACE PLANNING

Storage Requirements

The environmental requirements for different products vary, and this must be considered in planning for space. For example, some merchandise may require temperature or humidity control. Other products may be hazardous, and still others represent abnormal security risks. Some products must be stored in racks because packaging limits the ability to stack them. Some pallets must be stored in racks because of volume considerations, while others have both the quantity and packaging to allow a 20-foot free-standing stack.

Interaction of Storage and Handling Systems

The storage layout and handling system in a warehouse are inseparable. The handling system must fit the building and the storage job to be done. The storage plan must be adapted to the handling system and aisle width requirements. It is impossible to plan one without the other being considered.

Storage Space Calculation

Assumptions:
1. A portion of gross space must be dedicated to aisles and staging, leaving 60% net space.
2. Honeycombing losses further reduce net space, leaving only 80% of it typically in use.

Calculations
A. Each pallet of product contains five tiers with four cases per tier, or 20 cases total.
B. Each stack of three pallets contains 60 cases.
C. Each stack is the size of one pallet plus one inch overhang on each of four sides, or 42" × 50" = 2100 sq. in. = 14.6 sq. ft. rounded to 15.
D. Gross sq. ft. need per stack is 15 ÷ 60% = 25.
E. Sq. ft. per stack after honeycombing is 25 ÷ 80% = 31.25.
F. Sq. ft. per case is 31.25 ÷ 60 cases = .52083.

Figure 15-1

While the storage method aims at maximum cube use, the warehouse operator must recognize the ever-increasing cost of labor. There must be a balance between an effective storage system and an effective handling system. This balance changes if labor costs increase faster than storage costs. The goal is to plan a distribution center that operates at maximum cost efficiency today, but can be modified to keep in step with changes in the cost relationship between storage and handling.

Stock Location Systems—Fixed versus Floating Slots

Public warehouse operators are frequently required by their customers to keep all one customer's merchandise in the same area, even though it might be more efficient to share storage areas among several customers. A warehouse in which all storage locations are fixed will lose space because specific slots must be dedicated to items that fluctuate substantially in volume. When this loss is multiplied by the number of items stored in the warehouse, a totally fixed slot system will waste enormous amounts of space.

A floating slot system is one in which stock locations are assigned on a random basis. But such a system could result in the storage of products having incompatible storage characteristics. For example, a floating slot system that mingles grocery products and poisonous chemicals would cause needless danger. A floating slot system is effective only when inventory records are precisely maintained with location identified. When a floating slot is filled, the stock picker must be able to find another location in which the inventory can be stored. The floating system works best when inventory records are up-dated on a real-time basis.

Family Groupings

A compromise between fixed and floating slots is the grouping of items by family. One product family may be located in a fixed area, but members of that family will be in floating locations. The family grouping method allows the storer to handle products that have different storage characteristics. For example, a grocery wholesaler will

SPACE PLANNING

find it necessary to store all tobacco products in the same area. Soaps and detergents might also be stored as a family, and since these frequently have an odor, they are stored far from products which would absorb that odor.

A Locator Address System

Every location system needs an address designation system, preferably one having a sensible logic that anyone can readily understand. The illustration shown below is a basic address system. Warehouse bays are assigned a number to indicate their relative east-west position, No. 1 being the westernmost side of the warehouse. Letters are used to indicate relative north-south position, with "A" being the southernmost and "D" the northernmost. A storage address with a lower number is always reached by going west, while one with a lower alphabetic designation is reached by going south. A worker who must

	N						
	D1	D2	D3	D4	D5	D6	
	C1	C2	C3	C4	C5	C6	
W	B1	B2	B3	B4	B5	B6	E
	A1	A2	A3	A4	A5	A6	
	S						

SOURCE: The ABC's of Warehousing, © 1978, Marketing Publications, Inc.

Figure 15-2

move from B5 to D2 can quickly calculate that he or she must go to the north and west (*see Figure 16-2*).

A locator system can also help rotate the stock. If the date of receipt is noted at the same time as the location is recorded, outbound shipping instructions can send the order picker to the location with the oldest stock.

Cube Utilization

There are three ways to make better use of the cubic space available in your warehouse. The first is to increase the stacking height, the second is to reduce aisle width, and the third is to reduce the number of aisles.

It is easy to overlook the lowest cost space in any warehouse—that which is close to the roof. Most contemporary buildings are high enough to allow at least a 20-foot stack height, and some allow substantially more. Fire regulations typically require that high stacks be at least 18 inches below sprinkler heads, but with this knowledge you can and should calculate the highest feasible stack height in the building. Then determine whether or not it is being used.

For some warehouses, a mezzanine is a way to use the cube by allowing two levels of storage, or in a few cases even three. Mezzanine components are produced and sold by the same vendors who supply steel pallet racks. Like racking, the mezzanine can be readily moved and relocated, and it can be depreciated on the relatively short life allowed for warehouse equipment. Mezzanines have other advantages over conventional multi-story buildings. If the floor is of metal grid, you may avoid an underwriter's requirement for intermediate sprinkler heads. The elevating and lowering of materials from a mezzanine is usually done by lift truck, which is faster and more economical than a freight elevator.

Consider the use of higher lift trucks, racking designed to use the cube, and mezzanines. One or more of these options is likely to enable you to use the height of the building more effectively than it has been used in the past.

SPACE PLANNING

Lot Size

The typical quantity of each line item to be stored can critically affect space utilization in any warehouse. If storage spaces are planned to accommodate stacks four units deep, and the average inventory is two units deep, the result is either 50 percent wasted space or double-handling to store one line item behind the other.

Foods, pharmaceuticals and other controlled items must be completely segregated. Unless these lots are the same size as available storage slots, more warehouse space is wasted.

Reducing Aisle Losses

The width of aisles is dictated by the turning radius a lift truck needs to make a right angle turn and place goods in storage. For a 4,000-pound counterbalance fork lift, this is typically a 12-foot aisle. Increased lift capacities, carton clamps and slip-sheet attachments will further increase aisle width because of the space buffers needed to use the attachments as well as the extra weight caused by the attachment itself.

Aisle width can be narrowed if special types of lift truck are used. There are two types available. The more economical of these is the stand-up lift truck or reach truck which differs from a conventional fork lift, with the driver standing rather than seated. The stand-up lift truck makes a right-angle turn in about 2/3 of the space needed by a counterbalance truck.

Much narrower aisles can be planned by using a turret lift truck, which makes a right angle turn by swiveling the lift mechanism rather than the entire truck. Because the truck itself remains stationary, the turret lift truck can operate in an aisle which allows only a few inches of clearance on either side. Such trucks are typically wire-guided so that steering in a tight space is automated. But the turret truck is far more costly than a conventional fork lift truck, and it is a more complex machine which can be difficult to maintain. Two types of turret trucks are available, the man-aboard design and a conventional lift with the driver sitting down in the main truck body. The man-aboard type allows the operator to see exactly where the fork carriage is, which is not easy at extreme lifting heights and even lets the operator ride up and select less-than-pallet quantities without removing the pallet from the stack. These very narrow aisle trucks will permit aisle widths to be reduced from 12 feet to a little over 5 feet.

Reducing Number of Aisles

One way to reduce the number of aisles is a rack system designed for two-pallet-deep storage. The only disadvantage is the possibility that overcrowding and insufficient volume may cause one item to be blocked behind another.

Another means of improving space use is movable storage racks. These are roller-mounted rack installations which can be shifted from side-to-side to create an aisle whenever needed. This equipment is costly, but may be justified by the space it saves. It may also be impractical for a very fast-moving warehouse, because access to some facings must be blocked while the rack is rolled to create an aisle on the opposite facing. Mobile racks can be moved to close and open different aisles. A warehouse which is busy enough to require all aisles to be open at once cannot use the mobile rack.

Managing Your Space

While equipment is useful, the most critical element in improving use of space is good management and planning. How many cases of product will your facility hold? What variables in the operation will change its capacity? Finally, how close to capacity is your warehouse operation today?

Space planning starts as goods are received. When a truckload of new merchandise arrives at your receiving dock, who makes the decision about where that product will be stored? In altogether too many warehouses, the inbound location decision is delegated to the fork lift driver. It is only natural that the driver will make an expedient decision which saves time without saving space. Your warehouse should have at least one person designated as the space planner. The planner should know where currently empty storage locations are, where additional space can be created by re-warehousing, and what

SPACE PLANNING

inbound loads are expected as well as the identity of the items which will be on each of these loads. From this the planner develops specific storage instructions for every arriving load of freight.

You will probably do about 80% of your sales with only 20% of the stockkeeping units which you store. These few fast movers should be in those positions where they can be handled most inexpensively: that is, close to the door and close to the floor.[1]

---

1 From "Making Space Work Harder" by Richard E. Nelson, Pfizer Inc., in a paper prepared for the Warehousing Education Research Council.

16

PLANNING FOR PEOPLE AND EQUIPMENT

Before the warehouse doors ever open, management must run a simulation of the personnel and equipment requirements for the new operation. The number of people involved will affect parking requirements, as well as numbers of restrooms, locker and lunch room facilities. A projection of volume and nature of work-flow will dictate equipment requirements, and spaces needed for staging, shipping and receiving.

Personnel planning includes determining and quantifying the work to be performed, measuring the number of manhours needed to perform that job, and translating this calculation into numbers of people required. In addition to the touch labor associated with conventional warehousing tasks, the ability to handle office work and provide supervision should be included.

Productivity measurements are described in Part VII, although some rough productivity measurements must take place in the planning stages.

The Process of Making Flow Charts

Flow-charting helps the planner learn more about a job by forcing a description of that job, step by step.[1] Thus, flow-charting leads to work simplification, which in turn increases productivity. But first

---

[1] The flow charts and subchapter are taken from Vol. 7, Nos. 18 and 19 of *Distribution/Warehouse Cost Digest,* Marketing Publications, Inc., Silver Spring, MD.

you must go back to the what, where, when, who and how questions, such as:

Purpose:
What is done?
Why is it done?
Place:
Where is it done?
Why is it done there?
Where *else* might it be done?
Where *should* it be done?
Sequence:
When is it done?
Why is it done then?
When *might* it be done?
When *should* it be done?
Person:
Who does it?
Why does *that* person do it?
Who *else* might do it?
Who *should* do it?
Means:
How is it done?
Why is it done *that* way?
How *else* might it be done?
How *should* it be done?

Flow-charting will help you identify questions you may have overlooked. It imposes a discipline, makes you plan ahead and will provide a clue as to what a worker might do in a given situation when faced with a decision you may have failed to anticipate.

The flow chart in Figure 16-1 describes a simple materials handling job. Three different symbols are used. The first and the last are the same. This is the terminal symbol and shows where a job begins and where it ends. It is never used for any other purpose.

PLANNING FOR PEOPLE AND EQUIPMENT

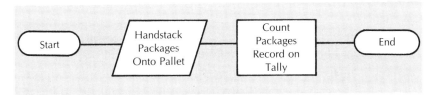

Figure 16-1

The second symbol is the physical process symbol, which identifies an element involving the use of energy or power. The third symbol, the clerical process symbol, identifies an element involving clerical work; in this instance, counting and recording the count.

This flow chart briefly and correctly describes the main elements of work involved in hand-stacking packages onto a pallet. It describes the basic job, the work elements and the sequence in which the elements are performed. However, it does not give a complete description of the job. We might want to know how the worker got to the job site, where and how he got a pallet, and what happens after the packages are counted.

Figure 16-2 gives us a more detailed picture of a job. It tells us what must happen before our worker can hand stack the packages onto a cart. It tells us what happens after the cart is loaded and makes us think about the sequence of the elements. It also may make us ask

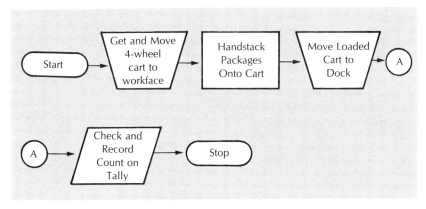

Figure 16-2

179

whether the elements are being performed in the most efficient sequence.

We have introduced three new symbols. Notice the arrowheads. As our charts become more complex, it will be necessary to use arrowheads to show the direction in which the work flows. The arrowhead is an important symbol as will become apparent later in this discussion. The second and fourth symbols denote input/output or transport. The fifth and sixth symbols are connectors.

The Best Sequence?

Refer to Figure 16-2 and prepare to use a little imagination. Where is the check and record count element being performed? In the carrier vehicle or on the dock? To find out, we locate the check and record-count element, refer to the transport element immediately preceding the check-and-record-count element, and then establish that

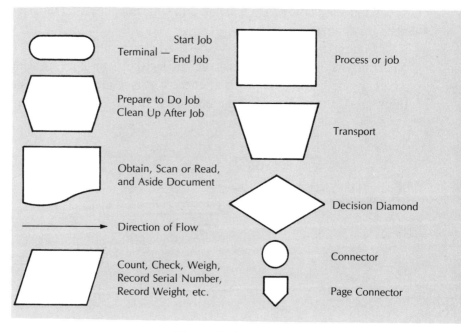

Figure 16-3

PLANNING FOR PEOPLE AND EQUIPMENT

the work is being done on the dock. Is this the best place to do the work? If the dock is well-lighted and the inside of the vehicle is not, the dock is the place to get the most accurate count-and-check. But if the carrier vehicle is well-lighted and the check and count can be made in the vehicle, then the transport move can be made directly into the warehouse and bypass a stop on the dock. Is this a good idea?

Shown in Figure 16-3 is a summary of all of the symbols commonly used in flow charts. Note the addition of new symbols to indicate documents and decisions.

Figure 16-4 shows flow charts used in some typical warehousing tasks.[2]

---

2   The flow charts and subchapter are taken from Vol. 7, 18 and 19 of Distribution/Warehouse Cost Digest, Marketing Publications, Inc., Silver Spring, MD.

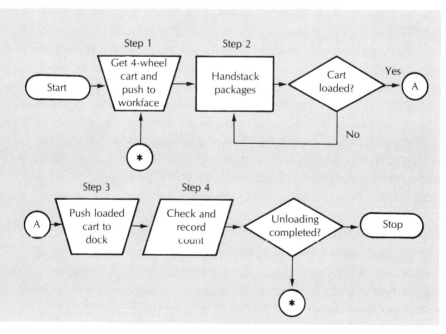

Figure 16-4

'What If' Questions

Flow-charting will force you to think and ask "what if" questions. For example, what will the order picker do if he or she arrives at the pick line stock location called for by the order and finds no stock? Remember, we made the simplifying assumption that stock would always be available, although this is clearly not realistic. Sooner or later, the picker will find the pick line location empty or with insufficient stock to satisfy the order. What will the worker do if he or she has an order pallet on the forks with a portion of the order already picked? How will the worker know where the reserve stock is and whether to take the order pallet to the reserve stock area, or set the pallet aside and proceed to the reserve stock to restock the pick line so as to continue picking? Should the picker be able to communicate with the office in the event that stock can't be located? What kind of communication system should be used? Should it be mobile or fixed? When you begin to cost out steps in terms of machine and labor time you may find you need to look into the communication problem.

One method of locating stock is to maintain records at the pick line location. When new stock is received, the receiving warehouseman posts its location at a designated point on the pick line. Thus, when the pick line is empty, the picker knows where to go to obtain reserve stock. This same system might be suitable for controlling withdrawals of full-pallet loads. The picker often has to pick the less-than-pallet-load portion of the order in the pick line and the full pallets from reserve stock. The picker would go to the pick line location to pick the less-than-pallet quantity, leaving full pallets until last.

Personnel

An important part of personnel planning is the exact description and definition of each warehouse job. Particularly in a plant governed by a union contract, the written job descriptions can have an important effect on the productivity of the work force. When personnel requirements are still in the planning stages, management has a valuable opportunity to ensure that these job descriptions are written to provide maximum flexibility to get the job done. If such descriptions are written narrowly, labor contracts may restrict the assignment of workers to secondary jobs when the primary job is completed.

Once jobs have been described, the number of manhours needed to do the work is determined.

Figure 16-5 shows a table published by the U.S. Department of Agriculture on work standards for a commonly-performed warehouse task.

The Critical Role of the Supervisor

Planning is absolutely essential for supervisors. It starts with time management. Better time management means planning, prioritizing, and scheduling. You should determine how your time is actually spent and then compare it with how it should be spent. One way to do this is to keep a daily time log to track your own day. Every half hour, list what you have done, including problems that required your attention during that time. One warehouse supervisor was surprised to learn from such a log that he had spent 60% of his time one week on grievances. He thought that limited time was his major headache, but his real problem was too many grievances.

There are two types of lost time: obvious and hidden. Obvious lost time is a fork lift driver leaning against his machine with nothing to do. Hidden lost time is more subtle, such as problems within operations. For example, one warehouse loses time at the beginning of each shift because work is not scheduled and waiting for the crew. Each morning, 15 workers gather while the first line supervisor sorts and distributes orders. Then they have to find equipment and pallets before actually getting down to work. A good supervisor solves the problem by having someone from maintenance arrive before the shift begins and prepare all equipment for the first orders of the day. Then the supervisor schedules the day well in advance, so that there are orders waiting for each operator when the shift begins.

Though the pace in many warehouses is hectic, the good supervisor uses planning to complete odd jobs that might otherwise fall by the wayside. Carry a notebook—a traveling "job jar"—in your pocket

TABLE IV—Labor required, per occurrence, for receiving elements when dead skids and low-lift platform trucks are used

Element description	Workers in crew	Base labor	Allowances	Productive labor
	Number	Man-hours	Man-hours	Man-hours
Position skid for loading: Begins when worker starts toward empty skid. Includes walking to empty skid, lifting, transporting an average distance of 24 feet, and placing skid in position for loading. Ends when skid has been released.				
1. Manually	1	0.0078	0.0020	0.0098
2. Manually	2	.0085	.0013	.0098
3. Manual platform truck (ends when skid is in position on floor)	1	.0069	.0010	.0079
4. Electric platform truck (ends when skid is in position on floor)	1	.0070	.0007	.0077
Enter carrier: Begins when transporter contacts bridge plate. Includes crossing bridge plate, transporting to pickup point, and aligning truck to move under loaded skid. Ends when truck is backed against loaded skid.				
1. Manual platform truck	1	.0028	.0004	.0032
2. Electric platform truck	1	.0023	.0002	.0025
Pick up loaded skid: Begins when truck is backed against loaded skid. Includes elevating skid for movement. Ends when loaded skid is clear of floor and ready to move.				
1. Manual platform truck, mechanical	1	.0021	.0004	.0025
2. Manual platform truck, hydraulic	1	.0067	.0013	.0080
3. Electric platform truck	1	.0027	.0003	.0030

Element				
Leave carrier: Begins when skid is clear of floor and ready to move. Includes moving skid clear of original position, maneuvering it when necessary in order to cross threshold of carrier and bridge plate. Ends when skid clears bridge plate.				
1. Manual platform truck	1	.0030	.0006	.0036
2. Electric platform truck	1	.0025	.0002	.0027
Position loaded skid for setdown: Begins when transporter starts maneuvering skid for setdown. Includes turning and pushing or pulling truck to align skid for setdown. Ends when transporter touches lowering control.				
1. Manual platform truck	1	.0072	.0018	.0090
2. Electric platform truck	1	.0038	.0003	.0042
Set down loaded skid: Begins when transporter touches lowering control. Includes lowering skid to floor, and moving truck out from under skid. Ends when truck clears skid.				
1. Manual platform truck, mechanical	1	.0014	.0003	.0017
2. Manual platform truck, hydraulic	1	.0032	.0008	.0040
3. Electric platform truck	1	.0011	.0001	.0012
Position truck to pick up load: Begins when transporter starts maneuvering truck for pickup. Includes turning and pushing or pulling truck, and positioning truck under skid. Ends when truck is backed against skid.				
1. Manual platform truck	1	.0043	.0006	.0049
2. Electric platform truck	1	.0015	.0002	.0017

Figure 16-5

to jot down one-time tasks. Use these for assignments when certain workers are idle. Here are some ideas for your job jar: looking for empty slots, counting products, sorting or repairing pallets, housekeeping, re-numbering aisles or slots, and re-packing damaged products.[3]

Planning for Equipment Use

In most warehouse tasks, as well as many office tasks, the number of manhours needed to finish the job will depend greatly on the equipment or tools available to the worker. The planner begins by considering the kinds of equipment available.

For example, the lift-truck is available with a wide array of attachments besides the conventional forks. A warehouse operation designed for carton-clamp handling will require significantly different planning for space and people than one designed for the receiving of loose-piled cases that will be palletized at the receiving dock. Use of slip-sheet handling devices requires still different planning. The best time to do this planning is before the warehouse is opened. (Part VIII will provide more information on materials handling, including the options available in warehousing equipment.)

At the planning stage, the warehouse manager should select the kinds of equipment that can best be used in the operation, taking into account lifting capacities and storage heights, and any environmental considerations that would restrict the design of the lift-trucks. For example, internal combustion engine trucks may not be acceptable in freezers or other confined areas.

Another determination the equipment planner must make is whether to buy or lease warehouse equipment. The present-value technique described in Chapter 12 can be applied with minor changes to the equipment decision.

Six basic items of information should be gathered before planning for manpower and equipment is completed:

---

3   This sub-chapter was adapted from "Supervising on the Line," copyright, 1987 by Gene Gagnon, published by Margo, Minnetonka, MN.

PLANNING FOR PEOPLE AND EQUIPMENT

Figure 16-6
Illustration courtesy of Gene Gagnon, from "Supervising on the Line."

- The average order-size provides a good indication of the amount of time needed to ship each order.
- Physical characteristics of the merchandise also may define the amount of effort needed to handle it.
- The ratio of receiving units to shipping units, and the question of whether goods are received in bulk or as individual units will greatly affect the effort involved.
- Seasonal variations and the intensity of these seasonal swings must be measured.
- Shipping and receiving requirements should be defined, including the times of day when goods will be received and shipped.
- Picking requirements must be defined. A strict first-in first-out

system is more costly from a labor, as well as a space, standpoint.

When this information is collected, a useful definition of manpower and equipment needs can be reliably projected.

17

CONTINGENCY PLANNING

Contingency planning is preparing in advance for the emergencies not considered in the regular planning process.[1] The best approach is to ask the question, "What If?" For example, what if our major suppliers are on strike? What if we are unable to obtain sufficient fuel to run our truck fleets? What if we are hit with an earthquake or a tornado at our biggest distribution center? What if we simply cannot find enough secretaries, order pickers, industrial engineers to staff our distribution facilities? What if we have to recall one of our major products?

Managers who fail to anticipate certain events may not act as quickly as they should, and in critical situations this will exacerbate the damage. Contingency planning should eliminate fumbling, uncertainty and delays in responding to an emergency. It also should bring more rational responses.

A contingency plan should have certain characteristics. First of all, the probability of an adverse occurrence should be lower than for events covered by the regular planning process. Secondly, if such a critical event occurs, it would cause serious damage. Finally, the contingency plan deals with subjects the company *can* deal with swiftly.

Both "How critical is it?" and "How probable?" must be considered. Some events have a low probability. For instance, the sudden loss of a distribution center could be disastrous if no plans existed to

---

[1] Based on an article by Bernard J. Hale, Bergen Brunswig Drug Corp. *Warehousing & Physical Distribution Productivity Report,* Vol. 14, No. 3, Marketing Publications, Inc., Silver Spring, MD.

meet that contingency. Even if such a distribution center is considered relatively fireproof and well-protected, possibly it could burn down. Contingency plans should be selected commensurate with the impact on the organization of the particular occurrence. The estimate of potential impact can be made in financial terms, competitive position, employee availability, or a combination of such considerations.

Strategies

Strategies in contingency plans should be as specific as possible. For instance, an organization may develop a contingency plan to be implemented in the event of a drop in sales below a certain level. Strategies such as reducing employment, or reducing advertising expenses, or delaying construction are simply too broad. They must be more specific.

As an example, "If sales drop by 10 percent below expectations, our net income will decline by X million dollars. In order to reduce this loss by a specific goal—say, 75 percent—it will be necessary to defer the program or expansion of plant X, to reduce the number of employees by 500 and to cut variable production costs in proportion to the sales decline." Comparable strategies can be prepared to offset income loss by 50 percent, 100 percent, or whatever is considered appropriate.

Wherever possible, the expected results of the actions taken should be calculated in financial terms, or in other meaningful measures, such as market share, capacity and available manpower.

Trigger Points

Contingency plans must specify trigger points, or warning signals, of the imminence of the event for which the plan was developed. For example, if a plan is prepared to deal with a 10 percent decline in sales below estimate, there will be warning signals. Sales may fluctuate daily, weekly or monthly. What will trigger the implementation of the plan? In some cases, such as fire, the trigger point is obviously the event itself. In others, however, the point is not so clear. But a trigger point should be specified. Thus, the contingency plan should

CONTINGENCY PLANNING

indicate the type of information to be collected and the action to be taken at the trigger point. Also, it is advisable to assign specific responsibilities to someone to collect relevant information and pass it on to the person who makes the decision to implement the plan.

Distribution managers should consider developing contingency plans for such potentially serious disruptions as the impact of strikes, energy shortages, labor shortages, natural disasters and product recalls. We will use as an example contingency planning for the impact of strikes.

Planning for the Impact of Strikes—Your Own Firm's as Well as Others'

As the first step in preparing to face a strike, be fully briefed on contract deadlines. It should not be hard to keep track of such deadlines in your own firm, although it is easy to miss them in the case of your suppliers. Many purchasing agents require each supplier to list contract expiration dates, with this kept as part of the master purchasing file.

Every distribution operation has traditional alternatives for dealing with a work stoppage. These include the use of public distribution facilities, non-union transportation companies, or even non-union private distribution facilities.

If your firm is preparing for a strike, the company's main objectives in bargaining should be outlined, since they are likely to affect decisions made during the work stoppage.

If public warehouses are used as a strike hedge, the manufacturer should begin the public warehousing operation before the strike starts—dispersing the product so that it represents only a small fraction of the total capacity of each warehousing supplier. If such precautions are taken, the warehouse operator can file a secondary-boycott charge with the National Labor Relations Board should illegal picketing of a public warehouse occur. In selecting a public warehouse as a strike hedge, it is important to appraise the labor stability and operational flexibility of the warehouse supplier. Properly applied, the "strike hedge" inventory can be a powerful bargaining tool in contract negotiation.

Strike Against Suppliers

The effect of other company strikes can be as disruptive as disturbances within your own organization. A first step in planning for this contingency is to determine which suppliers are critical to your operations. This may depend on whether they are sole sources or whether they supply a critical percentage of the purchasing volume.

Having identified the critical suppliers, the next step is to learn if those companies are unionized. If unionized, what is their labor relations record over the past decade? If there is a contract, when does it expire? Some information on contract expiration is publicly available, including the Negotiations Calendar issued by the U.S. Bureau of National Affairs.

The risk of a supplier's strike can be mitigated by developing an inventory hedge in the form of advance stockpiling of the critical items. At times, interim purchasing arrangements, such as the import of similar items, may provide a source of relief.

Strikes Against Customers

A strike against your customer also can have a serious effect on distribution operations. It is advisable to follow closely the contract expiration dates and negotiations affecting companies with which you have important business relationships. If a customer's strike seems imminent, consider accelerated deliveries, or such special accommodations as direct store delivery to bypass a strike in a retailer's distribution center.

Trucking Strikes

Transportation strikes require particular resourcefulness in planning. The traffic manager should be charged with finding alternative sources well in advance of the strike. The problem should be discussed with customers and suppliers, who may be able to use their own private trucking to move products in or out of your warehouse.

CONTINGENCY PLANNING

In times of trucking strikes, some small carriers will make a side agreement to settle on the master contract and agree to keep working during the general strike. The traffic manager should locate such firms and arrange to use them. Owner-operators of trucks also frequently ignore strikes. In addition to using customer and supplier private trucks, your own firm may have private transportation that can be used. Non-union transportation companies also can be used, though such firms naturally will serve their own regular customers first.

Similar steps should be taken in the case of strikes affecting rail unions or longshoremen. In such case consider diversion to other modes of transportation or to other routes that avoid the labor difficulty.

The approach outlined here for development of a contingency plan in the event of a strike can be applied to many similar emergencies.

Charting the Planning Process

Keep a record of planning actions taken, particularly on a complex plan in which several persons have been assigned tasks. Shown on pages 194-195 is a contingency planning action log that provides an example of how the plan can be charted.

Planning Steps

There are four steps in preparing a contingency plan:
- Identify the contingency and describe your objectives. For example, if the computer goes down, will your major objective be to get the equipment back in operation, or will you switch from computer to manual systems?
- Identify departments that will be affected by the contingency. For example, will the computer breakdown affect both office operations and warehouse operations? How will each of these departments cope with the situation?
- Determine what action should be taken to lessen the impact of

the event. For instance, would the maintenance and testing of standby manual systems ease the transition in the event of a computer breakdown?

- Determine what connection one event will have on another. For example, how will the computer breakdown affect shipping and receiving operations?

The increased attention to contingency planning is the result of

Contingency Planning Action Log

No.	Action to be taken	Assigned	Date	Due
1.	Determine expected length of labor stoppage.	AK	3/10/94	3/10/94
2.	Re-supply all using departments to maximum levels.	MM	3/10/94	3/17/94
3.	Notify key vendors of potential labor stoppage and advise shipping instructions if necessary.	PM	3/10/94	3/15/94
4.	Identify all available storage space at other locations	RS	3/11/94	3/14/94
5.	Move substantial reserve stock of critical high-usage items to available storage space and establish product locator system for same.	WS	3/14/94	3/18/94
6.	Determine probability of delivery vehicles crossing a picket line.	MJ	3/15/94	3/16/94
7.	Make up a work schedule for supervisors during the labor stoppage.	WS	3/21/94	3/21/94
8.	Continue to maintain maximum levels of inventory at using departments.	MM	3/21/94	3/31/94

Figure 17-1

CONTINGENCY PLANNING

a comparable increase in business uncertainty in recent years. Supply situations taken for granted in earlier years must be the subject of contingency planning today. In a time of increased government involvement in quality control, the contingency of product recall is more common. If the contingency event occurs—however unlikely it may be—the prudent manager will have available a plan of actions to be taken.

No.	Action to be taken	Assigned	Date	Due
9.	Notify key vendors to implement alternative shipping instructions as of 3/29/94.	PM	3/22/94	3/24/94
10.	Notify neighboring warehouses of potential labor stoppage and establish a "lending" program.	HA	3/23/94	3/24/94
11.	Notify all using departments what actions have been taken and what alternatives have been established.	WS	3/24/94	3/24/94
12.	Set up a security plan for the facility.	WS	3/24/94	3/31/94
13.	Specify locations for crossing picket line for personnel and vehicles.	JW	3/24/94	3/24/94
14.	Review probability of labor stoppage as of 3/31/94.	WS	3/25/94	3/25/94
15.	Meet with all using departments to review alternative plans.	WS	3/28/94	3/28/94
16.	Implement alternative work schedule.	WS	3/31/94	3/31/94
17.	Keep department supervisors informed.	WS	3/31/94	

18

POSTPONEMENT

"Postponement" in distribution is the art of putting-off to the last possible moment the final formulation of a product, or committing it to the market place. The theory of postponement is nothing new. It has been discussed in marketing and logistics textbooks for several decades.

But although the theory is not new, and the concept is simple, its full potential for saving money in the warehouse has not been reached. For example, you could make an inventory of 20,000 units replace an inventory of 100,000 simply by converting branded merchandise into products with no brand name. That's not only possible—it's being done right now.

Background of Postponement

Postponement was practiced long before marketing theorists started writing about it. One early example in industry—Coca-Cola—goes back to the early years of the 20th century. Postponement was described by a marketing theorist as the opposite of speculation. A speculative inventory costs more to maintain than the profit derived from using it to stimulate purchases.

Alderson was the first marketing expert to write about postponement. In 1950, he observed that "the most general method which can be applied in promoting the efficiency of a marketing system is the postponement of differentiation . . . postpone changes in form and identity to the latest possible point in the marketing flow; postpone changes in inventory location to the latest possible time."[1] The risk of speculation is reduced by delaying differentiation of the product to the time of purchase. Merchandise is moved in larger quantities, in bulk, or in relatively undifferentiated states, thus saving on the cost of transportation.

In *Marketing* James L. Heskett defines the principle of postponement as follows:

> A philosophy of management that holds that an organization should postpone changes in the form and identity of its products to the latest point in the marketing flow and postpone changes of the location of its inventories to the latest possible points in time prior to their sale, primarily to reduce the risks associated with having the wrong product at the wrong place at the wrong time and in the wrong form.[2]

Types of Postponement

Postponement can take any of several forms:

- Postponement of commitment.
- Postponement of passage of title.
- Postponement of branding.
- Postponement of consumer packaging.
- Postponement of final assembly.
- Postponement of mixture or blending.

A typical product moves through four channels: manufacturer to distributor to retailer to consumer. Movement through each involves costs of transactions, as well as the costs of storage, transportation, and inventory carrying-costs. The principle of postponement prescribes reducing these costs by delaying or bypassing some of the stages between producer and consumer.

Postponement of Commitment

To delay shipping a product avoids risking its miscommitment.

---

1   Alderson, Wroe — "Marketing Efficiency and the Principle of Postponement," Cost and Profit Outlook III,4 (September 1950).
2   Heskett, James L., *Marketing*. 1976, Macmillan Publishing Co., Inc., N.Y.

For example, using air freight enables some manufacturers to eliminate field inventories by keeping the entire stockpile at a factory distribution center located at the end of the assembly line. The product is moved into the market at speeds competitive with that of shipping from field distribution centers. This enables the manufacturer to avoid the potential waste of committing the wrong product to the wrong distribution center and later having to cross-ship products to correct an imbalance of inventory.

The most efficient distribution point is always at the end of the assembly line, since inventory at the point of manufacture will never have to be backhauled. If air freight is fast enough to allow all inventory to be kept at the factory, a manufacturer may trade off the premium costs of air freight by eliminating inventory losses from miscommitment or obsolescence.

With some merchandise, postponement includes dispensing with certain distribution channels. For example, an appliance retailer may have floor samples of each major appliance on the sales floor to show prospective customers. However, when this results in a sale, the actual appliance delivered to that customer may not even be in the same city. By using premium transportation from strategically-located distribution centers, the appliance distributor will avoid committing inventory to each retail store. In this case, postponement is particularly useful because of the cost of handling, storing and carrying an inventory of bulky and costly appliances. If the customer cannot take the merchandise home in a shopping bag, why move the item through the store?

Postponement of Title Passage

Delaying passage of title sometimes takes the form of "consignment." For example, a chain grocery retailer makes annual purchase commitments for canned vegetable items. The vendor's agreement calls for these items to be shipped to a public warehouse located at a point convenient to the retailer. However, the contract also specifies that warehousing charges will be paid by the manufacturer, and passage of title is postponed until the product is released from the warehouse.

Since the public warehouseman is an independent bailee, the manufacturer still has a measure of control over the inventory. However, the retailer has the power to release it upon call. By delaying passage of title, the buyer reduces the cost of carrying inventory. Yet, by positioning stocks at a convenient warehouse, passage of title is postponed without any sacrifice in customer service.

Postponement of Branding

Last-minute branding is another type of postponement. A vegetable canner may produce canned peas for his own house brand, as well as for the private brands of several different chain retailers. To meet these several brand requirements, the canner puts up all the canned peas "in bright"—blank cans that have no labels. Typically, these bright cans are held at the factory warehouse until orders are received for one of the private labels; then the order is run through a casing and labeling line. Thus, the manufacturer avoids the risk of putting the wrong label on a product and subsequently having to relabel it.

This type of postponement can be further refined if the manufacturer ships goods in bright to market-area distribution centers. By positioning bright cans (and casing and labeling lines) in distribution facilities close to the market, the canner may substantially improve customer service and still avoid the speculation of producing private label merchandise that may be slow to sell. Also, by postponing a commitment to perhaps six different private labels, one field inventory can do the work of six.

Postponement of Consumer Packaging

A fourth type of postponement is the delay of packaging. A pioneer in the manufacture of soft drinks, Coca-Cola, determined many years ago that its manufacturing strategy would be to produce a concentrated syrup that would be delivered to soda fountains or regional bottlers to be diluted with carbonated water and delivered in a package convenient and attractive for the consumer. Similarly, manufacturers of whiskey move the product across the ocean in bulk, then bottle it in the country in which it is to be sold. The risk of breakage and pilferage of glass bottles is eliminated, and a more favorable customs duty applies in some instances.

Postponement of Final Assembly

Postponement of final assembly is an accepted strategy in many manufacturing firms. For example, a manufacturer of office furniture may delay final assembly of desks or chairs until customer orders are received. Common components such as desk pedestals, seat cushions, frames and hardware can be produced and held in stock. When a customer orders a desk or chair of a certain style or color, the order is filled by assembling and finishing common components on an assembly line. By doing this, the manufacturer avoids the risk of overproducing certain styles which cannot be easily sold.

Postponement of Blending

A sixth type of postponement is to delay mixing or blending. One widely recognized example is the multi-grade gasoline pump developed by Sun Oil Co. By altering the mixture or blend of high and low octane gasoline, Sunoco offers its customers five different combinations of motor fuel at the time of purchase.

Some paint manufacturers offer a customer paint colors mixed at the store. This is a variation of postponement of final assembly, the final mixture or blending being made after the product is sold.

Increased Warehouse Efficiency

What does postponement do for warehousing efficiency? Basically, it enables one inventory to do the work of many. Where blending is postponed, the warehousemen can have a single stockkeeping unit that is convertible to many different products as the final brand is applied. To illustrate, in one line of automotive batteries, the only difference between private labels is a decal applied to the top of the battery. By delaying the application of the decal and the final consumer package until the last moment, the number of line items in the warehouse is dramatically reduced. (See Fig. 18-1).

Postponement can also reduce freight costs, which are usually the largest component of physical distribution costs. In the auto industry, the incentive for regional assembly plants has been to reduce freight costs. It is often cheaper to ship components than finished cars.

Another instance of postponement involves an Italian toy manufacturer that produces a light-density item. The item is shipped in bulk boxes to a warehouse at the port of entry where consumer packages purchased in the U.S. for the American market are also received, and the product is moved from bulk-shipping container to the final consumer package. Not only are freight costs reduced, but the con-

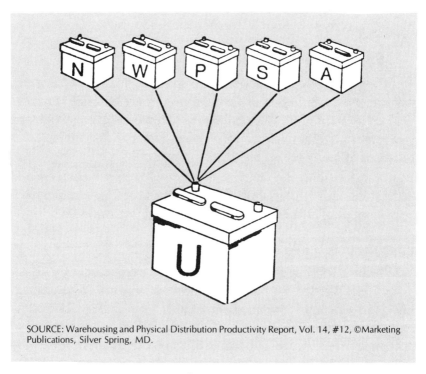

SOURCE: Warehousing and Physical Distribution Productivity Report, Vol. 14, #12, ©Marketing Publications, Silver Spring, MD.

Figure 18-1

sumer package is more attractive because it has not been subjected to the potential wear and tear of long-distance transportation.

Delaying final branding or packaging also reduces the warehouseman's risk of obsolescence. Certain brands may decline in popularity, reducing warehouse turnover and increasing the risk that the product will become obsolete from a marketing standpoint.

Potential Challenges

While productivity advantages in the use of postponement are substantial, the challenges are equally so.

Postponement at a warehouse level usually requires the warehouse operator to enter a whole new business—that of assembly and

---

A Manager's Questionnaire For Warehouse Postponement

1. Do we produce identical products under separate brands?
2. How is the final brand name applied?
3. Can the product be remotely assembled?
4. What is the cost-saving in shipping the product in a knocked-down (KD) condition?
5. What is the cost of remote assembly?
6. What are the quality control risks of remote assembly?
7. What are quality control risks of remote branding?
8. What are customer service levels for the product?
9. Does our competition have a postponement strategy?
10. What is the cost-advantage of bulk shipment of our product?
11. What is the cost of remote packaging of the product?
12. What is the quality control risk of remote packaging?
13. What is the cost of moving the product through each channel of distribution?
14. Can the product skip one or more of these channels?

Figure 18-2

packaging. This creates additional risks in quality control. In the course of packaging, the product may be damaged or contaminated. It could even by mislabeled or put in the wrong package. Quality control in final assembly may be compromised.

Ultimately, the widespread use of postponement could turn today's distribution centers into tomorrow's packaging and light-assembly centers. However, the management and labor skills involved at the warehouse level will need to be higher than those needed for receiving, storing and reshipping.

Postponement in the Warehouse

Uncertain money markets and high interest rates have greatly complicated the problem of carrying inventory. An escalation of fuel costs and deregulation have affected transportation. With both capital and transportation costs increasing abnormally, it is likely that the use of postponement as a distribution strategy will grow in popularity in the near future.

The future possibilities of postponement are limited only by the imagination of the warehouse operator. Any product that has multiple private brands, a fragile consumer package, a bulky configuration in final assembly, or a very slight modification from one form to another, is a candidate for postponement at the warehousing level. Products now packed only at the factory and frequently subject to breakage, obsolescence, or high freight costs may be shipped in an unbranded, unpackaged bulk state to a regional distribution and assembly center. Thus, there is no reason why the practice of postponement may not be far more widespread tomorrow than it is today.[3]

---

[3] This chapter is based on Vol. 14, No. 12 of Warehousing and Physical Distribution Productivity Report, Marketing Publications, Inc., Silver Spring, MD.

19

STRATEGIC PLANNING

Developing a corporate strategy is usually seen as a job for senior management or planning specialists. But though the warehouse manager may not have the last word in developing a corporate strategic plan, providing input from the warehousing standpoint is an important part of it.

Warehousing is a support activity, and in a larger corporation it may be subordinate to marketing or production. Sometimes the warehouse is used to store raw materials or work in process for production, as well as to distribute the finished product for marketing. Clearly, the strategic plans affecting both production and marketing will also affect the warehouse.

What Is It?

Corporate strategy is the development of company goals and the detailed policies for pursuing these goals. A corporate strategy statement may involve economic, social and even personal goals, as well as financial objectives. *See Figure 19-1 for a corporate strategy for a public warehousing organization.)*

The concept of a corporate strategy is not always popular. Some believe that organizations cannot have purposes—that only people can. Some anti-strategists insist it is the job of the manager to maximize profits and not waste time making policy decisions.

On the other hand, some persuasive business scholars argue that strategic planning is necessary for a corporation—and the managers themselves—to survive.

How to Start

If the warehouse manager is going to contribute to the company's strategic planning, a first—and obvious—step is to understand the strategic planning process occurring elsewhere in the company. The warehouse manager needs to know about the strategic thinking that takes place in marketing, production, and finance. Also, the changing business trends among suppliers and customers should be examined. The warehouse manager must also be sensitive to those

---

Corporate Strategy for a Public Warehousing Company

The following statements were designed to provide a better understanding of what kind of company we have, and what kind of organization we will build in the future.

We define our business as providing customers with warehousing and related services designed to offer 48-hour delivery time to any significant market in the continental U.S.

We will build, own and operate a chain of public distribution centers consistent with this objective. This will ensure the availability of efficient operating facilities at a real estate cost substantially below appraised value.

We will aim to fill a unique role in warehousing.

We will constantly seek innovative techniques in storage, handling, warehouse design, and construction. We will also seek to provide peripheral services, such as packaging and freight consolidation, to enhance both our capabilities in warehousing and the service package offered to our customers.

We will become the most competent, rather than the largest, company in our industry.

Figure 19-1

external changes that affect the way the warehouse will operate in the future.

How do you find out what's happening outside of your own area? A good way to start is by conducting a series of planning interviews with executives in customer and supplier organizations. Undoubtedly, the most important of these interviews will be with your customers, since changes in this sector will have the most immediate and profound effect on your company's future. Figure 19-2 gives a

---

This objective will be achieved by relating people to equipment to facilities to the job to be performed.

We will foster the development of a participative group of creative people, who will work in an atmosphere of teamwork and pride.

Compensation and benefits will be constantly upgraded, within limits allowed by the forces of competition. Opportunities will be available for stock ownership in the company, along with other means of sharing profits. Promotional opportunities from within will constantly be sought. We recognize a strong positive obligation on the part of management to see that all relationships between and among employees will eliminate any need for participation by outside entities.

We will, in our relationship with associates, customers and suppliers, maintain the highest ethical standards. We will conduct our affairs in an atmosphere of complete candor.

Policies will prohibit any "conflict-of-interest" situations, including personal ownership of assets used by the corporation. An outside board of directors shall continue to be active in reviewing the affairs of the corporation. Plans will be discussed with associates periodically. Financial affairs will be audited by an independent certified public accounting firm under the direction of an audit committee consisting of outside directors.

list of questions that might be asked of a customer using warehousing services.

Many of these questions are designed to stimulate a free flow of conversation about forward planning. During the planning interview, it's likely that the questioner will move into areas not covered in Figure 19-2, simply because discussions about planning will prompt the person being questioned to go into uncontemplated topics. Therefore, the listed questions should be considered as broad guidelines, not something to be followed verbatim.

Interview questions for suppliers generally will be quite similar,

---

1. In what way will current economic conditions change your business in the coming year?

2. How will the present price and availability of capital affect your company during the coming year?

3. How will current trends in government regulation change your product line, or your distribution and marketing methods?

4. Is your company planning any mergers or acquisitions that would change your warehousing plans?

5. Are you closing, selling, or otherwise removing certain divisions or products that will change your distribution program?

6. What new products are contemplated during the coming year, and how will these affect your warehousing plans?

7. Are there any changes in your sales plans that could change the location or number of warehouses now used to distribute your company's products?

8. What is the worst problem now affecting your business, and how do you anticipate the problem being resolved?

9. Do you have any plans within your department to streamline or otherwise change your current distribution pattern?

Figure 19-2

STRATEGIC PLANNING

but there may be significant variations in some instances. For example, suppliers of material handling equipment would presumably be asked many questions regarding anticipated changes in product line, as well as the reason for such changes. Questions about potential mergers should be posed as diligently to suppliers as to customers, since this kind of corporate change can affect the supplier's willingness or ability to serve you.

In handling these discussions with suppliers, do enough homework to anticipate some situations that might come out of the interview.

If you hold such interviews with the small group of customers and suppliers that normally account for most of your sales or purchases, you can obtain a reasonably accurate profile of the future planning taking place within each group. From this, a strategic planning profile can be developed—a summary of the planning ideas of both customers and suppliers. Obviously, a strategic plan is more than a reaction to the plans of others. However, understanding the changes in your environment will certainly help you develop a program to deal with those changes.

Corporate Strategy as a Morale Builder

In some companies, morale is low because managers are not sure about the future of the company. During a time of extensive mergers, sell-outs, and takeovers, absence of a known corporate strategy can cause fear for many managers.

If employees don't know what the company's development plans are for the next five years, they have a right to wonder whether management is thinking about the future or thinking about bailing out.

Some companies have a strategic plan but have not let their people know about it. Particularly in times of uncertainty, strategic plans can be a strong morale builder. When the plan is revised, let everyone involved in your company know about the revised plan for the near term future.

Production Changes

One way to make inventory more effective in a time of costly and scarce capital is to limit the number of options available to the consumer. If the number of different items in a finished-goods inventory can be reduced, the design and operation of the warehouse to hold that inventory will undergo significant change as well. If a production strategy of reducing the number of varieties is also accompanied by longer runs of each item, the need for finished-goods warehousing will certainly increase. Conversely, the reverse takes place if production runs are shortened.

Therefore, a warehouse manager should develop his strategic plan only after learning more about the production manager's plans. The production manager should be asked about the number of different items to be produced, as compared with the number from the previous year. Also, the production manager should be asked about the possibility of changes in the length of runs. As noted previously, if you have long production runs where you formerly had short runs, you have a completely different warehousing problem, one that you need to plan for.

Strategic Thinking in the Public Warehouse

The corporate strategy described in Figure 19-1 was designed to guide a public warehousing company through a period of rapid growth, but management's strategic thinking should also deal with the following topics:

- Operational expertise—Should the warehousing company specialize in certain products or product families, or should capabilities be developed to handle a wide range of products and services?
- Pricing policies—Should the public warehouse operator price aggressively, seeking a significant long-term financial gain, or should pricing be designed to cover costs and yield a very modest return? Are pricing policies in tune with the need of the customer?
- Assumption of risk—Will the public warehouse accept high-

STRATEGIC PLANNING

risk customers who want to store hazardous products or goods that have a high risk of theft or pilferage?

As the economy continues to change, strategic planning will become more and more important in corporate life, extending to the warehouse manager, as well as others in the corporation.

Financial Changes

Financial managers combat the high cost of capital by attempting to push inventory either forward or backward in the pipeline. When this kind of pulling or pushing takes place, the retailer tries to persuade the supplier to maximize inventory, positioning it at warehouse locations convenient to the retailer. On the other hand, the manufacturer tries to push its inventory down to the retailer, possibly by offering significant incentives in the form of discounts on off-season merchandise. For example, one producer of lawn and garden care products has long offered strong pricing incentives to persuade retailers to stock-up on the product during the mid-winter season when few consumers are thinking about gardens.

Since many distribution strategies evolve as a result of decisions from the finance department, it is important that you know what is happening in this area.

Government regulation also creates changes that affect warehousing. Figure 19-2 included some questions about government regulation in the proposed customer interviews. However, the warehouse manager also should look within the company for evidence of government activity likely to affect the warehousing operation. For example, a change in fire safety standards could force changes in current warehousing practices. More stringent fire regulations have forced many warehouse operators to add intermediate sprinklers in storage rack installations, even though such installations were considered safe without them just a few years ago.

The warehouse manager must understand the corporate strategies developed at the highest levels within the company. For example, if the strategy of the corporation is to aim for the largest market share in its industry, the warehouse manager may need to have ample inventory in many locations in order to create the most responsive customer service network money can buy. On the other hand, if corporate strategy is to produce the lowest-cost product in the industry without regard to market share, the warehouse manager may determine that the cheapest warehouse is one adjacent to the assembly line, and that no other warehouses are justified.[1]

Planning Warehouse Networks

Anyone who uses warehouses and needs more than one should ask at least these five questions about the number and location of warehouse facilities:

1. Do I have the right number of warehouses, and are they of the right size and in the right places to minimize total costs?

2. Should I stock every item in all locations, and if not which items should go where?

3. What will be the effect of any existing warehouse change on service levels to my customers?

4. What should my warehouse network be in three to five years to support the growth plans of my company?

5. What is the appropriate mix of fixed (owned or leased) versus variable (third-party) warehouses for my network?

What will stimulate management to study the existing warehouse network? Here are a few developments:

- Management becomes aware of costly practices, such as three separate warehouses in the same metropolitan area.
- Existing warehouse facilities seem to be outgrown.
- An industry leader alters warehousing strategy, causing competitors to seek ways to remain competitive.
- The sales organization urges establishment of a new warehouse location.
- A major shift in supply sources invalidates the current warehouse network.

---

1   Parts of this chapter are based on an article by Bruce Abels in the WERCsheet, Warehousing Education & Research Council, 1980.

- Customers are changing their inventory/purchasing policies (e.g., they pursue a "just-in-time" policy), forcing suppliers to change delivery patterns.

Sometimes computer models are used to answer this sort of question. A typical model problem is the following example: if your Atlanta distribution center will be out of capacity in 1996, consider all alternatives for moving product to your markets, recognizing the capacity constraints on all facilities. The solution might suggest opening a Charlotte warehouse, shipping farther distances from your New Jersey or Dallas warehouses or adding additional space to the Atlanta warehouse.

Reducing the number of warehouses is not necessarily a correct goal for every distributor. What you want to know is the *correct* number of warehouses for *your* company. That is where modeling becomes a valuable part of the strategic planning process.[2]

---

2 From "Planning Warehouse Networks," an article by T.E. Pollock, Warehousing Forum, Volume 1 No. 7, Ackerman Company, Columbus, Ohio.

20

SELECTING A PUBLIC WAREHOUSE

The purchasing process in warehouse selection follows the same steps involved in many industrial purchasing operations:

1. Assessing the need.
2. Learning about sources of supply.
3. Evaluating alternatives.
4. Final selection process.
5. Monitoring decision results.

First, the buyer must recognize the need for outside warehousing. Until there is general agreement about this, obviously no decision to purchase warehousing services will be made.

Once a need is recognized, the buyer must develop a list of qualified bidders and then evaluate each of these companies. Eventually one of the bidders is selected, but even then there is a possibility that the selection will not work out as planned. Therefore, the purchasing decision is closely monitored, particularly in its early stages.[1]

In both the final selection and monitoring of results, it is essential to refer continually to the goal which you expected to achieve by using a public warehouse. If your prime goal was improved customer service, how well has that been achieved? If the prime goal was cost reduction, your success is readily measurable. Consider whether the cost reductions are taking place in transportation, in warehousing, or in both.

---

[1] Much of this chapter is from "The Public Warehouse Selection Process," by Jim McBride, Affiliated Warehouse Companies, Inc.

The Preparatory Phase

The public warehouseman will not be able to prepare a thorough quotation unless he is provided with accurate specifications from which to bid. Your requirements will differ from those of other companies, and unless you are very specific in outlining those requirements, the bidder will probably play safe and make assumptions which will create a higher quotation than would be given if all facts were known. Above all, thorough preparation will eliminate misunderstandings later. It is important that all public warehouses asked to bid have the same data, because otherwise the bids will not be comparable.

Your first step in the selection process is to develop an accurate and detailed data summary to be distributed to the bidders. Figure 20-1 on page 000 shows an example of an account data summary.

If you have a procedures manual, it would be wise to allow bidders to review it. This will prevent nasty surprises once the operation has started.

The Investigative Phase

Once you have selected a city, you must then find the suppliers available in that city. There has been an increasing degree of specialization within public warehousing. Yet the great majority of suppliers are listed in telephone book yellow pages under the classification "Warehouses - Merchandise" or "Warehouses - Cold Storage." The buyer should be aware of the fact that telephone book publishers will take listings from anybody, and the listing of a company under these sections is no indication that the bidder is really qualified to handle your products. At least two trade magazines in the United States publish annual directories of public warehouses. Like the telephone book listings, these advertisements can presumably be purchased by nearly everyone. However, the magazine publishers do attempt to provide somewhat more detail about the qualifications of the firms listed. The American Warehousemen's Association continues its role of quality control in public warehousing, and the organization makes every effort to limit its membership to financially-responsible companies.

Many warehouse companies belong to sales organizations which represent a group of non-competing warehouses in various regions of the country. These organizations publish their own directories providing details about their members.

Probably the most widely used source of information is recommendations from contacts in other companies. You may wish to compare notes about warehousing companies with other users in your industry. It is certainly wise to talk to major customers of a bidder under consideration. Public warehousing is a "reputation" business, and the investigative phase should center on discovering the actual reputation of the bidder under consideration.

After you have compiled a list of qualified public warehouses, the evaluation process begins. Typically, it is handled initially by letter or telephone, followed by a visit to those companies which seem best qualified. Using a form for the evaluation can save time. Figure 20-2 on pages 220-223 shows an example of a public warehouse evaluation form.

Evaluating Quality

Once the evaluation form is filled out, some experience is needed in interpreting the results.

Your most important determination is the business stability of the potential supplier. The majority of public warehouses are privately-held or closely-held and have no requirement for divulging financial information. Those which are financially sound recognize the concerns of their customers and will provide ample proof of their stability. A check on financial stability must be made even after the relationship has started, since sometimes a warehouse that seemed sound at the time the decision was made will have difficulties at a later date.

Since management is probably the most important thing you are buying, both the quality and "depth" of management must be measured and monitored. Turnover can be a sign of bad management, so it is well to know how long various managers have held their jobs.

While the highest standards of sanitation are critical only for

PRACTICAL HANDBOOK OF WAREHOUSING

companies which produce food, health care and cosmetic products, housekeeping is usually a reliable indicator of good management and well-motivated workers.

Both condition and design must be considered in evaluating the facility. As you evaluate the buildings, consider your own requirements as well as the impression made by the structure.

Public warehousing has always been competitive, and cost of the service has always been important. But often the importance of

---

Figure 20-1
Account Data Summary for Warehousing Quotation

1. Cities/areas interested in _____
2. **Product specifications:**

Item & Pack	Length	Width	Height	Cube	Weight

Maximum no. items _____. _____ items will constitute _____ % of movement.
Value of product: Maximum $ _____ Minimum $ _____
No. units per pallet Maximum _____ Minimum _____
Pallet size used _____
Explanation/exceptions _____
Stacking height _____
Average inventory _____
Monthly thruput _____
Monthly thruput _____
Pallet weight: Maximum _____ Minimum _____
Inventory turns per year _____

SELECTING A PUBLIC WAREHOUSE

the actual prices charged is overstated. Unfortunately, a few public warehouse companies will quote a very low price to attract the buyer and then plan to adjust it once the operation has begun. For this reason, many buyers are rightly wary of "low-ball" quotations.

The transportation capabilities of the prospective warehouse are important. These include both the ability to handle freight consolidations and deal with various modes of transportation.

Exceptions for following items _____
Estimated square footage required:
   Maximum _____ Minimum _____
Term of requirement _____
Special characteristics/requirements:
   (Include temperature/humidity controls, food grade sanitation, label products, odors, recall, pick-and-pack, stock rotation, etc.) _____

3. Inbound information
Rail: Floor-loaded _____ Palletized _____ Slip sheets _____
Clamped _____ No. of cartons/pieces _____ No. units _____
Weight _____ Stretch wrapped _____

Truck: Floor-Loaded _____ Palletized _____
Slip Sheets _____ Clamped _____ No. of cartons/pieces _____
No. units _____ Weight _____ Stretch wrapped _____
Average inbounds: Day _____ Week _____ Month _____
Average no. items per inbound load _____ Maximum _____
If unitized, more than one item per unit _____

4. Outbound information
No. orders per day _____ Week _____ Month _____
No. of line items per order _____
Average weight per order _____
Are orders always in pallet quantities? _____
If not always, then what percentage of orders are? _____
Deliveries are to _____
Special characteristics/requirements:
(Include order entry system, credit approval, customer pickup, stenciling, document preparation, etc. _____

Figure 20-2
Public Warehouse Evaluation

Completed by: _____
Warehouse name _____
Warehouse address _____
_____

Telephone: _____
Contact: _____

Organization
Year established: _____
Investment _____
Owners _____
Is a financial statement available? _____
Own building _____ Lease building _____
If lease, term & expiration date _____

Management
Owners active & on site _____
Management quality _____
Ratio of supervisors/foremen to warehouse labor _____

Building Construction
Type of construction _____
Type of floor _____
Single or multi-story _____
If multi-story, number & capacity of elevators _____
Ceiling height _____ Floor load capacity _____ Square footage _____

Receiving & Shipping
Private siding _____ Railroad _____
Reciprocal switching _____ Switches per day _____
No. of outside rail docks _____ Inside rail docks _____
No. of outside truck doors _____ Sheltered truck doors _____
Automatic dock levelers _____

SELECTING A PUBLIC WAREHOUSE

Hours of operation (number of shifts, etc.) _____
Hours for receiving & shipping _____

Handling Equipment

Type & number of fork lifts _____
Clamp handling _____ Slip sheet handling _____ Drum handling _____
Maximum capacity _____

Storage Space: Ceiling Height   Square Feet

Unheated
Heated (temperature)
Air conditioned
60°F-70°F
Humidity control
Cooler 32°F-40°F
Freezer 0°F-32°F
Hazardous material space

Amount & type of racks _____
Paved & fenced yard space _____
Total square feet _____
Is warehouse in industrial or isolated area? _____
How close to residential area? _____
General appearance of storage area (neat, clean, messy, etc.) _____
Condition of recooperage area _____
Evidence of food grade sanitation _____

Types of accounts

Consumer products _____ No. _____
Food _____ No. _____
General Merchandise _____ No. _____
Other _____ No. _____
Adequate separation between food & non-food accounts? _____

Transportation-Trucking

Commercial zone cartage _____

(continued on next page)

PRACTICAL HANDBOOK OF WAREHOUSING

Beyond carriage & area served _____
Storage-in-transit capability _____
Consolidation program _____
Pick-up/deliver container (piggyback) _____
Pick-up/deliver container (pier) _____
Policy towards customer pick-up _____
Pool distribution _____

Administration/accessorial service

                                                  Available   Good/Poor, etc.

Desk space
Secretarial/clerical services
EDP services
Inventory procedures
Communications capability (TWX, FAX)
Record, report numbers
Lot/batch control
Deliver by serial number
Recall capabilities
Office appearance (neat, well-run, etc.)

Special Storage

Red Label _____ Yellow Label _____
Fertilizer _____ Poisons _____
Flammable or other label storage _____
Familiarity with Class B poisons _____
Is there an existing policy toward handling spills of hazardous material? _____
If so, what? _____
U.S. Customs bonded space _____ U.S. Foreign trade zone _____

Security

24-hour guard service _____ Uniformed guards _____
Night watchman _____ Fenced grounds _____
Fire alarm _____ Sprinkler _____
Security system (type) _____
Fire station—proximity to warehouse _____

SELECTING A PUBLIC WAREHOUSE

Warehouse Labor
Union _____ Non-union _____
If union:
what union _____
Contract expiration date _____
History of strikes/work stoppage _____

Insurance
Fire rate _____ Contents insurance rate _____
Warehousemen's legal liability, coverage per-building and total _____

Taxes
Freeport _____
Storage-in-transit exempt if destined for another state _____
Other exemptions applicable _____
Assessment date _____ Range of rates _____
Personal property taxes _____
State _____ County _____ City _____
Corporate sales tax _____

References
Bank _____
Customers _____
Trucking firms _____
Vendors _____

General
Does warehouseman appear aggressive, up-to-date, knowledgeable in warehousing/distribution business and genuinely interested in handling our account?
_____

Prominent services/qualities/characteristics of this company _____

Detrimental points or reasons for further research _____

Consider also whether the proposed supplier is experienced in handling customers similar to your company.

Insurance rates are an independent measure of quality, and should be carefully considered and compared. A supplier who has an unusually high rate may be a poor insurance buyer, or there may be a more serious reason.

Location of the building is absolutely critical, since it is obviously not possible to change this factor. If the location will not work for servicing your customers, you can save a lot of time by determining this quickly.

Labor is judged both in terms of quality and stability. A company with a history of frequent labor disputes has obvious disadvantages.

Finally, the warehouseman's presentation, his level of interest, and his reputation are perhaps the most telling points in the decision. If the company is not really interested in meeting your needs, nothing else in the proposal will be of great value.

The Decision

You may wish to establish a rating system based on the information collected in the evaluation form. However, the decision will be subjective even if a rating system is used. In essence, you will select a vendor who meets your "comfort level," and you should reject one who makes you uncomfortable.

Once a decision is made, business courtesy requires you to notify the unsuccessful bidders. Remember that you may be re-evaluating the process if things do not work out as expected.

Finally, the selection is only the first step in implementing a new warehousing program. Working with the warehouse selected and maintaining close communication in the early stages are the best ways to be sure that the company you selected performs the way you hoped it would.

PART V

PROTECTING THE WAREHOUSE OPERATION

21

PREVENTING CASUALTY LOSSES

The warehouse operator insures against (and aims to reduce) three kinds of casualty risks that can affect his business in different ways.[1] The first group of risks affect warehouse property. Losses here could destroy or render useless the building and land on which the warehousing operation is located. The second broad group of casualty risks affects contents and personal property—merchandise or other equipment stored in the building. And the third group of risks affects business operations, or the ability of the warehouse operation to continue in the face of a casualty loss of the building and/or its contents.

A serious fire or other major casualty loss affects all three—the warehouse, its contents and the business operation. However, since insurance carriers normally divide coverage in this way, it is important to be sure that all three risks are covered in purchasing insurance.

Types of Casualty Losses

Perhaps the most catastrophic of losses is fire. A warehouse fire, once out of control, is likely to destroy both the building and its contents.

Windstorm is another common threat to warehouse buildings. The modern warehouse is particularly vulnerable to windstorm damage because it has wide expanses of flat roof that can be readily dam-

---

1 From Warehousing & Physical Distribution Productivity Report, Vol. 18, No. 2, Marketing Publications, Inc., Silver Spring, MD.

aged by high winds. When such roofs are constructed with asphalt, insurance carriers may limit the amount of asphalt used in order to control risk of a roof fire. As the amount of asphalt is reduced to cut the fire risk, the risk of damage by windstorm is increased. (The asphalt serves as an adhesive.)

As well as the risk of flood-water damage, there also is a risk of a malfunction of the sprinkler systems. Sometimes ill-trained emergency fire brigades may set off the sprinkler systems, and then sprinkler system water may do as much damage as a fire.

Overloading the building, structural failure, or earthquake may all cause a building to collapse. (Note that the risk of collapse through earthquake is sometimes a separate peril that must be specifically covered in another insurance contract clause.)

The risk of explosion is clearly affected by the nature of the contents stored in or near the building.

Vandalism and malicious mischief clauses cover losses caused by sabotage or other acts of malicious destruction. A last and more general risk is that of "consequential loss"—the losses that may come as the result of an earlier event. One example would be a power failure disabling the refrigeration system in a temperature-controlled building, spoilage of stored products being a consequential damage.

Controlling Fire Risks

Perhaps the most important step in controlling fire risks is to purchase both insurance and loss-prevention advice from a competent insurance carrier. Some of the more progressive insurance companies have devoted vast sums to research and development on minimizing the risk of a fire starting in the first place. These companies believe that the best way to keep loyal customers and control insurance rates is to make every employee in a company aware of how to prevent fires from starting. Means of controlling the risk have varied at different times and in different countries, but in the U.S. today perhaps the most widely accepted means of reducing the risk of loss by fire is the automatic sprinkler system.

The automatic sprinkler is a series of pipes installed just under

PREVENTING CASUALTY LOSSES

the ceiling of a warehouse building, carrying sufficient water to extinguish or control any fire that breaks out beneath them. There are two widely used types of sprinkler systems—dry-pipe and wet-pipe systems.

The dry-pipe system has a series of heated valve houses, each containing a valve using compressed air to keep water back from the sprinkler pipes. The pipes throughout the warehouse are filled with compressed air, which is evacuated to permit water to flow when a sprinkler head is tripped. This is a system designed to function in unheated warehouses or in outdoor systems that operate in subfreezing temperatures.

Sprinkler piping has sprinkler heads at regular intervals. Each of these has a deflector to aim the flow of water and a trigger made of either wax or a metallic compound that will melt at a given temperature. When heat reaches that critical temperature, the trigger is snapped and the compressed air will escape and allow water to flow through the system and onto the fire. When this happens, most systems also include a water-flow alarm that will ring a bell and trip an electronic signal to summon the fire department. The electronic signal is particularly important for controlling a false alarm or an accidental discharge of the system.

In the wet-pipe system, water fills all the pipes right up to the dispenser valves, which means that there must be sufficient heat throughout the warehouse to keep this water from freezing. The wet-pipe system will respond faster since compressed air does not have to be released first. Thus, it is considered a safer system for fire control. Furthermore, the wet-pipe system is less susceptible to false alarms than the more complicated dry-pipe system.

All sprinkler systems are equipped with master control valves, which represent another potential risk in that these valves might be either accidentally closed or deliberately shut off by a vandal. Nearly a third of dollar losses from fires during one recent period occurred when the control valves had not been closed.

Fire extinguishers are an important "first aid" for fires. A portable extinguisher often prevents a small blaze from becoming a major

PRACTICAL HANDBOOK OF WAREHOUSING

fire. However, the best way to be sure that fire extinguishers will be effectively used is to have periodic training and drills in their use.

Most fire protection systems also include fire hose stations for use before the fire trucks arrive. When fire drills are held there also should be training in the use of this hose. Many major fires are prevented by maintaining a team trained to act effectively in the interim between discovery of a fire and the arrival of the professional fire department.

The warehouse operator's proficiency in loss-prevention is usually reflected in the insurance rate underwriters assign to his facility. Some public warehouse operators will advertise their low fire rate knowing that it reflects well on their success in loss prevention. Fire underwriters measure both management interest and employee attitudes toward fire protection. They recognize that loss risks can be in-

First aid for fires. Every employee should be trained in the correct use of fire extinguishers.

creased by careless smoking, poor housekeeping, or even poor labor relations.

Hazard Ratings for Stored Merchandise

The insurance inspector is far more concerned about the materials in your warehouse than the building itself. With few exceptions, most warehouse buildings constructed during the past few decades are very fire-resistant. If you are fortunate enough to be warehousing noncombustibles such as metals, concrete, or fire-proof chemicals, and if those materials are in bulk with no combustible packaging, then you have a very safe warehouse. However, almost every warehouse today is filled with products which at least contain some paper or corrugated packaging, plastics, wood, and combustible chemicals.

The insurance industry rates combustibles in four classes plus "special hazard." Let's look at the most hazardous first:

Training in handling high pressure fire hoses will prevent injuries to employees and improve fire fighting efficiency.

1. A special-hazard product is one in which the plastics content of packaging and product is more than 15% by weight or more than 25% by volume. Expanded plastic packaging, such as the foam boxes which are used to protect computers and television sets, are particularly hazardous.

2. A Class IV commodity can be any product, even a metal product, which is packaged in a cardboard box if non-expanded plastic is more than 15% of the product weight, or if expanded plastic is more than 25% of the product's volume. An example item would be a metal typewriter with plastic parts which is protected by a foam plastic container.

3. A Class III item is one in which both product and packaging are combustible. An example might be facial tissue in cartons.

4. A Class II product is a non-combustible item which is stored in wood, corrugated cardboard, or other combustible packaging. While the contents will not burn, the package will. An example might be home appliances such as refrigerators, washers and dryers.

5. A Class I is the least-hazardous commodity, and thus would be non-combustible products which are stored in non-combustible packaging on pallets. The pallets themselves of course will burn, but if everything on the pallet is non-combustible, it is rated as Class I.

The fire inspector will consider the quantity of material in your warehouse which falls into each of the five classes and the manner in which you store materials. From a protection standpoint, solid-pile storage, free-standing stacks which are snugly against each other, provide the least-hazardous storage arrangement. The exception would be a commodity that is in itself quite dangerous, such as a highly flammable chemical.

Fire inspectors consider rack storage to be the most hazardous, simply because a pallet rack layout allows a great deal of exposure to air which will feed the fire. For this reason, an increasing number of underwriters are demanding that storage racks have sprinkler heads dropped down into each rack position. In-rack sprinklers make the warehouse operator's job more difficult. It becomes almost impossible to remove or rearrange rack installations because the sprinkler sys-

PREVENTING CASUALTY LOSSES

tem would have to be moved as well. Furthermore, there is a danger of striking and breaking sprinkler heads as merchandise is moved.

You can do certain things with racks to make them safer from the underwriter's point-of-view. One is to allow a vertical space between rack structures which will permit sprinkler water to run down into the racks. Pallet racks or installations which use metal grates are considered to be safer than solid-steel shelving, simply because the openings allow water to run down through the levels in the event that fire breaks out.

The fire inspector will also look at the number and width of your aisles, and the height at which goods are stored. Nearly all underwriters request that goods be at least 18 inches below sprinkler heads, but with more hazardous materials, an even greater space between product and sprinklers will increase the possibility that the sprinkler system can control the fire. The problem with high-piled storage is a chimney or flue effect. If you have a 30-foot pile of combustible boxes with a 6-inch space between rows, a fire in that space will move up the 30-foot flue like a blow torch. The flame at the top is extremely hot and violent, and this heat can expand the steel framework of the building and cause a collapse.

Fire safety for the most dangerous materials is improved with ample aisle separation. If the hazardous products are separated from the rest of the merchandise by wide aisles, the chance of controlling a fire is increased. In some cases, extremely hazardous materials are removed from the warehouse and put in a special-hazard building which is far enough away from the rest of the facility so that a fire in that building will not threaten the larger warehouse structure.

Wind Storm Losses

Certain parts of the country are recognized as having a higher hazard from storms. Gulf Coast areas are notable for hurricanes and parts of the midwest have repetitive problems with tornadoes. Probably no economical warehouse design could withstand a direct hit from a tornado or the full force of a hurricane. However, certain kinds of construction have proven more resistant to such perils. A pre-en-

PRACTICAL HANDBOOK OF WAREHOUSING

Types of doors that cannot be opened or broken without triggering an alarm, and infrared beam sensors will detect a break-in.

gineered metal roof, for example, has greater wind-resistance than a built-up composition roof, simply because the metal roof does not rely on asphalt or any other chemical adhesive. Overhanging truck canopies should be avoided in high-wind areas, since they are particularly susceptible to wind damage.

Flood and Leakage

One way to control the risk of loss from either flood or leakage is to be sure that no merchandise is stored directly on the warehouse floor. The use of storage pallets or lumber dunnage to keep all cardboard cartons at least a few inches from the floor can provide a measure of protection in the event of minor leakage.

Mass Theft

Any electronic burglary protection system designed by one human being can be overcome by another. For this reason, warehouse managers can never assume that a protection system is guaranteed to prevent mass theft. Furthermore, the system is only as good as the people who operate it.

Periodically, prudent warehouse managers should deliberately breach the security system just to learn how fast the police will re-

PREVENTING CASUALTY LOSSES

spond. But holding this kind of burglary drill is potentially dangerous, and the manager who does it must be sure that he or she does not cause an accident in the course of running the drill. One way to reduce this risk is to inform the local police chief that a drill is taking place, but not the alarm company. The police chief need not tell everyone expected to respond to the alarm, but steps should be taken to assure that nobody is accidentally hurt.

Since the technology for electronic protection is constantly being improved, the alert warehouse operator should always look for new equipment better than that currently installed. But all such equipment should be "fail-safe"—meaning that it will ring an alarm if wires are cut or power is interrupted.

Outside lighting is an important deterrent to mass theft. Intense outside light will make a thief feel conspicuous. So rather than run the risk of being seen, the thief probably will select a less well-lit building.

Good outside lighting will make a thief feel conspicuous. He will probably choose another less well-lit building.

Vandalism

Vandalism, sabotage or other deliberate destruction is a very difficult risk to control. In the case of fire, the most serious risk is that sprinkler valves might be deliberately closed. There are electronic devices to counter this. Also, it is possible to apply a seal that shows at a glance if a valve has been turned by an unauthorized person. The risk of vandalism increases during a period of labor strife and management should be especially alert at those times.

Plant-Emergency Organizations

A carefully-defined plant-emergency plan can greatly reduce the risk of major loss. For this reason, most major casualty underwriters strongly encourage the development and training of plant emergency teams.

The job of the emergency team is to fill the gap between the time that the emergency is discovered and the time when a fire department or other professionals are able to arrive at the warehouse. The emergency organization should always include representatives of plant maintenance or engineering departments, since these are the ones most familiar with the protection systems, control panels, or circuit breakers. The plant emergency team should know how to operate all protective equipment and be trained to react with confidence and accuracy.

The most important aspect of training is practice drills. Such drills should include sounding the alarm, moving to a prescribed emergency station and handling fire hoses, extinguishers and sprinkler system controls. The last of these is the most complex, since a large sprinkler system usually has many valves as well as a booster fire-pump. A good emergency organization also is schooled in first aid, since the first priority is always saving lives.

Emergency teams need regular drill and preparation. When emergency plans are made and then forgotten, the team rapidly becomes disorganized and useless.

PREVENTING CASUALTY LOSSES

New Developments in Fire Protection

In the United States, fire protection in warehouses depends primarily upon the automatic sprinkler system, a technology which has been used in industry for more than a century. Fire walls are considered to be a second line of defense, though some underwriters tend to minimize the value of a fire wall in a major disaster.[2]

The current general-commodity warehouse sprinkler system is designed to cover a specified control area. A typical control area is 3,000 square feet, normally covered by 30 sprinkler heads. A sprinkler system to cover this area will require 30 gallons per-minute for each sprinkler head, which means that if all sprinklers are released by a fire, a flow of 900 gallons per-minute (30 sprinkler heads x 30 GPM) is needed.

The system also allows the use of two fire hydrants for manual fire-fighting, each of which theoretically draws 250 gallons per-minute. This means that controlling a fire in that 3,000 square-foot area could require a total water-flow of 1,400 gallons per-minute (900 + 250 + 250). This flow should be at a minimum pressure of 30 pounds. Flow and pressure to meet such a need must be available, and it is commonly found in either a city water system, or by boosting pressure with a water tower or mechanical pumps.

The current common expectation is that a warehouse fire in its initial stages will be controlled by between 20 and 40 automatic sprinklers triggered by the first alarm.

The state-of-the-art today includes many sprinkler systems which have a trigger designed to release at 280° F, when a fusible link melts and activates the system. Unfortunately, we store products today which can create a raging fire before temperatures reach 280° F at the sprinkler head. In one ignition test of a fire at the base of storage racks, it took about two minutes before the sprinkler head reached 280° F and actuated the system. By that time, the updraft from the

---

[2] This sub-chapter is based on information received from Monfort A. Johnson of Peterson/Puritan, Inc. Johnson is author of "The Aerosol Handbook" and a recognized authority on fire protection.

fire had reached speeds of nearly 30 miles per-hour, and air temperature at the ceiling was over 1,300° F,. When the fire reached this stage, it was too far advanced to be controlled by the sprinkler system.

The fire protection industry's answer to this problem has been the development of two newer types of sprinkler heads. The products are referred to as High Challenge and Early Suppression Fast Response (abbreviated as ESFR). The response time of both sprinkler heads can be reduced by using a metallic trigger with a lower melt point. Both heads normally activate at 160° F. With this lower setting, the response time is significantly faster, which is critical in controlling a warehouse fire. In a storage rack, the fire updraft speed is greatly damped and ceiling temperature can be kept under 400° F.

If your water-supply limits force you to change the coverage of the control area, this would require major reconstruction of a sprinkler system in an existing building. To install higher flow heads, either the water supply must be made adequate for the greatly increased water consumption of the new heads, or the sprinkler system must be reconstructed to provide for the smaller control areas described earlier. In most cases, ESFR heads will require a booster pump for adequate pressure and flow. When booster pumps are installed, diesel-powered units are better than electric pumps because they could function even if electrical lines are broken in a fire, or shut down by the fire department for the safety of fire fighters. The pump is typically housed in a small building separate from the main warehouse.

The Importance of Management

While the prime goal of a warehouse manager must be to produce a profit, insurance underwriters tend to evaluate management interest in reducing fire risk as much as any other factor in rating a facility.

The best insurance policies will not adequately cover all the losses in a major casualty disaster. The ill-will from customers and suppliers is difficult to measure and nearly impossible to recover.

When the chief distribution executive is not overly interested in protecting property, it is unlikely that his subordinates will be either.

On the other hand, when senior management interest in property conservation is strong, this attitude will pervade the organization. The cornerstone of a program to control casualty losses is senior management's commitment to it.

22

'MYSTERIOUS DISAPPEARANCE'

Protecting merchandise from "mysterious disappearance" requires a careful blend of the physical techniques of security with personnel and procedural precautions.[1]

The physical techniques—the easiest to put into practice—too frequently are the only ones used in a warehouse. However, the best security procedures in the world cannot defeat collusion between two dishonest individuals—one employed inside the warehouse, one outside. And while your goal is to hire only honest, trust-worthy employees, even a warehouse with 100 percent-honest employees is vulnerable to dishonesty from outsiders. So procedural checks become important to ensure that honest employees remember to use security systems. Thus, a combination of all three—physical, personnel and procedural precautions—must be considered when you look at warehouse security.

Electronic Detectors and Alarms

A first line of defense in theft security is electronic alarm systems, which generally are more reliable than a watchman. But it is important to remember that an electronic system can always be defeated. Any system designed by one human can be overcome by another, especially since a sophisticated theft ring may include a former employee of an alarm company. Furthermore, some alarm companies

---

[1] From Warehousing and Physical Distribution Productivity Report, Vol. 17, Number 9, Marketing Publications, Inc., Silver Spring, MD.

side electricians as sub-contractors to install equipment. And one dishonest electrician can defeat most systems.

However, it's harder to defeat the electronic systems being installed today, thanks to the increasing sophistication of the new equipment.

An example is the use of closed-circuit television cameras. These are most effective when combined with videotape, which give a reviewable record of an occurrence seen by the camera. Some closed-circuit TV systems are activated by the opening of a dock door, so that the camera operates only when a loading door is open. The tape systems can be set to operate through an evening or weekend when the warehouse is closed, with high-speed review of the tape by supervisory personnel after the warehouse has reopened.

In many warehouses, closed-circuit TV is used primarily as a psychological deterrent. However, when properly installed, it also can be a material aid in both detecting theft and convicting those involved.

What About False Alarms?

One of the great drawbacks of electronic alarm systems is the frequency of false alarms. A too-sensitive system tied directly to police stations can set off so many false alarms that it quickly becomes a nuisance. While most law enforcement officials encourage electronic systems, they naturally want the devices to be reasonably reliable.

In a warehouse, the most sophisticated electronic systems are also those most likely to send the most false alarms. Heading the list would be ultrasonic systems, which are designed to detect noise or movement. The problem, however, is that most warehouses are located in or near railroad yards, and the normal switching of freight cars may be sufficient to trip the alarm. Some warehouses have an interior rail dock which must be accessible to railroad crews for switching cars at night. Under these conditions, use of ultrasonic alarms is most difficult, if not impossible.

'MYSTERIOUS DISAPPEARANCE'

Door Protection Alarms

The least-costly electronic alarms are those that provide protection for doors only, but they are not adequate for most warehouses. With current construction techniques, the outside walls of a warehouse are seldom load-bearing, and therefore are relatively thin. Whether the walls are constructed of masonry or metal, a determined thief will have little trouble cutting a large enough hole in the wall to allow a mass theft.

In one such case, thieves backed a truck against a wall of the warehouse. Working from inside the van, they removed concrete blocks from the exterior wall to gain entry, and loaded the van with the highest-value product in the building. Fortunately, one of the thieves tripped an electronic beam in the rail dock area and sent in an alarm.

Because of the vulnerability of nearly all warehouse walls, the best electronic alarms include a "wall of light" surrounding most of the storage areas. With this, the thief who manages to penetrate an exterior wall will still sound an alarm when entering the storage areas.

Skylights and roofs are also easily entered, primarily because of light-weight materials used in modern construction. Protecting an entire roof area with electronic beams is not practical. Fortunately, however, it is equally impractical to take out any significant volume of freight over a roof.

One answer to the compromises necessary with alarm systems is to install sectional alarms, providing one standard of security for most inventory of normal value, and a high-security storage area for high-value items. Even the ultrasonic system can be practical in an isolated room designed for a small portion of the inventory.

The typical alarm system used in warehouses usually must be switched on or off as an entire system, so if there is a breach it's difficult to pin down where it occurred. Warehouse supervisors may have difficulty "setting up" the system at night when one of many doors is partially ajar, or one electronic beam is blocked. However, more sophisticated systems have an "annunciator panel" to show exactly which doors or beams have been breached.

Such a system may also permit you to keep certain doors on alarm, but take others off. This enables you to activate the system for portions of the building not in daily use. In a very large or lightly-staffed building, this system may be worth its additional cost.

Fire codes in most communities require many safety pedestrian doors along the walls of the warehouse to allow workers to escape from the building in case of fire. Since such doors can also make it easier to pilfer merchandise, you should stop any casual use by having an alarm on each personnel door which rings when it is opened.

Grounds Security

The physical arrangement of grounds and access to your warehouse will either invite or discourage theft. While it would be ideal if you could restrict access to the property surrounding the warehouse, this often is not practical. Because most warehouses are involved in high-volume movements of freight, common carrier truck drivers and railroad employees must come on the property to spot or remove vehicles. Such access usually is required at night or after working hours.

High fences and gates help control access to the property. After-hours patrol of the grounds by a guard dog and trained handler is an excellent psychological deterrent. Some guard dog services operate on a roving basis, with protection provided for a number of facilities by the same guard team.

Where you locate employee and visitor parking can also deter or encourage theft. Persons entering or leaving a warehouse should all use a single entrance, with other personnel doors available only for emergency. Parking should never be permitted adjacent to warehouse walls or doors.

Seals

A seal is a thin metal strip (with a serial number) which is fastened so that it can't be broken without its being evident. Properly used, the seal is a good physical deterrent to cargo theft. Because theft in interstate commerce may be investigated by the FBI, for years the mere presence of a seal on a box car was enough to discourage unauthorized entry. More recently, however, bolder thieves have attacked the box cars themselves.

If a box car is broken into while spotted at the shipper's or receiver's plant, the railroad has a legal right to disclaim responsibility, as the car was not under its control. For this reason, the use of an inside rail dock may be the only means of protecting against theft from a railroad car. A parked truck-trailer presents similar problems. You can buy electronic alarm systems that extend coverage to spotted trailers.

To use seals effectively, you must have procedures that are carefully followed, as well as good communication between shipper and receiver. You can't allow seals to be either applied or removed by unauthorized personnel if you expect them to serve their purpose.

Storage Rack

One way to reduce losses of highly-vulnerable products is to place them on the upper levels of storage racks where they can be readily removed only by a trained lift-truck driver. You can thus use storage racks to improve security as well as to improve storage productivity.

Restricted Access

Only warehouse employees should have access to storage areas. Bank customers aren't offended by barriers that prohibit them from walking behind the counters where the money is stored. Similarly, no user of a warehouse would be offended if his or her movement is restricted in the same manner. Even warehouse employees should understand that they are authorized to enter only those areas involved in their own work.

Yet many warehouses have no physical restriction that prevents visitors from wandering at will. Furthermore, many warehouses don't prevent visiting truck drivers from walking or loitering in storage areas. Posting signs or painting stripes may not be good enough. Look into how fencing, cages or counters can be used to prevent unauthorized persons from entering storage areas.

Marketability of Product

In analyzing your theft security, study the marketability of each product in the warehouse: Some merchandise may be valuable, but not easily sold; other products may be of lower value, but easily sold. Thieves generally will consider marketability of product more closely than value. For example, basic food products, paper products and other products that do not have high value but do have a ready market have been the object of mass thefts. If you're not sure of the marketability of certain products, law enforcement officials can give you a good idea of how frequently various products are stolen for resale.

After you determine marketability, apply selective security standards, with the highest security measures used for the most marketable merchandise.

Hiring Honest People

Even the most elaborate physical protective devices can often be defeated by one dishonest warehouse worker who is in collusion with an outside truck driver or some other regular visitor to the property, such as a trash collector. The only way to control this kind of theft is to try to hire only honest workers. This has become increasingly difficult, particularly since the use of polygraph or lie detector tests in the U.S. is now banned by legislation passed in 1988. Use of detailed credit checks is discouraged by some legislation in certain states.

A newer method of testing a job applicant's honesty is pre-employment screening through voice stress analysis. Some claim that the results are superior to the polygraph. When the person being questioned intends deceit in his or her answer, the voice will contain stress which may not be audible to the human ear, but which can be measured by a trained analyst using a special recorder. While most interviews are conducted in the presence of the speaker, it is also possible to analyze a tape recording or a telephone conversation.

The 1988 ban on polygraph caused renewed interest in handwriting analysis, a personnel technique which is more widely used in Europe than in the United States. The effectiveness of this technique has been debated, yet many claim that handwriting analysis will reveal a great deal about the personality of the writer.

Using Supervision for Security

A good way to defeat collusion is to have a "second-count" check. Ideally, the person who makes it should be a foreman or supervisor who is a member of management.

There are several ways you can use a foreman as a first line of defense against collusion theft.

- First, the placement and breaking of all seals should be the responsibility of a foreman, accompanied by the truck driver whose vehicle was sealed.
- Second, upon receipt of any load, the foreman as well as the receiver can verify merchandise in the staging area. In a busy or crowded warehouse, this verification may take place in a storage bay, since the load is taken directly from the vehicle to storage.
- Third, as only a random check of certain loads can be made, that check is best done by a member of management.

This does not mean that a collusion ring could not extend to management, or could not include more than two persons. However, experience has shown that few collusion thefts involve more than two individuals.

Procedures to Promote Security

Personnel policies also can be structured to encourage honesty. Strict policy on the acceptance of gifts is one example. Truckers report that getting a convenient unloading appointment at many busy grocery warehouses depends on a "gift" to the receiving clerk. When this kind of corruption exists, the moral atmosphere that encourages pilferage of merchandise is also likely to be present.

Even honest persons and good security systems can be overcome by creative outside thieves. Therefore, strict adherence to security procedures must be carefully enforced.

For example, if only foremen are to break and apply seals, you must create—and use—paperwork that certifies a certain foreman performed the task.

Every empty container is a potential repository for stolen merchandise. An empty trailer left at a dock could be the staging area for a theft. The same is true of empty box cars, and even trash hoppers. One approach is to restrict access to such containers. Empty freight vehicles should either be kept under seal, or inspected as they leave the warehouse. Furthermore, this inspection must be covered by a written record that it took place.

Customer Pick-ups and Returns

Customer pick-ups create an additional risk of theft. One way to control this risk is to have merchandise for pick-up pulled from stock by one individual, with delivery to the customer made by a different worker. Again, this procedure is based on the fact that there usually are only two people involved in collusion thefts.

Customer returns present an unusual security problem—particularly if the merchandise is returned in high volume or in non-standard cartons. If this is the case in your warehouse, be sure customer returns are checked thoroughly as soon as the merchandise arrives at the dock. If you don't keep your returned-goods processing current, you're inviting pilferage.

Control of documents can be as important as control of freight. Since fraudulent parcel labels can be used to divert small shipments, the best way to prevent this is to supervise the issuing of labels.

Many warehouse crews break for lunch or coffee at the same time, with no one available to inspect exposed docks or entrance doors. It's a good idea to stagger rest breaks, with a few individuals always on duty to protect against unauthorized entry.

Security Audits

Good management must actively combat theft, focusing on the three aspects of defense: the physical, personnel and procedural. No one of the three will be effective without the other two, and a "Maginot Line" complex can be as fatal in warehouse security as it was in

'MYSTERIOUS DISAPPEARANCE'

Security Checklist

1. When was your alarm system last reviewed and checked?
2. Have you considered sectional alarms?
3. Are restricted areas defined in writing?
4. Are the boundaries of the restricted areas obvious to employees and visitors?
5. Are the means of restriction adequate (do you use fences, counters, cages as necessary)?
6. Does the warehouse have a reputation for tight security?
7. Are rules enforced strictly and impartially?
8. Does management set the proper example for loss prevention.?
9. Do you have the necessary checks to ensure that procedures are followed correctly?
10. Is employee/visitor parking located away from the building?
11. Do you make spot checks to ensure that watchmen and security guards do their jobs properly?
12. Are walls and fences checked periodically for evidence of entry?
13. Are employees and visitors required to use one door only, with all other doors equipped with alarms?
14. Are docks and doors never left unattended?
15. Are rest breaks staggered?
16. Is your returned goods processing up-to-date?
17. Do you deliberately overship goods to see if the overshipment is reported by receiving personnel?
18. Have you determined which products are most attractive to thieves?
19. Do you have different levels of security for different value products?
20. Do you emphasize randomness in your security checks?
21. Do your supervisors look for patterns that might indicate pilferage or collusion?
22. Is access to empty freight cars and trailers restricted?
23. Do you check trash containers randomly?
24. Do you have a systematic, rigorous audit of your security procedures and policies?

Figure 22-1

military history. The greatest enemy of security is complacency. An outside security audit is a good way to guard against management complacency.

A periodic spot check of outbound truck shipments may be the best way to impress both warehouse employees and truck drivers of your vigilance. Under this audit procedure, perhaps one truckload per month leaving the warehouse is stopped, returned to the dock and thoroughly recounted to check loading accuracy. Common carrier management will cooperate with such a procedure when its purpose is understood, and you demonstrate to employees that you are serious about discovering any "errors" that could cause mysterious disappearance.

23

SAFETY, SANITATION AND HOUSEKEEPING

In 1970, a law that became known at the Occupational Safety and Health Act (OSHA) cleared Congress. This piece of legislation had a substantial effect on the warehousing industry, as well as on many other industries where accidents are a serious risk. For their first decade, the OSHA laws were enforced stringently, with the warehousing industry frequently found to be in violation. In one survey, nearly one-third of responding public warehouse operators indicated that they had received recent OSHA inspections, while more than half of those inspected received some sort of citation for a safety violation. Although no figures are available on private warehouses alone, it is doubtful if their safety procedures were any better.

Perhaps the most frequent safety problem in warehouses is unstable storage stacks. Other safety hazards are insecure dock plates and cluttered aisles.

One result of OSHA activity has been increased interest in safety training and equipment. Use of hard hats and safety shoes in warehouses has increased. Training courses for fork-lift operators to improve safety awareness are now commonplace.

The establishment of employee safety committees is a good way to foster greater awareness of safety, as well as improving personnel communication.

Improving Safety in Lift Truck Operations

Accidents involving lift trucks can usually be traced to the failures of operators, warehouse layout, the machines themselves—and management.[1] The role of managers is perhaps the most important.

EXAMPLE: who's responsible for the bent racking in a warehouse with 102″ aisles and counterbalanced lift trucks specified by the manufacturer as suitable for right-angle stacking in 110″ inch plus clearances—the operator who hit the rack or the manager who said making the aisles narrower will allow room for another row of racking?

EXAMPLE: who's responsible when a truck runs through a three-day old puddle of oil, the wheels skid, the operator loses control and runs over the foreman's toe—the operator or the supervisor, who, for three days running, having walked by, through and over the puddle, was waiting for a break in the action before scheduling "non-productive" housekeeping?

The examples might go on indefinitely. The absence of lift truck accidents is a characteristic of a well-managed warehouse, which is another way of saying that any listing of causes of accidents must include managers.

Action Steps to Improve Safety

Of course all of us are safety conscious—when asked. Good managers, however, not only *talk* safety, they *act* safety. Here are a few examples of management safety action:

- Planning and establishing traffic patterns, lanes, aisles, etc., that contribute to safe operations as opposed to ones that create "booby traps."
- Providing for pedestrians by routing traffic flow away from work stations.
- Using the full range of safety equipment and techniques to support the plan; marking the aisles, ensuring they are wide enough, hanging fisheye mirrors at restricted-visibility inter-

---

[1] From *Lift Truck Fleet Management and Operator Training* by Leon Cohan, International Thomson Transport Press, Washington, D.C.

SAFETY, SANITATION AND HOUSEKEEPING

sections, posting speed limits and stop signs, establishing one-way aisles and eliminating cross-dock traffic.

- Discussing safety and traffic flow with the operators and identifying potentially dangerous areas and situations, encouraging their participation and getting their ideas.

- Establishing specific (to-the-job/to-the-location) safety rules based on operator consensus and participation. Most plants and warehouses have safety rules, safety committees, safety signs and safety posters. Often they are all but ignored. When workers have helped set the rules they are more likely to follow them.

- Enforcing safety rules to both the letter and the spirit.

- Combating management apathy (demonstrated as an unwillingness to do what it takes to enforce safety rules). This will be quickly picked up by the workers. Good managers have the courage to insist that safety procedures must be followed even when it appears that this will, in the short term, unfavorably affect productivity.

- Training operators in safety procedures. This includes instruction in both how and why. Instructions not to put a wrench on top of a battery are more likely to be remembered when the operator understands that the gas generated during charging is explosive and may be set off by a spark. Training programs should include, at a minimum, instruction on (1) company safety rules pertaining to powered industrial truck operations; (2) basic material handling techniques; (3) the specific piece of equipment to be operated; and (4) written and practical tests to ensure that the training has been absorbed.

- Specifying safety equipment.

Lift truck capacity must be adequate for the job. This means that management must be fully capable of selecting equipment that is appropriate and suited for the task. A piece of equipment may be suited for, and perfectly safe for, one job, but totally wrong or unsafe for another. For example, batteries *can* be changed by using a chain hanging from the forks of a second lift truck; it is safer and easier, however, to use a crane or roller stand.

In the final analysis, management must follow talk with action. Good managers set the example and tone for an operation. Those whose actions demonstrate their sincerity will find that the workforce complies. A professional manager working in full cooperation with trained lift truck operators will find that a safe, accident-free workplace, once achieved, fosters productivity.

Sanitation

Although the Federal Food, Drug and Cosmetic Act was passed in the 1930s, strict warehouse inspection by representatives of the Food and Drug Administration (FDA) was not a source of serious concern until the 1960s and beyond. Everyone involved in warehousing took notice when the chief executive of a major east coast food chain was convicted of a felony in connection with extreme contamination in one of his company's distribution centers. This case, which went to the U.S. Supreme Court, certainly provided general recognition of the power of FDA to enforce its citations, targeting its prosecution on the senior management of a company if necessary.[2]

The result was greatly increased interest on the part of food warehousemen, in particular in ensuring their warehouses maintained conditions that would prevent similar litigation by the FDA. Some food companies hired professional inspection services, firms that could function as independent auditors of sanitary conditions in both private and public warehouses.

FDA inspectors may hold the warehouseman responsible for any evidence of contamination, even when there is some doubt as to whether it originated within the facility. Rodents are not the only sanitation problem, though they have generally caused the greatest concern. Insects and birds also are a source of contamination. And, because they fly, they are more difficult to control.

Any sanitation program must consider the following:

---

2   74-215 (June 9, 1975) United States, Petitioner vs. John R. Park.

SAFETY, SANITATION AND HOUSEKEEPING

- Building design and construction.
- Rodent exclusion and extermination.
- Bird control.
- Insect control.
- Inbound inspection.
- Cleanup methods and equipment.

Other types of contamination can occur by odor transfer from an odoriferous product to one that has absorption qualities. Leakage, staining and chemical reaction can be caused by incompatible items stored in close proximity.

Many companies have developed internal sanitation guidelines listing quality standards to be followed. Some companies retain an inspection service even more rigorous than the inspections typically performed by FDA or local health inspectors. The American Institute of Baking is one of these inspecting organizations, and they have listed conditions which will automatically rate a warehouse "unsatisfactory" (Figure 23-1).[3]

Housekeeping

What are the consequences of a major breakdown in good housekeeping in a warehouse? Bad sanitation and safety are the consequences that concern the government. Many others should be of even greater concern to warehouse management. The first and foremost important sign of poor housekeeping is a drop in productivity.

Poor housekeeping is often the result of overcrowding and one symptom is the habit of dropping staged merchandise in aisles rather than in the staging area. In extreme cases, merchandise may be permanently stored in aisle space. Obviously, using part of the aisle for storage will slow traffic in those aisles, occasionally requiring goods to be moved to get at a picking face.

Another symptom of poor housekeeping is storage of the same product in many locations, often without a usable location system.

---

3 From a publication of the American Institute of Baking.

> **Conditions that Automatically Rate a Warehouse 'Unsatisfactory' to the F.D.A.**
>
> 1. No sanitation program or pest control in effect.
> 2. Presence of live rodents or birds inside the distribution center.
> 3. Any significant insect infestation, internal or on the surface of the stored product.
> 4. Rodent gnawed bags or rodent evidence on pallets of stored stock, or numerous excreta pellets on the floor-wall junction of the food distribution center.
> 5. Cockroach infestation in the food distribution center that could cause regulatory action or could adulterate product.
> 6. Stored product contaminated with poisonous or hazardous chemicals (i.e., pesticides, motor oil, etc.)
> 7. Perimeter floor-wall junctions and storage areas completely restricted, preventing cleaning, inspection and pest control procedures.
> 8. No internal rodent control program. As an example, mechanical traps, snap traps or glue boards not present or evidence of dead and decomposed rodents in traps, or an insufficient number of traps present.
> 9. Rodent bait stations used inside the stock storage area of the food distribution center with evidence of dead rodents or spilled bait present.
> 10. Dock doors and pedestrian doors not rodent proofed and significant rodent evidence noted inside the food distribution center.
> 11. Severe roof leaks that may have contaminated product in storage.
> 12. Insect or rodent-infested transport vehicles.
> 13. Freezers and coolers, without thermometers, and temperatures above required minimums except during periods of heavy traffic.
> 14. Poor personnel practices, such as eating and smoking in undesignated areas, that could contaminate food in storage.
> 15. Illegal pesticides used in the distribution center or transport vehicles.
> 16. Non-certified persons using restricted-use pesticides.
> 17. Transporting food ingredients and product in the same trailer or rail car with toxic chemicals except when they are physically separated on the load. Separation can be by partition or location provided that leakage or spillage could not contaminate stock.
>
> From a publication of the American Institute of Baking.

Figure 23-1

SAFETY, SANITATION AND HOUSEKEEPING

As locations become mixed, order pickers may spend more time searching than picking. Stockkeeping units may be mixed on a pallet or in a stack, and the mixed stack may not be discovered until it is time to ship or after the wrong product has been shipped.

Finally, disorderly work in itself will slow productivity. Employees can work most effectively in a relatively neat and clean environment. A breakdown in neatness is likely to cause a slowdown in work pace.

Damage Spreads

A second consequence of sloppy housekeeping is an increase in damage. When merchandise is staged or stored in aisleway areas, it is more likely to be struck by materials handling equipment.

One of the signs of poor housekeeping is the presence of damaged or leaking cartons in storage areas. Unfortunately, damage, like a contagious disease, tends to spread. Leaking liquids will stain other packages when they come in contact. Damage and leaks will cause packaging to lose its strength, and the result can be a collapse of a storage pile. Because of the tendency of damage to spread, one of the first rules of housekeeping is to isolate and repackage any damaged product immediately.

Personal Appearance—Part of Good Housekeeping

Most discussion of warehouse housekeeping overlooks the importance of personal appearance of the people who work in the warehouse. Do the people who handle the freight in your warehouse, and the people who manage that warehouse always maintain the appearance of professionals?

Some companies deal with this by issuing uniforms to everybody working in the warehouse. However, uniforms are sometimes not as acceptable in our culture as they are in other parts of the world. Americans are by nature individualists, and some people feel stifled by the requirement to wear a uniform. Others feel that it is a divider, since the manager wears "civilian" clothes but the hourly worker wears a uniform. A few companies have addressed this problem by issuing uniforms to *everyone* in the company.

There is a variety of ways to handle the cost of uniforms. Only a little over half of the companies which have uniforms pay all the cost of this service. About a tenth of those using the service ask their workers to pay all the cost, and over one third using such services have a cost-sharing arrangement between company and employees. Uniform companies generally avoid dealing directly with employees, even when they pay all the costs. The company guarantees return of garments, and a person leaving the company must return uniforms at the time that the final paycheck is delivered.

A warehouse worker who presents a sloppy appearance, such as a flapping shirt tail, unpressed trousers or dirty clothes surely does not give the appearance of a professional. Hairstyles as well can create a good or a bad impression. Yet many people will take offense if their personal appearance is criticized. Given this sensitivity, the manager must set his own personal example in making appearance a positive factor in the housekeeping in the plant. Consider the use of at least a partial uniform for management as well as for hourly people. Offering positive reinforcement in the form of public praise for those workers who exhibit a professional appearance will help remaining workers get the message.

Effect on Employee Morale

Perhaps the most destructive result of extremely poor housekeeping, however, is its effect on employee morale. Workers do not function productively in a sloppy environment. In fact, they may develop a "don't-give-a-damn" attitude. In one recent inspection of a warehouse suffering from extreme housekeeping disorder, an employee asked about his job said, "It takes a strong stomach to work in this place." His disgust with the environment was evident.

The loss of morale is pernicious, since it in turn creates further housekeeping hazards. One is a tendency to abuse the building and equipment. This is usually a subconscious act, resulting from an attitude problem. Basically, the worker assumes that if management is not interested in keeping the place neat, there is no reason the worker should feel differently. If anything is damaged, whether it is merchandise, handling equipment, or the building, there is a tendency to leave it alone since the place is a mess already. Furthermore, the worker may reason, if things are already sloppy, what difference does a little more damage make?

Effects on Customer Service

Another consequence of bad housekeeping and bad morale is a tendency to ship unsaleable products to customers.

Any carton in storage for more than several months may look a little dirty, but a warehouse with good housekeeping practices will remove such dirt before the box is shipped to the customer. However, when housekeeping slips to a certain level so does all interest in neatness and cleanliness, and the result is the shipment of stained or dirty cartons, which may be rejected by customers.

Most warehouse inventories face a problem of obsolescence, with certain older product become unsaleable either because the product has been replaced by a newer model, or because its maximum shelf life has been exceeded. But product may also become obsolete because some merchandise was buried in a mixed-pallet or mixed-stack. If the product is "lost" within the warehouse, it may become obsolete simply because it cannot be found.

A Source of Shame

Finally, the most devastating result of poor housekeeping is the fact that the entire warehouse operation becomes a source of shame rather than of pride. In many companies, the distribution operation is a showcase for visitors, a highlight of any company tour. A look at a well-run warehouse is impressive, a reflection on the management of the company and a showcase for the company's products. Neat and well-kept stacks of beautifully packaged products will impress any visitor, which is why the distribution center is so popular on most plant tours. Conversely, a warehouse with extremely poor housekeeping will elicit a strong negative reaction from anyone who sees it. As a result, firms with sloppy warehouses usually hate to show them to visitors. Something that should be a showcase becomes just the opposite.

Impact on Safety

Not all warehouse accidents are caused by bad housekeeping, but a significant percentage of them are. A frequent cause is a fall, which can be the result of slippery floors caused by product spills. When leakage or spillage causes a stack to collapse, there is a risk the worker will be struck by the falling stack. If materials handling equipment is poorly maintained, equipment failure can cause a personal injury accident. Every safety inspector looks at housekeeping, and management's interest in housekeeping will be viewed as a reflection of management's interest in employee safety.

A Housekeeping Checklist

Figure 23-2 is a checklist which can be used for periodic inspection of your warehouse.

Figure 23-2
A Housekeeping Checklist

Grounds   Yes   No

1. Do any paved areas need repair?
2. Does vegetation growing in the 18-inch gravel strip around the perimeter of the building need weed killer application?
3. Do grass, shrubs, weeds growing around the warehouse need mowing?
4. Do splash blocks need to be repaired or repositioned?
5. Is there stagnant, ponding water anywhere on the property?
6. Are there rodent burrows which need to be destroyed?
7. Does the waste disposal area need cleaning; does any waste need to be placed in covered containers?
8. Are there any unessential materials lying about the premises?
9. Is there any equipment stored outside the building which is closer than 10 feet from the building?
10. Do pallets stacked outside need destruction or restacking?
11. Are there any damaged pallets in stacks?
12. Is the grass around the office in need of water, fertilizer, or treatment of disease?
13. Do parking lot bumper blocks need repositioning?

Equipment

1. Does the materials handling equipment require cleaning/repairing?
2. Is there evidence of hydraulic oil leaks from any fork lifts?
3. Is equipment clean?

Housekeeping—Building Interior

1. Do any floors need to be swept or scrubbed?
2. Is there trash which should be removed from the warehouse?
3. Are there any areas under lockers or around vending machines which need to be cleaned?
4. Does the maintenance area/shop need cleaning/organizing?
5. Do any windows need to be washed?

SAFETY, SANITATION AND HOUSEKEEPING

(Housekeeping Checklist continued from previous page)	Yes	No
6. Do rail pits need cleaning?	___	___
7. Do pallets need to be cleaned or restacked?	___	___
8. Are any personnel, loading dock, or riser house doors blocked?	___	___
9. Do any bins, shelves, or closets need to be cleaned?	___	___

Product Handling and Storage

1. Are there any leaning stacks or damaged products in stacks?	___	___
2. Are there any instances in which pallets are stored too high as indicated by the bottom carton being crushed?	___	___
3. Are any products stored in aisleways?	___	___
4. Are any storage rack bars or uprights bent or damaged to the extent that replacement is necessary?	___	___
5. If supports are used in storage rack, are any broken boards causing a potential hazard?	___	___

Damage Hold Area

1. Does any damage need to be segregated from regular stock?	___	___
2. Does the recoup area need cleaning/organizing?	___	___

Maintenance—Building Exterior

1. Are there cracks in any of the exterior walls which need to be repaired?	___	___
2. Is there any indication that gutters or downspouts are clogged or damaged?	___	___
3. Do any canopies need to be repaired?	___	___
4. Do any of the dock leveler pits need to be cleaned or repaired?	___	___
5. Do any security lights require maintenance?	___	___
6. Is snow removal from grounds, sidewalks, door overhangs, canopies, etc. required?	___	___

Maintenance—Building Interior

1. Are there any cracks or crevices in walls or floors?	___	___
2. Are there any openings at the top of concrete block walls which should be capped?	___	___
3. Are there any inside walls which should be cleaned?	___	___
4. Are there any holes of greater than ¼ inch in size beneath doors or dock plates which should be filled?	___	___
5. Do steel door tracks or dock springs require adjustments?	___	___
6. Do unused pedestrian doors need securing?	___	___

SAFETY, SANITATION AND HOUSEKEEPING

(Housekeeping Checklist continued from previous page)   Yes   No

7. Are there any areas which require additional lighting? ____ ____
8. Are there any burned out bulbs which need to be replaced in the warehouse? ____ ____
9. Do any floor drain covers require a seal? ____ ____
10. Are any circuit breakers to riser house heaters turned off? ____ ____
11. Do any fire extinguisher hangers need to be remounted? ____ ____
12. Is insulation needed at the bottom of truck, rail, overhead doors to reduce heat loss and to restrict rodent entry? ____ ____

Toilet Facilities

1. Do toilet or hand washing facilities need repair, cleaning, or supplies? ____ ____
2. Do toilets/dressing rooms need additional lighting/ventilation? ____ ____
3. Do toilets require disinfectant? ____ ____
4. Are there any leaks in toilet facilities? ____ ____

Eating Area

1. Does the eating area need cleaning? ____ ____
2. Do floors, tables, countertops, ranges, refrigerators, microwave ovens need cleaning? ____ ____
3. Do wastecans/lids need cleaning/emptying? ____ ____
4. Do water fountains need cleaning/repair? ____ ____
5. Is additional lighting/ventilation needed? ____ ____

Pest Control

1. Is there any evidence of rodents, birds, or insects within the building? ____ ____
2. Were insecticides found other than in a locked cabinet or room? ____ ____
3. Are there any traps which need new bait? ____ ____
4. Do toilet and eating facilities need to be sprayed with pesticide for insects? ____ ____

Additional Areas to Be Checked

1. _____ ____ ____

2. _____ ____ ____

3. _____ ____ ____

4. _____ ____ ____

24

CONTRACTS AND LIABILITY

A public warehouseman can be legally described as a bailee-for-hire. The responsibilities of a bailee-for-hire have been described in common law for centuries. Yet the contractual responsibility of a public warehouseman is frequently misunderstood.[1] Since the user wants maximum protection from loss, the operator must be sure that his or her liability insurance is not compromised.

Any warehouse user dealing with public warehousemen must have a clear understanding of the liability of these suppliers, and of generally accepted contracts which control that liability.

In the U.S., the responsibility of a warehouseman is specifically outlined in the Uniform Commercial Code *(see Appendix I)* This code is used in every state except Louisiana, which uses the Uniform Warehouse Receipts Act. Under the U.C.C., the warehouseman is responsible for issuing a warehouse receipt for stored goods, for exercising due care while the goods are in his custody, and for delivering the goods upon appropriate demand. The UCC section describing the warehouseman's duty of care states that a warehouseman is liable for damages for loss or injury to goods if such damage could not have been avoided had he or she exercised that quality of care that a reasonably careful person would have exercised under like circumstances. The UCC also provides that a warehouseman cannot impair this duty of care. For example, he cannot provide that he will not be

---

[1] The author is indebted to Richard R. Murphey, Attorney, Murphey, Young & Smith, for help in preparation of this chapter. William H. Towle, General Counsel for A.W.A., reviewed and revised the chapter in 1988.

responsible for loss due to fire, or any other cause. One might well conclude that the only contract needed in the U.S. between the warehouse operator and his customer is one in which the operator agrees to be responsible to the full extent provided by law.

However, many warehouse users are not satisfied with such a simple agreement. In a desire to combine legal contract and operating procedure, some corporate attorneys have developed complex agreements. Seeking to achieve simplicity and standardization, the American Warehousemen's Association has developed a set of contract terms and conditions. They have been constantly revised since the first version was issued in 1926. The latest revisions, published in 1976, are reproduced as Appendix I.

Any warehouse user dealing with public warehousemen must have a clear understanding of the liability of these suppliers, and of generally accepted contracts which control that liability.

The Changing Role of the Warehouseman

The user's interest should be in the business justification and discipline underlying the contracts, rather than in the form of a document. Originally the only function of the warehouseman was that of a custodian. He delivered goods placed in his care to a depositor or to a designated third party. Storage and protection were almost his only duties. This limited function of caretaker has long since been expanded to a variety of warehousing services. The development of the warehousing industry, particularly advances since World War II, have plucked warehousemen from the historic role as custodian and placed them in the center of the complex pattern of today's distribution system. Warehousemen have assumed responsibilities, on behalf of the depositor, to a host of entities with whom they have no direct contractual relationship. They now are frequently responsible for seeing that goods are delivered to third party consignees following their period of storage. They must deal with common carriers. At times, the creditors of the depositor take an interest in the goods that are stored. Insurance companies are always extremely interested in their activities. And today, various agencies of government monitor cleanliness, health and safety as they apply to the goods stored or shipped.

The Insurance Problem

The warehouseman's ability to perform this expanding role at a reasonable cost is to some extent dependent on the availability of liability insurance coverage. There is a vast difference between the typical net worth of a public warehouseman and that of most customers served. Few public warehousemen could prudently be self-insurers. Yet for many years, a warehouseman's liability insurance policy was difficult or impossible to buy. The standard warehousemen's liability policy currently does not provide coverage if the warehouseman's duty of care with respect to goods in his or her charge goes beyond the standards set out in section 7-204 of the Uniform Commercial Code. Therefore, both warehouseman and customer must be vigilant to see that no contract requires the warehouseman to assume obligations that go beyond the responsibility described in that section. On the other hand, since many users will insist on a contract, the warehouse operator must determine whether the contract is one which he or she can sign without the danger of compromising insurance coverage.

Few public warehousemen will hesitate to sign the AWA contract forms. If the customer presents a new agreement, the prudent public warehouseman will refuse to sign it until it has been checked to be certain it does not endanger insurance coverage.

Customer's Contracts

Some warehouse users insist on their own contract, perhaps one that would expand the liability of the public warehouseman. In dealing with such situations, the warehouseman should ask customers whether they really want to deal with an uninsured supplier who could not pay a major claim if it occurred. It is to the mutual benefit of both parties to have an agreement providing service at a cost that makes economic sense. If the agreement would endanger insurance coverage, any economic benefits of the arrangement are not worth much.

In examining customer-drafted contracts, the warehouseman should be alert for clauses likely to create problems. Some contracts are so absolute with regard to assumption of liability that they quickly surpass the Uniform Commercial Code. Clauses such as "any and all loss," or "assumes absolute liability for," or "warehouse shall have the burden of proving," or "any and all claims," are the types of clauses that frequently are signs of a potentially troublesome contract.

Liability Limitation

In complying with section 7-204 of the UCC, the warehouseman should remember that liability may be limited for loss or damage to goods by an agreed-on amount. That limitation is very flexible, with nothing in the law stating that the limit should be 200 times the base storage rate, or any other figure. This figure may properly be negotiated in each contract.

A Meeting of the Minds

Some users would have a contract that describes in detail the jobs the warehouseman will perform. Others have left such description to a procedures manual, which becomes part of the agreement between user and operator. Each change in the procedures should be approved by both parties to ensure that it is noted. The services provided by the warehouseman are very broad and become more so each year, and describing them clearly in a contract becomes more difficult.

Most misunderstandings arise from a failure of the parties to discuss fully and understand the implications of their responsibilities. If the parties really do not have a meeting of the minds, the writing of a contract may not be a practical solution to the problem. Few warehouse users would want the intervention of a court of law in developing their distribution program.

Since public warehousing arrangements are customarily subject to termination on 30 days' notice, the user can always remedy a service failure by moving to another warehouse. And since public warehousing has traditionally been a business service without long-term agreements, one might again ask why a detailed contract is really nec-

CONTRACTS AND LIABILITY

essary. The warehouseman holds the customer by giving good service every day, and knows that is the only way to keep the customer.

The most successful relationship is one in which the warehouse operator is not just a physical resource, but a distribution planning consultant. If this relationship exists, then the contractual understandings should provide for written communication of control documents, to eliminate potential for misunderstandings, adequate notice of changes in operating procedures of either party, and a recognition by both parties of the physical capacity of the warehouse in terms of its size, shipping and receiving capabilities.

A contract between warehouseman and depositor usually attempts to reflect the understanding between the two as to services that will be performed. The user and the operator must perceive each other as partners in accomplishing a distribution objective instead of as adversaries bound by a contract.

25

VERIFICATION OF INVENTORIES AND CYCLE COUNTING

If there is any warehousing job that is usually taken for granted, it is that of taking physical inventories.[1] This job is considered a "necessary evil," yet it involves a considerable investment in terms of resources—people, equipment and time. While there are no magic solutions, certain systems, most notably bar coding, can help improve the accuracy and shorten the time of physical inventories.

A newer alternative to taking physical inventory is cycle counting, a technique which many people believe to be significantly more efficient that the traditional approach.

Purpose of Inventories

Why even take physical inventories? An obvious answer lies in the impact that inventory has on the corporate financial statement. Finance people are certainly concerned about accountability, and public accountants are retained to verify the accountants' work. The very fact that we need to take physical inventories is a symptom of our failure to receive, store, ship and record inventory correctly.

Annual physical inventories are similar to annual medical ex-

---

[1] Most of this chapter is based on an article by W.G. Sheehan, President, Distribution Centers, Inc., in Warehousing and Physical Distribution Productivity Report, Vol. 16, No. 11. Marketing Publications, Inc., Silver Spring, MD. Revised with the help of Bruce A. Edwards, Distribution Centers, Inc.

aminations— they provide a look at what is potentially wrong and may provide the opportunity to correct past errors. Considering the impact that physical inventories have on operations, they deserve more attention and emphasis than we normally give them. Inaccurate inventories can impair a company's ability to plan materials requirements, and cause uncertainty in financial planning. Inaccurate inventories lead to lost sales, lower fill rates, and lost productivity in the warehouse.

How often should the warehouse take physical inventory? Only you can decide for your company. But bear in mind that loss of assets is not the only loss suffered by the company when the physical inventory and the book inventory do not balance.[2]

Inventory Carrying Cost

Inventory carrying cost is one of the major costs of doing business—anywhere from 30 to 60 percent of the value of the inventory carried. Inventory cost as a percentage of sales is largely regulated by the rate of turnover. For example, if it costs 30 percent of the value of the goods to carry the inventory and you are turning the inventory over at a rate of 10 times each year, inventory carrying costs are about three percent of sales. Failing to turn the inventory can erode profits.

Benefits of Improvement

Have you measured the investment in resources you make each year in performing physical inventories? How much time is spent on your physical inventories in planning, counting, recounting and reconciling?

What would be the potential savings for a 10 to 20 percent reduction in the hours required for physical inventories? In many cases the savings can be substantial.

---

2 From Distribution Warehouse Cost Digest, Vol. 12, No. 18, Marketing Publications, Inc., Silver Spring, MD.

VERIFICATION OF INVENTORIES AND CYCLE COUNTING

Problem Areas

Neglect may be the biggest problem. Too many things are taken for granted. Too many potential problems are overlooked. Too much time is spent "fighting fires." Many counts are not accurate, hence have to be redone. Too many counts are impossible to trace, including who did what, where, when and how. There is a lack of communication between the warehouse, the office and supervisors. This is often compounded by inaccurate counting on the part of those persons actually taking the physical inventory.

How Can a 'Physical' Be Improved?

Remember that each department or function in your company needs the results for different reasons. Finance needs the numbers for accountability. Marketing needs to be sure what is available to sell and ship. Everybody is concerned about accuracy!

The primary purpose of the physical inventory is to determine the correct quantity and location of material in storage. Accuracy in the location system alone will provide tangible improvement in warehouse productivity.

Fiscal responsibility for stored goods is an inherent responsibility of the warehouse manager. In addition to confirming the accuracy of inbound and outbound shipments, completing a reliable physical inventory is a measure of management's success in protecting stored merchandise.

To achieve a reliable count, it is best to conduct two separate and independent counts. Reconciliation of the two counts is paramount in proving the reliability of the physical inventory without regard to differences between the physical and "book" balances. Reliability is enhanced through proper scheduling of the inventory. If the warehouse manager can be completely sure a thorough and reliable physical count has been taken, much more credence is provided to subsequent reconciliations with the book inventory.

In order to be sure that the counts are truly independent, one or more persons should be assigned to different areas for the first and second counts. If the inventory is to be effective, it must be accurate, fiscally responsible, reliable.

Efficiency in inventories is achieved by minimizing resources required. Not only do we want an accurate count, but we also want the inventory conducted within a reasonable or minimum time.

Incentives

How can efficiency and effectiveness be combined? One suggestion is to provide for an incentive system. During the actual count it is vital to achieve a balance between effectiveness and efficiency. There is little to be gained by rushing through a physical count and then doing an inordinate number of recounts. Nor is it sensible for the counters to take so much time ensuring an infallible count that taking the inventory becomes extremely time-consuming.

In order to achieve the proper balance it is not only appropriate, but necessary to establish a goal for the number of required recounts or third counts. While this goal will fluctuate, a goal of 98 percent accuracy between the first two counts is certainly realistic. Considering the total number of manhours invested in a physical inventory, it is feasible to provide some incentives to workers who do the count for meeting performance goals of quantity and quality. One incentive program provided money prizes to each member of the team that counted the most items or turned in the greatest number of inventory tickets. Another prize went to team members who achieved the most accurate count. One criterion, however, was that the winners in each category had to rank in the upper third in the other category. The team that counted the most tickets also had to have a reasonable level of accuracy, with the most accurate team also having to achieve a given level of productivity.

Achieving efficiency and effectiveness through motivation—the incentive system—will assure the reliability necessary for the reconciliation process. Reducing the number of third counts can materially reduce the hours devoted to the inventory-taking process.

VERIFICATION OF INVENTORIES AND CYCLE COUNTING

Managing the Job

In order to improve the taking of physical inventories, we draw on the classic management principles of planning, staffing, organizing, coordinating and controlling. Plan for physical inventories. An efficient physical inventory starts weeks or even months before the actual inventory date.

The first step is to assign an inventory coordinator for each of the physical inventories to be conducted. This appointment pinpoints responsibility and establishes a sole source for information flow. An annual wall calendar can be posted and marked with dates (whether tentative or firm) for inventories. The calendar should include specific dates for preparation for the inventory.

Involve the Whole Team

After suffering several excruciatingly painful physical inventories, a group of warehouse and office employees met to share ideas on the experience. It was interesting to note the enthusiasm generated simply by including employees in the planning process. And good standard inventory procedures can be developed as a result of this type of meeting.

Getting Organized

The next step to take prior to the actual physical inventory date is to make every effort to get rid of obsolete items, waste, or excess material that can only lengthen the time required to take the count. Then, two weeks before the inventory date, get organized by bringing key personnel together to touch base on responsibilities, availability of workers, equipment requirements, supplies required, etc.

Advance Preparation

Decide whether the inventory should be pre-tagged, or tagged during the first count. Some computer systems can easily produce inventory tags preprinted by location. You may want to consider filling out inventory tickets in advance—but not the quantity!—in order to facilitate the actual taking of the physical inventory. Damaged goods and items in the repack area should be counted in advance. Employees must be trained how to fill out tags and who to contact regarding problems with the tags.

The Importance of Cutoff

A cutoff date for receipts and shipments must be established for an accurate reconciliation after the physical count has been completed. Certain areas should be zoned off (and not included in the actual physical inventory) for material received after the cutoff date, or material shown on the inventory records as shipped, even though it has not left the shipping dock. Documents for merchandise received or shipped after the cutoff date should be stamped so they are identified as *not* part of the reconciliation process.

Zones

A useful technique is to zone the inventory area into sections that require similar amounts of time to count. Teams should then be assigned to those sections. After the physical inventory coordinator has developed the plan, zoned the area to be counted and assigned inventory teams by zones for first and second counts, conduct a physical inventory briefing for all persons concerned. This briefing should be for the counters in the warehouse, as well as clerical personnel who will work on the physical inventory and subsequent reconciliation. Highlight any difficult-to-count items, potentially difficult product codes, or other unique identification problems. Have examples of such items available for everyone to see before starting a count. The meeting should cover any unique requirements, team assignments, designated zones, an expected time schedule, how the tickets are to be filled out, and requirements to record serial numbers, lot numbers, or production date codes.

This meeting also is the time to warn personnel conducting the physical inventory about possible differences in pallet patterns and the difference it can make on the actual quantity on hand. This is particularly important since so many companies have standard unit-load programs.

VERIFICATION OF INVENTORIES AND CYCLE COUNTING

Arithmetic Aids

If bar code readers cannot be used, encourage your warehouse personnel to use hand-held calculators. Issue guidelines on how or where counters are to make calculations. Don't allow them to write such calculations on the carton or, worse yet, on the inventory ticket, because if they do, it becomes impossible to ensure the independence of separate and distinct physical counts. Provide inventory teams with scratch paper and have them make their calculations on the back of *the ticket that is turned in*. Keeping these calculations assists in providing an audit trail when reconciling discrepancies.

Controlling the Count

Having a supervisor present during the count is valuable in ensuring an effective and efficient inventory. This individual should constantly spot-check procedures, read inventory tickets to ensure they have been properly filled out, and then sample some counts for accuracy.

We can't over-emphasize the necessity for proper control of inventory tickets. This control begins with each inventory ticket having a unique serial number. The tickets should be handed out in sequence, according to zones, and should be placed on the product, also in sequence, so that an audit trail is maintained. This also expedites any third counts needed, since the counter will know precisely where to find the inventory ticket in question. The coordinator must ensure that inventory teams turn in all tickets, then verify that every ticket is accounted for at the conclusion of the inventory. One simple way to provide for this is to void the first and last tickets for each inventory.

Reconciliation

Remember that the end of the count is not the end of the inventory. Office procedures are just as important as warehouse procedures. Exercising proper control over the inventory in the warehouse is no more important than exercising proper control over the inventory within the office. The inventory is not complete until all counts have been reconciled and inventory records updated.

An important last step of inventory-taking is a review among team leaders, including representatives from the warehouse and office, which allows participants to cover problem areas and to solicit suggestions for improving future inventories. Notes should be kept concerning the number of hours required for the inventory, the number of persons, the equipment, the items counted and the quantities of merchandise involved. This information is useful in improving planning for the next inventory. You can also use this session to recognize effective performance.

Eliminating Inventory Errors by Cycle Counting[3]

The goal of every warehouse and distribution center manager is inventory accuracy. The best way to achieve this is to count a small percentage of the inventory on a regular cycle basis. This sample of the total inventory can be easily compared to the inventory records, and since you are continually counting, when errors are identified, you have an excellent opportunity to define the cause of the error and take corrective action. This process of regular cycle counts, reconciling with the inventory records, identifying and correcting these errors, is called cycle counting.

It is important to realize that the immediate objective of cycle counting is not accurate inventory; it is the identification and elimination of errors. A by-product of identifying and eliminating errors will be an accurate inventory.

To respond to the question, "What to count?" one must realize that this question is really two questions in one. The first question is, "How often should an item be counted?" and the second is, "When should each specific item be counted?" The answer to the first question lies in Pareto's Law. Pareto's Law, or the A-B-C Concept, has many different applications but the A-B-C's of cycle counting are:

- Approximately 80% of a warehouse's dollar throughput is typically attributed to 20% of the stockkeeping units (A items),

---

[3] This subchapter is by James A. Tompkins, Ph.D., President of Tompkins Associates, Inc., Raleigh, NC.

VERIFICATION OF INVENTORIES AND CYCLE COUNTING

- Approximately 15% of a warehouse's dollar throughput is typically attributed to 40% of the stockkeeping units (B items), and
- Approximately 5% of a warehouse's dollar throughput is typically attributed to 40% of the stockkeeping units (C items).

The obvious cycle counting application is that the A items should be counted much more frequently than the B items, and the B items much more frequently than the C items. A typical scenario would involve counting six percent of your A items each week, four percent of your B items each week, and two percent of your C items each week.

The second question, "When should each specific item be counted?" is relatively straightforward. An item should be counted when it is easiest and cheapest to get the most accurate count. When is this?

1. When an item is reordered.
2. When an inventory balance is zero or a negative quantity.
3. When an order is received.
4. When the inventory balance is low.

Thus, the answer to the question, "What to count?" is that you should count on an A-B-C basis and count when it is easy to count.

The number of people who should do cycle counts depends upon the number of items in inventory, the desired count frequency, the number of storage locations for each item, (and the number of count irregularities such as number of recounts), accessibility of items, and physical characteristics of the items. A realistic standard is that a cycle counter can count forty items per-day. The cycle counters should be very familiar with the stock location system, the warehouse layout, and the items being counted. Cycle counters should be assigned to the job on a permanent basis. This does not mean that cycle counting is necessarily a full-time job, only that the persons assigned to cycle counting should be part of a permanent team.

Cycle counters must recognize the probability of crossovers. A well-established cycle counting procedure will include checks of items which are likely to be misshipped. When a cycle count reveals an overage in one item, there is an immediate check of those items which would normally be confused with the one which is not in balance.

Some auditors are concerned with the "proper checks and balances" within a warehouse, and believe that only people from outside the warehouse should be assigned full-time to cycle counting. In fact, to the contrary, it has been shown that true inventory management will begin when management realizes that the job of maintaining accurate inventory is not that of accountants and auditors, but that of warehouse and distribution managers. Thus, the whole "checks and balances" discussion is irrelevant and it doesn't matter to whom the cycle counters report.

Having determined what to count and who should do the counting, the remaining question is "What are the procedures for cycle counting?" Valid cutoff controls are critical. As cycle counting should take place without affecting normal operations, a very careful coordination of counting and transaction processing must be planned. Any transaction which takes place after the inventory balance is reported and the actual count is made must be isolated so that an accurate inventory reconciliation may take place. This can be done in many ways:

- Record all inventory transactions and cycle counts in "real time."
- Record the time when each location is counted and report transaction times.
- Coordinate transaction with the count process so that counts are taken when transactions do not occur.
- Don't process any transaction for items scheduled to be counted until after the items are counted.

Once the cutoff controls have been set up, the next procedure is count documentation. The information which should initially be recorded on the cycle count document is:

- part number
- part location

VERIFICATION OF INVENTORIES AND CYCLE COUNTING

The cycle counter should then record:
- count quantity
- counter's name
- date and time of count

The cycle counting document should then be forwarded to a reconciler. Counters should not reconcile their own counts. Reconcilers should compare the cycle count to the inventory record and determine if it was a good count. For this, one needs to have established a cycle count tolerance limit. Cycle count tolerance limits should be based upon the value of the items being counted. For example, if a $1,500 item is counted and a cycle count of 490 is made, and this is reconciled against an inventory record of 500, it is clear that this is a bad count. But if a $.02 item is counted and a cycle count of 490 is made, and this is reconciled against an inventory record of 500, it is clear that this is a good count. A typical count tolerance limit would be $50. That is, if the cycle count is within $50 of the inventory record, the count is considered to be a good count. If the cycle count indicates a discrepancy of $50 or more, it is considered a bad count. When a bad count occurs, the item should be recounted. The recount will verify

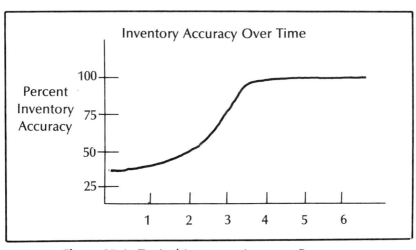

Figure 25-1. Typical Inventory Accuracy Progress

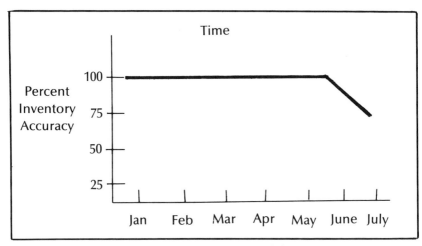

Figure 25-2. A Problem Indicating a Need for Corrective Action.

if the original count was truly a bad count or if an error was made and the item count is within the tolerance limit. Whether the recount variance is within the count tolerance limit or not, the reconciler should adjust the inventory record to agree with the verified recount. If the variance of the recount is outside the tolerance limits, further investigation is necessary.

The further investigation of a bad count is required to determine why the inventory record is incorrect. If the counted quantity is verified as less than the inventory record, there are two easily checked possibilities:

1. An outstanding allocation of this item may have been filled without recording the transaction, or

2. A recently completed sales or production order may have been incorrectly subtracted from inventory.

If the variance shows that the count quantity is higher than the inventory record, replenishment orders should be checked to see if any product was received and not properly recorded, and recently completed sales or production order should be checked for erroneous subtractions from the inventory records. It is important to recall that

VERIFICATION OF INVENTORIES AND CYCLE COUNTING

the objective of cycle counting is the identification and elimination of errors. Thus, the error investigation process is critical in cycle counting.

At the end of each month, a cycle counting accuracy plot should be given to management. Cycle counting accuracy is defined as:

$$Cycle\ Counting\ Accuracy = \frac{Good\ Counts}{Total\ Counts}$$

Figure 25-1 presents a graph illustrating the typical progress made once cycle counting is implemented. If a problem occurs, such as that shown in Figure 25-2, it is crucial that either the error investigation identify the problem and resolve it, or that the cycle counting quantities be increased until the errors can be identified and eliminated.

Cycle counting should be implemented in two phases. The first phase:

1. Identify approximately 100 items. These items should be A, B and C items.

2. Repeat counts on these items. Count frequency should be whatever is necessary to identify errors.

3. Eliminate errors.

Once the first phase results in 100% accuracy for several weeks, the full-cycle counting system should be implemented. But the original 100 items should continue to be counted for at least two months. The special counts on these 100 items should be continued until a 98%+ inventory accuracy is achieved.

Some have questioned whether or not public accountants who are responsible for audits will accept a cycle count. The fact is that cycle counting was developed because of frequent and severe problems in the accuracy of traditional physical inventories. As cycle counting has matured, auditors have recognized that the procedure is usually far superior to the traditional physical inventory. As long as the cycle counting is cleared with the audit team, a cycle count is recognized as a superior means of checking inventory.

What Does the Future Hold?

Considering where we stand today with computer technology and automated systems, it is not difficult to project where we might be two or five years hence with respect to taking physical inventories. By using universal product codes, as well as optical character readers, the job of taking physical inventories will become less difficult and less time-consuming, as well as more accurate.

If inventory changes can be recorded and records updated on receipt of material and shipment (or sale of material), then it is a natural extension of that concept to incorporate existing technology, equipment and systems into a truly automated physical inventory system.

Product codes or item identification can be input to computer through the "wand" or readers now being used with several distribution systems. With portable keyboard/wand equipment, it is possible to input the count accurately either manually or on the basis of a standard unit-load program.

If a warehouse handles items in a standard pallet pattern, and this information is preloaded in the computer, the simple reading of a universal code with the standard pallet quantity will eliminate many mathematical calculations now done by the inventory team.

PART VI

THE HUMAN ELEMENT

26

ORIENTATION AND TRAINING

Some companies do little more for the new employee than to explain who the boss is and where the restroom is located. Since every new employee will be "filled-in" on the company by somebody, management should see that this orientation is directed by somebody knowledgeable. Orientation should include:
- Description of the industry.
- History of the company.
- Unique advantages of the company and its industry.
- Information about company management and ownership.
- Identification of company customers and products handled.
- Information about unions and labor agreements (if any).
- Procedure for resolving problems or misunderstandings.
- Location of restroom, lunch facilities, parking, time clock.
- Details about pay and benefits.
- Recommended work clothing.
- Work and safety rules.

This information is often best provided by the employee's supervisor. This allows the new employee the chance to become acquainted with the supervisor, and it establishes him or her as the best source of information about the job and the company.

Figure 26-1 shows excerpts from an employees' handbook used by a public warehouse company. The full handbook gives details about benefit programs, work rules and company history.

PRACTICAL HANDBOOK OF WAREHOUSING

Figure 26-1
Excerpts from Personnel Handbook

Personnel Relations Policies

Personal honesty, responsibility, enthusiastic cooperation, loyalty and productivity are expected of everyone.

It is the policy of the company to treat everyone with respect, provide fair treatment for everyone, good, safe working conditions, superior benefits, and wages comparable with other companies in the public warehousing industry.

Anyone should feel free to discuss his work or work relationships with his supervisor or, through his supervisor, with any other member of management.

Whenever possible, promotions are made from within the company thus providing the opportunity for all to advance to the limits of their capabilities.

Work Schedule—The normal work week begins at 12:01 a.m. Sunday and ends at 12:00 p.m. the following Saturday. Work schedules vary at each location due to local business hours or shift schedules. Our normal warehouse hours are 8:00 a.m. to 4:30 p.m. with a one-half hour lunch period and a morning coffee break.

Overtime—Overtime, in excess of forty hours per week (or in accordance with local labor agreement) for hourly employees, will be compensated at the rate of time and one-half. Supervisors and management trainees will work some overtime.

Pay Categories—Hourly warehouse and clerical employees are paid at a specific rate per hour and are paid time and one-half for all hours over 40 per week. Hourly employees use the time clock and time cards.

Weekly salaried employees are paid by the week. The hours worked, lost time, vacations, etc. are recorded on a weekly time card.

All semi-monthly salaried employees are paid twice per month. They have no time cards and overtime is not paid.

Time Card Regulations—Each employee is required to punch his or her own time card not more than 12 minutes earlier than the scheduled start of the shift. At the end of the shift the time card must be punched immediately. When an employee leaves the premises for lunch or other personal business, the employee must clock out upon leaving and in upon returning.

Adjustments—All pay adjustments are made on the basis of $1/10$ of an hour. You must work a full $1/10$ hour (6 minutes) to earn $1/10$ hour of overtime pay. If you are late from 1 through 6 minutes, you will lose 6 minutes pay. If you are late 7 minutes, you will lose 12 minutes pay, etc.

Holidays—The entire company observes holidays as established between management and the collective bargaining unit in local collective bargaining agreements. Each full time employee will receive eight hours pay for established holidays which generally include:

ORIENTATION AND TRAINING

New Years Day
Memorial Day
Independence Day
Labor Day
Thanksgiving Day
Christmas Day
Additional day in conjunction with Christmas or New Years

To be eligible for holiday or shift pay, you must have worked both your last scheduled work day or shift prior to the holiday and your first scheduled work day or shift following the holiday. If a holiday falls on Sunday, warehouses will be closed on the following Monday. For holidays falling on Saturday, warehouses will be closed on the preceding Friday. You will not receive holiday premium pay unless you actually work on the calendar day on which the holiday falls. Holiday premium pay or additional holidays shall be stipulated in the collective bargaining agreement at each location.

Vacations—All full-time employees earn paid vacations on the anniversary date of hire as a full-time employee in accordance with the following schedule or as indicated in the local collective bargaining agreement.

After the 1st year—
   5 days at regular rate (1 week)
After the 2nd year—
   10 days at regular rate (2 weeks)
After the 10th year—
   15 days at regular rate (3 weeks)

A list will be posted each April for you to request your preferred vacation period. Your supervisor will make every effort to allow you to take your vacation at the time requested. However, the company reserves the right to limit the number of employees on vacation at any one time to insure adequate service to our customers. Eligible employees must take at least one week vacation each year. If a holiday occurs during your vacation, you will be paid for it or your vacation may be extended one day.

Managers must take their full vacation each year. Since managers, supervisors and the corporate clerical staff often work more than a 40 hour week, they receive 2 weeks vacation for one to seven years' service, and 3 weeks after seven years.

Unused vacation may not be accumulated from one anniversary year to the next, except by written approval of the general manager.

Call-in Pay—Hourly employees will receive at least 4 hours pay whenever called to report to work, unless the plant is shut down by an emergency (storm, power failure, etc.) which is beyond the control of management.

Death in Family—Any employee having a death in the immediate family will be given up to a 3 day paid leave to attend the funeral. The immediate family is considered to be wife or husband, child, parent, brother, sister, or as stipulated in the local collective bargaining agreement.

In nearly every organization there are three sources of information on how the company operates: the supervisor, the shop steward and the "grapevine." Because these information sources are competitive, evidence of unreliability in any one of them will discredit that source for the future. If the supervisor is to be credited as the best information source, that individual must be most supportive, well-informed and honest in dealing with subordinates.

Training

There was a time when the average warehouse manager hired employees by seeking people who already had the desired skills. If fork-lift operators were needed, an advertisement was placed for experienced lift-truck operators. The warehouse manager sometimes evaluated applicants by what they said they knew about operating lift trucks, not how much skill they actually might have. In more recent years, some of the best warehouse managers have come to realize that a well-motivated person who wants to learn will be a far better warehouse employee than the applicant who claims that "I can drive anything!"

Training aids have now been developed by warehouse operators, lift truck manufacturers, and trade associations. They are designed to help inexperienced people become skilled warehousemen.

Additional job training following first-day orientation should be conducted by an experienced employee, a group leader, a supervisor, a member of higher management, or a professional trainer.

Because training is a never-ending process, it should not be restricted to newly-hired people. New procedures, different equipment, and new customers create conditions where re-training is needed. Most people appreciate an opportunity to be involved in a continuing training and education program, since it gives them a chance to learn to do the job better. On the other hand, serious job dissatisfaction can arise from lack of training opportunities.

Some Training Examples

One warehousing company has a four-week program for trainees, most of whom have never worked in a warehouse before. The first week includes orientation, including a detailed tour, an explanation of lift trucks, and a viewing of safety films provided by lift truck manufacturers.

During the second week, workers are trained "live" with the equipment which they will use under the supervision of a training coordinator. The third week is spent in the warehouse, where the new worker is instructed while actively involved in a variety of warehouse tasks. The fourth week is also spent in the warehouse doing typical warehouse jobs. After four weeks, the new worker's progress is evaluated. Those who need additional instruction are given remedial attention.

Another warehousing company has a trainer checklist designed to guide the people responsible for training. Figure 26-2 shows a typical page from this checklist.

One company which offers contract warehousing services has an extensive training course which includes obstacle courses to test the operating skills of trainees after they have received orientation and training in the use of mobile equipment. Figure 26-3 shows three courses designed to develop maneuvering skills for lift trucks.

NHAW (Northamerican Heating & Airconditioning Wholesalers Association) has prepared an 80-page home study program titled *"Fundamentals of Materials Handling."* One of the unique features of this publication is a glossary of terms used in warehousing. A sample exhibit from the publication is shown as Figure 26-4.

Retraining for New Equipment

Many warehouses are making the transition from fork-lift and pallet handling to the use of slipsheets or other palletless handling methods. Training for this transition represents a specific example of how a warehouse training program can function.

Since the maneuvers involved in handling a truck equipped with either a carton clamp or push-pull attachment are more complex than conventional fork-lift handling, the lift-truck operator must learn new skills. Furthermore, the probability of hiring drivers who already possess these skills is slim, since specialized attachments are used less frequently than forks in most warehouse operations.

When palletless handling is introduced to a warehouse previously palletized, workers may react to the change with fear or uncertainty. Therefore, a key to a successful training program is the positive motivation of the warehousemen whose skills are needed to implement the new system.

Trainer Checklist

Effective training programs must be carried out by people who have developed a thorough understanding of lift trucks and have devoted the time necessary to prepare a program that will help them effectively accomplish these training objectives.

About Lift Trucks

How lift trucks differ from other vehicles. How lift trucks are built generally and how they operate. What uprights and forks do and how they are different. Uses and types of attachments and how they affect driving and load capacity. Matching lift trucks to loads and operating conditions. What happens when a truck "lifts."

About Using Lift Trucks

Good practice for starting the days' work. Using maintenance reports. Safe practices in driving, turning, climbing and descending, parking and stopping. Safe practices for picking up and depositing loads. Safe practices before entering truck and railroad cars. Safe practices for the unexpected situation. Watching the truck's "health" during the day. The real test of productivity—steady, careful operation. Safe practice of installing an LPG tank on trucks so equipped. Proper methods of battery care on electric trucks.

Supervisor's Relations with Operations

Do operators understand and practice daily maintenance and emergency service procedures? Do they understand and observe limitations on loads? Do they understand and observe limitations on speed, turning, stopping, and accelerating? Do they understand and accommodate clearance requirements in operating areas? Do they understand and practice best load positioning, release and pick-up techiniques? Do they understand the effects of health and attitudes on safety of truck, load, fellow workers, and themselves?

Courtesy of Distribution Centers, Inc.

Figure 26-2

ORIENTATION AND TRAINING

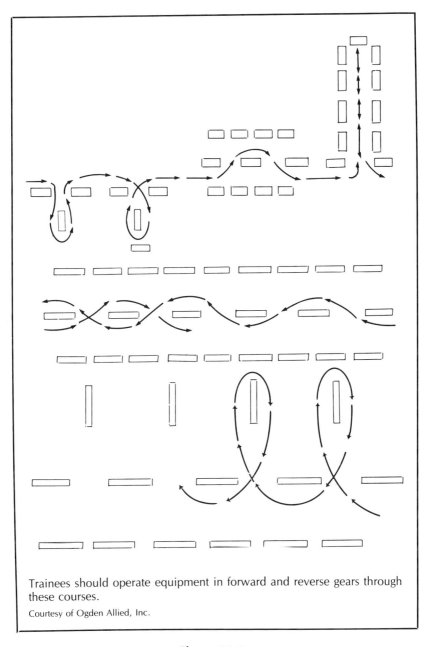

Trainees should operate equipment in forward and reverse gears through these courses.

Courtesy of Ogden Allied, Inc.

Figure 26-3

PRACTICAL HANDBOOK OF WAREHOUSING

One grocery products warehouseman has devised a simple method to help individuals train themselves in slipsheet handling. A worker takes a stack of eight or ten pallets and uses them as a dummy load. The stack of pallets is placed on top of a slipsheet, and the driver practices picking up and pushing off this load until he gains confidence in controlling the equipment. The same technique also is used to train the lift driver to handle unit loads high in a stack, where vis-

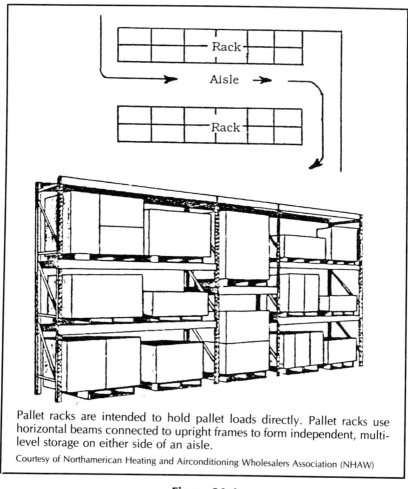

Pallet racks are intended to hold pallet loads directly. Pallet racks use horizontal beams connected to upright frames to form independent, multi-level storage on either side of an aisle.
Courtesy of Northamerican Heating and Airconditioning Wholesalers Association (NHAW)

Figure 26-4

ibility is marginal. Using a stack of pallets instead of a real load of merchandise enables the operator to practice with little danger of property damage.

Films and Video Tape

The Clark Equipment Company has developed an interactive video training program which has been used by many warehouse operators. The video enables the trainee to study the program at his or her own level of comprehension. The program uses a video cassette recorder tied into a small printer which records the success of the student in taking short quizzes. Clark offers three films designed to be used with the interactive video. The first is called "The Great Betrayal," the second is called "The Great Deception," and the third is called "Live to Tell About It." The emphasis of the films is safety. Photography and acting in the training films is superior, far better than most training films which we see. Furthermore, the tests are not easy. A multiple-choice test may have as many as six options, and sometimes four of them are correct. Unless the trainee keys in all the correct options, the answer is counted as incorrect, and as remedial training the trainee is required to watch that section of the film again until a correct answer is given.

Some firms have developed their own training materials. A wholesale grocery company commissioned a nine-minute television tape designed specifically to train its own order pickers. During the first week on the job, the order picker is asked to look at the tape at least twice. Still more viewings are available if a person seems in need of additional instruction. The tape includes warnings about safety practices and disciplinary information as well as ideas about how to do the job correctly.

Sometimes equipment companies have teamed up with their customers in developing training tapes. One video training tape in the use of push-pull attachments is a cooperative effort by the attachment manufacturer and a leading health-care products manufacturer. While it naturally shows its sponsor's products in the illustrations, the tape is not commercial. It emphasizes safety and shows details of equipment maintenance and correct operation of the attachment. It is designed for managers and lift-truck operators who are using this equipment for the first time.

Training films have many operator tips to make the job easier, such as the suggestion that the lift platens should be highly polished with steel wool and wax to improve the operation. One film suggests the use of stops to limit the travel of the push-pull bar and prevent damage. A technique known as "scooping" is demonstrated, in which the platens are chiseled under the load without using the bar to pull the load onto the forks. Scooping can be done only when the load is resting against another stack of merchandise or a wall that will stabilize the load. The film points out that scooping is faster than pulling the load with the attachment. It also prevents damage to the lip of the slipsheet.

Operator's Guide

A printed operator's guide has been published by the Cascade Corp. to deal with load-push and push-pull attachments. Shown in Figure 26-5 are steps for pickup of a slipsheet load from the floor.

The manual points out the importance of keeping the platens at a downward angle when the load is picked up. It also notes that handling of sloppy or poorly distributed loads should not even be attempted.

Mentoring

Mentoring is a system of pairing a new employee with an experienced one as a tutor. The mentor is a manager or executive who is completely familiar with company policies, but is also a recognized "people person." Each mentor has only one new employee at a time. Under a formal mentoring program, a six-month schedule is developed. The mentor visits his "pupil" every two weeks during the first three months of employment, followed by one visit per-month during the final three months. This formal schedule should not prohibit informal visits at any time, since the chance to ask questions on a random basis can be an important feature of the program. In all programs confidentiality between employee and mentor is essential. The role of the mentor is to give advice on company policy and procedures, but to provide personal advice only when asked.

But mentoring is a time-consuming process. It requires frequent interaction with the employee and could result in a short-term drop in productivity. If mentoring does not involve all employees, it could also cause resentment among those not chosen to participate.

Record Keeping

Comprehensive training files should be maintained for all trainees.[1] These records include "exam" and project performance data, as well as general evaluations by the course instructor. A standardized form to evaluate the trainee on such points as attendance, participation, examination scores and class project is used. The form and the file are periodically reviewed by the trainee's supervisor. The files are effective in charting the employee's development, noting strengths and weaknesses, and for evaluating the trainee for potential management positions.

A course instructor evaluation should be administered at the conclusion of each training session, using a standard evaluation form.

Goals of Training

As you create or revise your training program, a final check of it should be a review to be certain that everything in it is related to the goals of the training program. While those goals will differ from one company to another, most warehousing organizations would include at least these six training goals:

- to improve safety, sanitation and security,
- to help every employee make decisions which improve productivity and quality,

---

1 From an article by Professor Thomas W. Speh, Miami University (Ohio) and Albert D. Creel, consultant, from Vol. 16, No. 2 of *Warehousing and Physical Distribution Productivity Report*. Marketing Publications, Inc., Silver Spring, MD.

PRACTICAL HANDBOOK OF WAREHOUSING

Load Push/Pull Operating Instructions —Pickup from the Floor

1 Line up the platens squarely with the load.

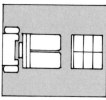

2 Raise the platens about 3" above the floor.

3 Tilt the mast forward until the tips of the platens touch the floor. *DO NOT* try to pick up a load with the platens flat on the floor. Doing so may rip the slip sheet and/or gouge the bottom layers of cartons.

4 Drive the truck forward until the platen tips are under the slip sheet lip.

5 Extend the pusher plate so that the slip sheet lip fits into the gripper channel opening.

6 Retract the pusher plate. The gripper bar will automatically clamp the slip sheet lip.

7 Move forward slowly as the load is being pulled onto the platens.

8 As the weight of the load is transferred to the platens, they will deflect downward slightly. Raise the carriage about 1" to prevent the platen from digging into the floor. Slowly tilt the mast to a vertical position as you scoop up the load.

9 Tilt the mast back, and raise the load 3"—4" above the floor. You are now ready to transport the load.

Load Push/Pull Pickup from a Stack

1 Line up the platens squarely with the load.

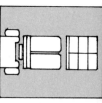

2 Raise the platens to just above the slip sheet of the load you are picking up.

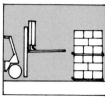

3 Tilt the mast forward 3° to 4°.

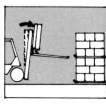

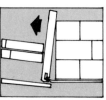

ORIENTATION AND TRAINING

Load Push/Pull Discharge

Drive the truck forward until the platen tips are under the slip sheet opening.

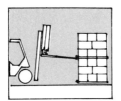

1 Carefully position the load exactly above where it is to be discharged.

Extend the pusher plate so the lip of the slip sheet fits into the grip-channel opening.

2 Tilt the mast forward 3° to 4°.

Apply the truck brakes and retract the pusher plate. The gripper bar will automatically clamp the slip sheet lip.

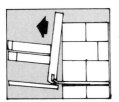

3 Lower the carriage until the platen tips are about 1" above the discharging area.

Pull the load onto the platens. As the load reaches the platens they will deflect downward. Raise the carriage slightly to compensate for this. DO NOT allow the truck to be pulled forward since the platen tips may gouge the load underneath.

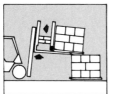

4 Place the truck's transmission in neutral and start the pushing operation. The gripper bar will automatically release the slip sheet lip. As the pusher plate moves the load off, the truck will move backwards, (with light loads drive backwards slowly). As the weight is withdrawn from the platens they will deflect upward slightly. Compensate for this by lowering the carriage slightly.

NOTE: Some operations use two slip sheets between the loads. In this case, insert the platens into the opening between the slip sheets and drive the truck forward as the load is being pulled onto the platens.

NOTE: If your truck does not operate in neutral, place the transmission in reverse and drive backwards slowly to coordinate the pushing motion with the reverse movement.

Tilt the mast back and lower the load to 3"—4" above the floor. You are now ready to transport the load.

5 Back away, tilt the mast back and raise or lower the platens to 3"—4" above the floor for traveling.

COURTESY: Cascade Corporation

Figure 26-5

- to reduce damage to warehouse equipment and stored merchandise,
- to give trainees a feeling of accomplishment,
- to motivate trainees to do a better job,
- to ensure that all new employees understand company rules and procedures.

Training builds teamwork. It can and should be a morale builder. At times when qualified workers are scarce, an effective training program can be a vital part of the growth of your warehousing organization.

27

LABOR RELATIONS

The risk of strikes or other labor disputes is a key consideration in warehousing. In some industries, work stoppages or other disputes among production workers are a likely occurrence, and therefore it's important to have a strategy for continuous distribution of finished products, even when one or more production plants is on strike.[1]

This strategy may be carried out in many ways. Some companies separate distribution facilities from manufacturing locations specifically to cushion the effect of work stoppages. In some cases, the distribution center is located in a rural area with a favorable labor climate so that its non-union status is easier to maintain.

Although such distribution centers can legally be picketed by manufacturing workers engaged in a lawful strike, factors of cost and distance may make such picketing impractical, allowing orderly distribution to take place despite the strike. And if the distribution centers are not unionized, there is little danger of a strike by their employees, either out of sympathy for striking manufacturing employees or in support of economic demands of their own.

---

1 The author is indebted to Theodore J. Tierney, a partner in the law firm of Vedder, Price, Kaufman & Kammholz, Chicago, for extensive research and advice in the preparation of this chapter.

Creating a Union-Free Environment in the Warehouse

There are many ways to create an environment which is not conducive to unionization.[2] One retailer provides a dignified and relatively luxurious setting for warehouse workers. In that company, every worker, from president to trainee, enters through the same front door and lobby. Furthermore, everyone on the payroll uses the same cafeteria for lunch. There is no "back door," and every hourly worker eats in the same room with the chief executive. This egalitarian environment is even more pronounced in a non-union manufacturer where everyone in the plant, from the president to factory worker, wears a white shop coat when on the production floor. The white uniform hides the pinstripes of the corporate executive or the blue collar of the factory worker. The result is a removal of the "we-they" symbols of business dress.

In contrast to such egalitarianism, unionism usually flourishes when a company has insecure or highly authoritarian line supervisors.

For the hourly worker, the line supervisor is the symbol of management. Yet, the typical line supervisor was promoted not because of any proven ability in managing people, but because he was a good "doer," or an above-average "producer." There is nothing wrong with rewarding productivity, but rewarding the productive worker with supervisory responsibility is not always appropriate.

Unfortunately, most managers are blinded by certain pre-conceived ideas about the hourly worker. These include the impression that the hourly worker is only interested in money, and the supposition that unionism is a grass-roots movement which, once under way, cannot be stopped.

Surveys of hourly workers suggest that money is relatively low in their priorities. They are more interested in prestige, security, re-

---

[2] From Volume 19 Number 8, Warehouse and Physical Distribution Report, ©Marketing Publications Inc., Silver Spring, MD. Some of the ideas here were based on lectures and writings of Charles Hughes, Center for Values Research; Fred Pryor, Fred Pryor Seminars Inc.; Alfred T. DeMaria, attorney at Clifton Budd Barlach and DeMaria.

spect, and a feeling of belonging. These desires can be thwarted by insensitive or poorly-trained line supervisors.

The professional union movement is in fact anything but a democracy—it is more like a closely-held franchise business with involuntary franchisees. Any unionized worker has the right to quit the company, but he or she probably will find it difficult to quit the union without also quitting the job.

A labor problem is often at heart a management problem. The franchise operators who solicit union membership are seldom able to market their product in a happy workplace. Workers who are fairly treated will respect both line supervisors and senior management. They are not susceptible to the appeal of the union organizer. While low pay is occasionally the cause, the motivation to unionize is usually non-financial. Typically, the worker who supports a union drive is reacting to either the reality or the perception that he or she has been treated unfairly in the workplace.

An Example

One medium-sized retail organization has a chain of small shops operating all over the country. The organization is also vertically integrated and owns plants that manufacture some of the products it sells. Many of its production facilities are unionized, as are some of its retail stores. In spite of this, the company has carefully preserved the independent status of its distribution operation. Some of its products are distributed with private trucking, but a stevedoring company is retained to hire and control the truck drivers. A group of private distribution centers is maintained on a non-union status.

Some time ago, the company endured a long strike to keep its largest distribution center non-union. More recently, a regional distribution center suffered the loss of a union election by just a few votes. The company signed a one-year labor agreement, and its employees subsequently petitioned for a decertification election. At the expiration of the contract, the union was rejected by the great majority of the distribution center workers.

Management of this company feels that distribution is the one branch of their business that can and should be kept non-union. Yet they accept the presence of unions at the manufacturing level, and even at the retail level.

Public Warehousing Alternative

Some manufacturers have used public warehouses rather than their own distribution centers to offer insulation against the impact of strikes in the distribution area. In order to use a public warehouse successfully as part of a strike-cushioning plan, the manufacturer must observe three ground rules:

- Rule 1. The manufacturer should enter the warehousing relationship well before the strike begins. Any agreement entered into shortly before the strike and terminated just shortly thereafter would be viewed as legally suspect.
- Rule 2. A manufacturer should avoid using more than a limited percentage of a public warehouse's total capacity. In *Samoff v. Local 8-732, OCAW, 307 F. Supp. 434 (D.Del. 1969)*, the court denied a struck employer injunctive relief against picketing, in part, because its goods occupied 85 percent of one warehouse's total available space. However, in *Auburndale Freezer Corp. v. NLRB, 434 F.2d. 1219 (5th Cir. 1970)* the court determined that an employer was entitled to injunctive relief to stop picketing where the employer's goods occupied only 5% of the total capacity of the warehouse.
- Rule 3. The warehouse's management must have full control over the placement of goods in the facility. To avoid lawful picketing, the customer should not rent any particular physical space at the site. The NLRB has held that, under the secondary boycott provisions of the National Labor Relations Act, a public warehouse cannot be picketed by a union having a dispute with one of its customers unless that customer's employees,

trucks, or equipment are on the warehouse premises.[3] In another ruling, the court pointed out that primary picketing at site of a neutral employer may be lawful where primary company's employees are on the premises.[4]

Public Warehouse Strikes

What happens if the public warehouse goes on strike? In most states the user can remove goods through the picket line by filing an action for replevin—a legal move to recover goods unlawfully withheld from their rightful owner. Under this action, the court issues an order to the marshal or sheriff compelling removal of the goods from the warehouse and their placement in the owner's hands.

Using Subcontractor Warehouse Services

A warehousing subcontractor arrangement involves the use of an independent warehouse operator's personnel within the manufacturer's facilities. In using the subcontracting option, the user must beware of creating a "joint-employer" relationship with the subcontractor's employees. To avoid this possibility, the warehousing contractor must be in full and sole charge of determining how many employees are needed, who they will be and the method used to perform the work. A subcontractor must also be in sole control of hiring, training, disciplining and discharging workers, as well as all other aspects of the employment relationship.

Transportation Strikes

The transportation industry has a long history of union activity and many warehouses are designed to deal with carrier strikes. For example, a distribution center with a railroad spur can be used in the event of trucking strikes. From a practical standpoint, strike activity at a warehouse seldom impedes a rail movement of goods in and out

---

3   Local 8, Steel Fabricators & Warehousemen (Fein Can Co.) 131 NLRB 59 (1961); Local 6, Int'l Longshoremen's & Warehousemen's Union (Hershey Chocolate Corp.), 153 NLRB 1051 enf'd 378 F.2d 1 (9th Cir. 1967).
4   *NLRB v. Ironworkers Local 433, 850 F.2d. 551, 554, fn. 2 (9th Cir. 1988).*

of the facility. Railroad companies will usually provide switching service with supervisory personnel. It is often physically impossible for pickets to block industrial railroad tracks because they are all on private property. Therefore, rail movement is frequently an effective hedge in the event a strike disrupts trucking operations.

Inventory Hedging

Once a manufacturer sets up a distribution system that provides an effective buffer in the event of work stoppages, the creation of a "strike hedge" inventory becomes possible. This can be a powerful bargaining tool in contract negotiations. Unions can be warned that the company has vast stockpiles strategically-located around the country, and that it thus is able to absorb a strike for a significant period without losing its position in the marketplace.

Turning from Union to Union-Free

Turning a unionized warehouse into a non-union warehouse is a difficult and delicate project. The most peaceful way to do it is by a decertification election. One such election took place because the chief executive of the warehouse company suggested to several workers that the union really had not done them any good. The group had been organized for about twenty years, and union activity had been virtually unknown during the last ten years. The workers agreed with the executive and a peaceful decertification election was held.

The details of decertification are closely regulated by law, and the most important thing to remember about the process is the need for close consultation with a lawyer who specializes in this field.

Other ways to remove a union are more difficult, such as closing a plant, or subcontracting the warehouse work and closing it. These matters may also be limited by law, though recent court decisions have greatly enlarged management's right to cease unprofitable operations.

Managers should remember that only the company creates warehouse jobs—the union does not create any. In the heat of organizing campaigns, some workers may get the idea that the union provides security—but in actual fact the opposite is often true. Ultimately, job security is based on the customer's satisfaction with the warehouse services, maintained only by a well coordinated labor-management team. No union assures that—only good worker-management teamwork does.

Pre-Employment Screening

Just as the old saying "do good and avoid evil" serves as a guide we can all live by, so does "hire good employees and avoid bad ones" give the most effective method by which an employer can avoid legal problems and maximize productivity. While living by either tenet can be rather difficult, the employer has many acceptable approaches in selecting good employees.

Figure 27-1 lists typical pre-hiring inquiries, with typical instances of when such an inquiry may be lawful. In addition, such inquiries must be job-related and administered in a uniform manner. Use this information only as a guide and always check with local counsel regarding the law in your own state.

Evaluating the applicant's previous employment record can be a useful tool. While previous employers are understandably skittish about giving out too much information about former employees, any information they provide in this area can be helpful. For example, an employee who had attendance problems on one job is quite likely to have them on another. Employers may be more inclined to honor requests for objective information, such as a former employee's attendance record, than giving subjective appraisals or conclusions. At the very minimum, verification of job duties and length of service is critical.

Ask the job applicant for an authorization for release of data by former employers. Obviously, such a step must be taken without regard to the race, sex, religion or national origin of the applicant. Moreover, be aware that when you do check references, you may have an obligation to furnish the applicant with the results of your questioning.

Figure 27-1
**Acceptable and Non-acceptable
Pre-employment Questions—
What Can and Cannot be Asked
In Most Situations**

? **Education**—Acceptable.

? **Work Experience**—Acceptable.

? **Arrest Record**—Acceptable in most states: Have you been convicted of a crime within the last two years? Not acceptable: Have you ever been arrested?

? **Language**—Acceptable: Inquiry into language applicant speaks and writes fluently, if job-related. Not acceptable: What is your native language?

? **Photograph**—Not acceptable.

? **Financial status**—Generally not acceptable: *But:* Have your wages ever been garnisheed? Do you own a car? Do you have a telephone? Have you ever filed bankruptcy? Have you ever been refused credit? (Some of these questions could be asked of applicants for sensitive financial positions.)

? **Reliable means of transportation to work**—Acceptable.

? **Applicant can supply a phone number**—Acceptable.

? **Marital status**—Not acceptable.

? **Religious holidays observed by applicant**—Not acceptable.

? **Days on which applicant is unavailable for work**—Acceptable.

? **Relatives**—Acceptable: Names of applicant's relatives, other than spouse, already employed by you. Not acceptable: Names, addresses, ages, number or other information concerning applicant's spouse, children or other relatives not employed by you.

? Applicant has any religious objection to paying union dues—Not acceptable.

? Age—Generally acceptable, but rejection after inquiry could create age discrimination problems; application should set forth specific disclaimer regarding this inquiry.

? Applicant has ever filed a Worker's Compensation claim—Not acceptable in states that have statutes on this subject.

? Applicant has a valid driver's license—Not acceptable (unless the position involved includes driving duties.

? Military experience—Acceptable: Applicant's experience in the U.S. armed forces or in a state militia. Not acceptable: What type of discharge did you receive?

? Citizenship—Acceptable: Are you a citizen of the U.S.? If not, do you have the legal right to remain permanently in the U.S.? Not acceptable: Of what country are you a citizen?

? Reasons for leaving former employment—Acceptable.

? "Have you had a physical examination in the last three years?" —Acceptable.

? "To your knowledge, do you have any of the following ailments (specific medical problems listed)?"—Acceptable, provided space is allowed for an explanation of the circumstances.

? Date of the occurrence and the treatment regarding any ailment checked—Acceptable.

? "Do you have any ailments not listed that might interfere with the performance of the job for which you are applying? If so, explain."—Acceptable if job-related and space is allowed for explanation.

Employee Complaints

Closely related to good communications is the proper handling of employee complaints. Several suggestions concerning this important management function follow:

1. Appreciate that the employee registering the complaint is paying management a compliment.

2. Listen carefully to the complaint and get all of the details.

3. Even if the complaint sounds frivolous, appreciate that it may be a real problem to the employee.

4. Avoid sarcastic responses and do not attempt to poke holes in the complaint.

5. Investigate the complaint carefully. If it is valid, follow through.

6. Where the complaint calls for a decision, do not hesitate to make it.

7. Realize that complaints may be symptomatic of deeper concerns, so be sure to explore them fully.

One of the keys to good supervision is follow-up. After a full and careful investigation, an employee's complaint must either be dealt with or the employee given a full explanation as to why action was not taken. Simply telling an employee, "You're wrong," or "You don't know what you're talking about," will only create further problems.

The cardinal sin in this respect is for a supervisor to fail to make any response. It is equally wrong for the foreman to "pass the buck" or to say simply that "the front office has turned you down." If the employee was concerned enough to bring the complaint to the foreman, he or she is entitled at least to a complete and meaningful response.

The Role of the Supervisor

In many respects an employee's supervisor *is* the company to that employee. A supervisor often has more impact on morale and perceptions of the company than any other single factor. So the selection and continued training of supervisors is vital.

Selecting supervisors should involve more analysis than merely promoting the most efficient workers. While a worker's productivity certainly should be considered, it does not guarantee that he or she has the qualities necessary to be an effective supervisor of other workers.

The credibility of your supervisors is critical. It is essential that you do all you can to enhance and maintain that credibility. Supervisors should be kept well informed and, in turn, should be the origin of most information related to the workforce. Supervisors should also, whenever possible, be given credit for working-condition improvements.

Arbitration Standards Pertaining to Discharge

No matter how carefully management screens its applicants, and no matter how effective its employee relations program is, there are some occasions where it simply must remove some workers from its workforce. In those situations, nothing can be much more harmful to employee discipline or supervisory morale than to have the discharged person reinstated by an arbitrator or court. Here are some suggestions as to what management can do to minimize the risk that its discharges will be challenged . . . or at least to increase the likelihood of a successful defense!

1. Establish written rules and communicate them to your workforce. If you decide to crack down on the enforcement of a rule, notify your employees in advance. rather than suddenly invoking stiffer or more frequent penalties.

2. Specify those rules violations which will result in immediate discharge.

3. Enforce the rules fairly and consistently.

4. Use progressive discipline, including lost-time suspension prior to discharge, except for the most serious offenses.

5. Conduct a thorough investigation that includes, wherever possible, written statements from all witnesses, including the disciplined employees, prior to discharge.

6. Carefully communicate the full reason for the discharge to the employee. If his past employment record was a factor, say so.

7. Maintain careful, complete records documenting all disciplinary action.

Two of the most aggravating offenses are chronic absenteeism and chronic tardiness. Because arbitrators and government agencies often fail to appreciate the seriousness of chronic absenteeism and tardiness, "building record" is essential. Offenses must be documented, penalties must be progressive—including at least one suspension prior to discharge. And management *must* be consistent.

One of the most common mistakes management makes in this area is to be too lenient by merely making oral reprimands—possibly because manpower needs often make that the easiest route, until management's patience is exhausted and the employee is suddenly discharged. Arbitrators often overrule these discharges on the grounds that the employer has lulled the employee into a false sense of security. Furthermore, because such sudden discipline often is not uniform, government agencies often find it to be discriminatory.[5]

The Essence of Union-Free Management

Union-free management is really a state of mind. It comes out of a shared confidence that teamwork builds a company, and that everyone in the organization is on the same team. It is nurtured by elimination of traditional "perks" which created corporate royalty and corporate commoners. This includes eliminating reserved parking places as well as private restrooms and lunchrooms. In the union-free company, it is common to find everyone on the payroll referred to as an "associate" rather than to have managers and subordinates, officers and employees. In one warehousing company, supervisors are called "team leaders" and hourly workers are team members.

In effect, union-free management is not a power play, not a strat-

---

5 From Vol. 15 No. 4, Warehousing and Physical Distribution Productivity Report. 1980 Marketing Publications, Inc., Silver Spring, MD.

egy for management to defeat a union. Instead, it is the maintenance of an environment in which nobody in the workplace needs a third party to feel secure.[6]

Summary

Many warehouse users have gone to great lengths to design a distribution system that is practically immune from labor problems. But there is an element of risk in anything connected with labor relations. The practice of good employee relations within your warehouse is one of the best ways to ensure against the kinds of labor problems that may result in work stoppages.

---

6   From Volume 19 Number 8, Warehouse and Physical Distribution Report, Marketing Publications, Inc., Silver Spring, MD.

28

MOTIVATION

Most jobs in a warehouse require a substantial amount of self-motivation. Since it is difficult to supervise warehouse work closely, the worker who is not motivated to do a quality job can do a great deal of damage before the sloppy performance is discovered.

In order to develop competence and good performance the following conditions should exist:

- Employees should know if they are improving.
- Good performers should get favorable attention.
- Poor performers should be punished.
- Employees should know what is expected of them.
- Performance feedback should come in a timely fashion.
- Employees should know how to track and measure their own performance.[1]

Output Criteria

Developing output criteria involves determining which of the quantity, quality and/or financial requirements are most relevant for each desired performance result. For a warehouse worker, we may be interested in the requirements of picking rate (quantity), accuracy (quality), or cost-per-case handled (financial). We could select one

---

1 From "Are You Creating Incompetence in Your Warehouse?" by Frank Petrock, Ph.D. and George Dahm, General Systems Consulting Group, Ann Arbor, MI. Published in Warehousing and Physical Distribution Productivity Report, Vol. 15, No. 5. Marketing Publications, Inc., Silver Spring, MD.

of these as being the most meaningful, a combination or any two, or use all three.

When the requirements have been specified, then the measures for each requirement are developed. When measures are developed, they should be meaningful to the person who is doing the work.

The next step is to develop standards of desired performance for each measure. Not having a goal, target, or standard makes a measurement system meaningless.

Performance results are analyzed first for specificity and appropriateness, even before measures are developed or changes in the job are developed. It doesn't make sense to set up a measuring system if you are measuring the wrong thing in the first place. In one warehouse, a target for improvement was based on cases shipped by each branch warehouse. However, in one branch more than half the cases shipped were moving to other distribution centers, rather than directly to the customer. This bulk shipping went at a faster rate which created an artificial decrease in cost per-case handled.

Feedback

The results of performance must be relayed back to the person in the job, if motivation is to be maintained.

One warehousing manager tested positive feedback on a project that required warehousemen to wear safety glasses. This was a long-term problem, so the manager felt if feedback could make an improvement here, it could be used to improve productivity as well. In the past, discipline had failed to increase the number of warehousemen wearing safety glasses. Even buying safety glasses with attractive frames didn't work.

The manager told all foremen to use verbal praise as feedback. Every time a foreman saw someone wearing safety glasses, that person was praised. In 10 days the number of persons wearing the glasses increased from 36 percent to 84 percent, simply as a result of using verbal praise as feedback.[2]

---

2   Ibid.

MOTIVATION

In another warehouse, supervisors designed and implemented a major performance improvement project. Each warehouseman was provided with feedback on his or her productivity in terms of "cost/cases handled" in relationship to a budgeted goal. The warehousemen could calculate their own productivity, as at the end of the shift the foremen provided them with their results for the day. Thereafter, "cost/case" feedback was summarized into shift performance. The "cost/case" productivity for all three shifts was continuously posted on a large scoreboard. The foremen insisted that the feedback be on a three-shift, 24-hour basis so that inter-shift cooperation would be encouraged (they did not want each shift's productivity posted on the scoreboard since they felt that this would cause destructive inter-shift competition).

Without feedback, personnel in work organizations tend to be self-correcting, which means that on-or-off course deviations from a goal cannot be detected. Feedback on results or outputs leads to performance improvements and behavior, especially if the feedback is positive. Best of all, feedback systems are relatively inexpensive in comparison with the performance improvement pay-offs.

Too often, feedback on performance results goes to executives, managers and, sometimes, supervisors. Rarely is a system set up to give warehouse employees feedback on their results and accomplishments. Yet, it is the warehouse workers who must change their behavior to improve results. Without feedback there is no way they can improve their own on-the-job performance in a systematic way.

Failing to give feedback to warehousemen is similar to asking them to bowl blindfolded. Even though we provide them with the best job situation (bowling alley, bowling balls, lighting, shoes, etc.), and know exactly what we want in terms of performance results, the most skilled bowlers will not do well. If we go a step further, and provide the bowlers with a supervisor to direct their activities, scores will be improved, but first-line supervision will be very difficult. If we take the blindfolds off, however, the bowlers get immediate feedback on results, can keep their own scores, and modify their behavior to improve those scores.

Feedback makes the employees more independent of the supervisor and sets up a situation where they can manage their behavior in a positive direction. This makes the job of warehouse supervision easier, since it is usually impossible to monitor the activities of workers in storage areas where they cannot be observed.[3]

Skills

In warehousing the most basic skills or lack of them often have the greatest impact on productivity. For example, one warehouseman seemed to have a chronic problem with order accuracy. Rail cars and trucks were either loaded with too much or too little product, and there was no apparent pattern to the overs and shorts. At first, it seemed that the problem must be in the warehouseman's bad attitude, i.e., he just didn't care. Later it was discovered that he couldn't do basic math and, therefore, was not able to keep track of what was or was not loaded. He needed training, or a different job assignment.

Everyone is interested in improving warehouse productivity. In the final analysis, however, we are not improving warehouse performance, but rather improving the performance of people who work in warehousing. If this distinction is made, then "motivation" becomes the key word for a performance system.[4]

Balance of Consequences

Managers prefer technological changes over behavior change, because technological change is seen as predictable, measurable and controllable. By contrast, human behavior is seen as random, hard to appraise, beyond comprehension and elusive. It is much easier for managers to deal with lift-trucks, storage layouts and picking patterns than with human behavior.

There is a psychological principle that, in order for improved behavior to last, it must be reinforced. Both desired and undesired behavior is learned and maintained through reinforcement. Which

---

3   Ibid.
4   Ibid.

MOTIVATION

type of behavior is displayed depends on the balance of consequences between the two.

An example of how balance of consequences works is illustrated in the following imaginary dialogue between a warehouseman supervisor and employee:

Supervisor: "You are the best warehouseman I have when it comes to loading orders with multiple brands. From now on, I'm going to give you the tough orders to fill."

Warehouseman (under his breath): "The only way to win around here is to do a lousy job."

The goal for analyzing the balance of consequences is to ensure that workers with desired performance results get immediate, reinforcing consequences. Conversely, the goal is to prevent the reinforcement of undesired performance results or behaviors. Balancing the consequences helps ensure that the person's behavior will be motivated in the desired direction. This results in a different type of dialogue between supervisor and warehouseman:

Supervisor: "You are the best warehouseman I have when it comes to loading orders with multiple brands. So you can leave the warehouse two hours early on Friday."

Warehouseman: "Thanks!"

Participatory Circles[5]

Also referred to as "quality circles," these groups provide a way to organize workers to talk about ways to improve the job. Quality circles were used in the U.S. just after World War II. Although they had some success, they did not become widely popular. Ironically, the idea has now been identified with Japan, even though the Japanese borrowed the idea from this country.

The goal of the circles is to create a concept of psychological ownership. Psychological ownership is an attitude anyone can possess. Sometimes an hourly employee may have it to a greater degree

---

5 From *Participatory Circles — Capitalizing Human Assets* by Saul Pilnick and Jo Ellen Gabel of Human Systems, Inc., Morristown, NJ.

than a member of management. Psychological ownership is achieved through mastery of the job. It does not depend on title or position within the organization. No promotion can create it, since it is a feeling of being in command of one's job.

It is participation rather than legal ownership that creates a feeling of psychological ownership. The fact that you are a part owner of the U.S. Post Office or the Army does not give you a feeling of psychological ownership. However, involvement in a participatory circle can do so. No one is ever pressured to join—membership is entirely voluntary.

For example, on one participatory circle, a formal meeting is held each week. Membership consists of six to 10 persons who are there voluntarily. The first rule is that every problem coming up must be important to the workforce and to the business. What the circle does is identify the problems, find solutions, distribute responsibility for undertaking them and, then, hold the members accountable for follow-through.

The circle itself is a simple roundtable. It consists of a chairperson and someone to take notes. And it is held on company time. An important condition of circle success is that no item is ever dropped from the agenda until it has been resolved. Members should be able to call on anyone from within the company for input, information or advice. Meetings should be limited to 90 minutes in order to prevent the session from deteriorating into an irrelevant discussion that wastes time. An essential to circle success is that the meeting is closed by going around the table and reviewing person-by-person who is committed to do what—and by when. This makes each person accountable for his or her part in the success of each project that the group is working on. They begin to feel psychological ownership when they are able to play a role in decision-making that affects their job.

Developing a High-Performance Work Culture

There are four principles which are essential to developing a high-performance work culture.[6]

- *Leadership must show the way.* Each manager and supervisor must have a planned strategy to behave every day in a fashion that supports the procedures, policies and goals of your warehouse organization.
- *Accountability for performance must become a way of life.* A formal system of performance accountability should be maintained in your warehouse. If goods are lost or damaged for any reason, everyone knows who is accountable for each occurrence.
- *Each team must practice "what we say is what we do."* Team meetings are conducted in a professional manner. Agendas are published ahead, and commitments made in the meeting are recorded. The next meeting begins with review of the team's past commitments and its results in meeting them.
- *Reward positive behavior and confront negative behavior.* Outstanding performance in the warehouse should be rewarded and praised. At the same time, rule violations, failure to maintain housekeeping, tardiness or discourtesy are negative behaviors which must be recognized. Failure to confront negative behavior is not a neutral act but is a statement of support for the negative behavior.

A Case Example of Participative Management in the Warehouse[7]

In one public warehouse, workers are asked by management for input before bidding on major new accounts. Small groups of warehouse workers are asked by a member of management what they think

---

6   From an article written by Saul Pilnick and Jo Ellen Gabel of Human Systems, Inc., Morristown, NJ.
7   Adapted from Vol. 19, No. 4, *Warehousing and Physical Distribution Productivity Report,* Marketing Publications Incorporated, Silver Spring, MD.

performance levels would be on a potential account, and whether they think productivity levels could be met. The manager talks in terms the employee can understand, such as "we have one-and-a-half hours to unload a truck" or "we need to fill these orders in an average of ten minutes per order." The workers' comments and suggestions are then taken and quantified, determining the rates for the prospective customer.

Because workers have provided input, and in effect help set the standards of performance, their input brings about productivity goals set by workers and management alike. If the account is sold and performance levels are not met, it is then possible for a manager to approach employees and say, "We planned on unloading a truck every 70 minutes, and it's taking 80—any idea what turned out to be different?" A worker may spot a problem or solution that management did not see. For example, one employee might point out, "This product is too light, and we're jerking the slip sheets. Maybe the customer could glue the bottom cases like XYZ company does."

The key to this participative management approach is threefold:

1. Workers have to be approached in small groups, such as a group of three or four people who are comfortable with each other. A large group, or one with unfamiliar people, will make workers hesitate to speak.

2. The person who leads the discussion needs to be at management level in order to bring credibility and a heavier commitment to the goals being discussed.

3. Feedback on whether targets are being met or missed needs to be given with regularity.

Warehouse Incentives

Any incentive system begins with a measurable and attainable goal, along with a known and worthwhile reward. In a warehouse operation, the variety of work and productivity conditions sometimes make the creation of an individual incentive system difficult. Group incentives directed toward a general goal are easier to set up and can result in improved effort and efficiency.

Incentives are sometimes structured on the basis of pounds or units handled per manhour. They also can be based on goals other than productivity, such as reduction of damage, elimination of errors, improved housekeeping and safety. Money is not the only reward that will stir interest. Individual prize awards, banquets and public note of accomplishments are also motivating elements.

But it is difficult to create an equitable incentive system for warehouse work, because the work content in most warehouses is subject to significant change over time. If you are considering an incentive system:

- Keep it simple. If the program is not readily understood by the average worker, it could fail as a motivator.
- Be sure the system is equitable. For some operations, a group incentive may be the only way to keep the system fair. Under a group incentive, rewards are divided among a work team rather than awarded to individuals.
- Maintain ample engineering support. Should anyone question the fairness of the program, it is necessary to have ample data to show how the system works and to prove that the system is fair and equitable.

The Importance of Listening

Successful warehouse managers communicate clearly.[8] The clear communicator has a genuine understanding of what he/she is trying to convey, uses appropriate words to phrase the ideas, chooses an effective way to put them across—and listens effectively. Many managers enjoy talking, telling, informing, teaching and judging. Listening—active attending to what another is saying—is frequently an undeveloped skill, particularly in those who talk a lot. Successful managers who learn to listen effectively often achieve new insights. Sometimes "generous listening" is used as a way to reward others.

In a warehouse, the continual movement of material in an effi-

---

8   From "Listening: A Warehouse Manager's Tool" by Richard E. Rogala; Volume 3, No. 6 of Warehousing Forum, Ackerman Company.

cient, profitable manner requires close cooperation and teamwork. Teamwork is based on good listening skills. Choose a place which is relatively quiet and comfortable to give people time to share information. The effective listener minimizes the probability of interruptions from telephone calls and secretaries. You should try to listen in physically comfortable surroundings.

One new warehouse manager was promoted to a new facility from another location in the company for which he worked. He was also promoted to general manager of the operation after serving as administrative assistant to the general manager in his old location. He was nervous and highly self-conscious of his age and minimal experience. The people who worked for him—foremen, customer service representatives and hourly workers—knew nothing about him or his reputation previous to his arrival. When he made his first appearance, he called everybody together and lectured for forty-five minutes to prove his knowledge and competence. His audience reacted, however, in a very noncommittal, reserved way. They had not been given any opportunity to raise questions, voice concerns or make their observations about the way the new location was operating. The new general manager's intent was to prove his competence by talking about his experience. In fact, he alienated the people he was trying so hard to please.

Had he taken the opportunity to spend time with individuals, listened to their experiences, heard their observation of how well the warehouse was operating and then commented on the things he saw they were doing right, he would have been in a much better position to elicit their confidence, loyalty and motivation.

Effective listening is only one part of the complex task of communicating. A manager needs to share information, teach, train, coach, talk at and inform people—and practice the art of listening. The art of listening is a fundamental management tool but it requires practice. Becoming a better listener will make you become a better warehouse manager.

29

MAINTAINING AND IMPROVING PERFORMANCE

Both feedback and recognition are important factors in *maintaining* performance. Here are some key points which you can use to maintain and even improve performance in your warehouse:

- It makes better sense to spend more time with your "winners" than your "losers." Winners tend to improve more than the lower performers.[1]
- The best recognition is *symbolic* versus monetary. Some managers find it helpful to memorize "planned spontaneous recognition" . . . .

Planned: We rarely *plan* to recognize and, hence, tend to lose those opportunities to even notice performance which is "above and beyond." Some warehouse operators hold monthly meetings to discuss coming things to recognize.

Spontaneous: The best recognition is *unexpected*.

Recognition: Whatever you do, somebody won't like it. Thus, the solution is lots of different types of recognition: from soft drinks and pizza, to plaques, to banners, and so on.

- One final key to maintaining performance is actually to *enhance* performance; there is no way to just maintain. Either you get better or you get worse.

---

[1] From an article by Stephen J. Mulvany, President, Management Tools Inc., Orange California.

Development

A business theory known as the Peter Principle says that managers in large organizations are promoted steadily to their level of incompetence. In the warehouse there are times when, say, a superior lift-truck driver is promoted to foreman primarily because of performance in his old job. Skills in operating a lift-truck may not be transferable to supervisory situations. When an error in promotion takes place, the honest but difficult thing to do is reverse the step as quickly as possible.

At the same time, a development system will at least reduce the number of times when the Peter Principle applies. Management seminars represent one way in which individuals can be encouraged to grow in personal skills. Individual efforts to gain outside education should be encouraged.

Discipline

Discipline is part of the recognition system. If a warehouse employee performs poorly, your duty as manager is to let that worker know of that performance failure. Likewise, when that individual changes his or her behavior and does the job correctly, it is vitally important that the supervisor recognize this positive change as well. Most employees will tolerate strict discipline as long as it is accompanied by recognition of improved performance.

Peer Review—A New Approach to Discipline

Traditional methods of disciplining employees sometimes foster an "us against them" mentality which can create the perception that the company is not committed to resolving workplace disputes fairly.[2] Some employers deal with this by using a dispute resolution process known as "peer review."

Peer review systems replace the traditional grievance arbitration process by allowing an employee to appeal disciplinary actions to a

---

[2] This subchapter is from an article by Theodore J. Tierney; Warehousing Forum, Volume 3, No. 4, Ackerman Company, Columbus, Ohio.

committee comprised, at least in part, of co-workers. Normally, all disciplinary actions are subject to peer review. In addition, the employee may challenge either the discipline itself or its severity; although some companies have chosen to narrow the scope of their peer review system to cover only specified actions, such as involuntary terminations, overtime, or the proper application of a company's layoff procedure. Where peer review systems are limited in their coverage, companies usually maintain an "open door" for other review issues.

Generally, the peer review process commences with the lodging of an appeal by the aggrieved employee. An employee typically is afforded five to seven days within which to lodge the appeal.

The next step in the appeal process often is the employee's supervisor, who reviews the discipline; hopefully, the matter can be resolved at this stage. However, if the employee remains dissatisfied following the supervisor's review, he or she may take the appeal to the Peer Review Board.

Although the composition of the Peer Review Board varies from one company to the next, a typical Board is comprised of five members: the aggrieved employee selects two members; the plant manager or personnel director also selects two members; and the fifth member is selected by the other four members.

Within a few days following its formation, the Board is convened. The matter is presented to the Board in an informal, non-legal setting. Each party—the aggrieved employee and his or her supervisor—presents his or her side of the dispute to the Board. The Board may ask questions or call additional witnesses and, in addition, may review the employee's personnel records. Once the Board has collected all the necessary information, the supervisor and the employee are asked to leave the room while the Board deliberates.

In considering the employee's appeal, the Board only decides whether there has been a violation of company policy. The Board is *not* empowered to change company policy. The Board may uphold the discipline, reduce or increase its severity, or overturn the discipline entirely.

Usually majority rules and the Board's decision is final. Some companies, however, require a unanimous decision. If such a decision cannot be agreed upon, a majority and minority report is given to the plant manager or personnel director who then renders a final decision.

Upon arriving at a decision, the Board immediately informs the supervisor and the employee. All aspects of the decision-making process remain confidential.

In companies that use some form of peer review, both employees and management comment upon it favorably. Employees are satisfied with peer review systems largely because such systems allow employees to participate in management decisions that have a direct effect on them. Employers have found that Board members take their responsibility seriously and do not seek to abuse or undermine the process. In addition, employers recognize that peer review reduces employee discontent. Generally speaking, where discipline is imposed properly in the first place, employees hesitate to invoke their appeal rights under the peer review system. In fact, a complaining employee's peers on the Board often are harder on him or her for certain rules violations, such as those related to absenteeism, than is the company.

Summing up, peer review programs:
- Increase employee morale by providing employees with a voice in disciplinary matters directly affecting them, thus building a bridge between management and employees;
- Thwart attempts at unionization by removing a common source of employee discontent (i.e., the absence of employee participation in discipline);
- Help to avoid expensive litigation and arbitration;
- Help to support the company's position in any subsequent, related litigation, such as EEOC charges; and
- Help to maintain company control over employee relations by keeping workplace disputes in-house.

MAINTAINING AND IMPROVING PERFORMANCE

New Personnel

If we define management as "achieving results through others" then a successful manager must have competent associates to work with. A critical step in finding people from outside is an effective interview. Before the interviewing process begins, there are three pre-interview steps that must be taken:

The first of these is the development of a concise description of the job you are filling. From that description, you can then make a list of the traits that are essential in a candidate. Typical qualities might be "organized, sets a fast pace, results oriented," all necessary traits in a management candidate.

A second list should itemize traits which are desirable but not absolutely necessary. Then list those traits which are absolutely unacceptable. The lists of positive and negative traits might cover as many as 15 or 20. When the actual interview is held, the manager should determine the presence or lack of these traits.

The next pre-interview step is gathering background information. Before you spend valuable time interviewing, you need to know whether prospective employees have the background to justify further consideration. Is their prior work experience comparable? What educational level has the person attained? Does the applicant have friends or relatives employed in your work force? This might help tell you whether or not you are likely to hire a malcontent or a team player. What kind of personal credit does the individual have? As a generalization, people with credit problems tend to be a poor risk. Learn why the applicant left former employers. While past behavior is not always an indication of future performance, if a pattern of absenteeism has developed with former employers, it is reasonable to assume that you will have the same difficulty.

Pre-Employment Investigation

Since warehouse employees are custodians of property, their integrity is as important as their ability to handle and store merchandise. Background investigation of prospective workers is, therefore, of great importance. And since warehouse employees must be physically capable of performing the work safely, physical fitness should be verified through both investigation and examination.

Use an application blank which allows you to collect such information. Before a final decision is reached, information supplied by an applicant should be verified by reference checking and the use of private investigation services.

Establish tests which an applicant should pass before employment with your company. The application blank itself is the first test. Always insist that the application be filled out in your presence. Record the time a prospect took to complete your application and compare it to previous interviews.

A math test is also appropriate in warehouses.

The ability to operate materials handling equipment is essential. Some applicants claim that they know how to drive a forklift, but not many can really do it well. Set up a test obstacle course and a series of lift maneuvers to see how well the applicant does.

Another key test is a meaningful physical examination. Urinalysis tests for drug use should be employed. The physician should be asked to carefully check backs, knees, and shoulders—these are high workers' compensation claim items.

Pre-interview checks may eliminate a high percentage of the applicants. No need to waste interview time on applicants who cannot pass your programmed tests or don't have the appropriate backgrounds.

The Interview

Try to observe the applicant from the time he or she arrives on the property. Is the applicant on time? How do applicants conduct themselves during the initial greeting? Do they look you in the eye? How are they dressed? Are they clean, neat and well groomed?

Some interviewers make the mistake of showing the job description to a candidate and then taking him on a tour of the facilities. In doing so, the interviewer reveals too much about himself and the job. Naturally, the candidate is going to tell you that he can do the job as outlined in the interview. In addition, he will probably learn enough about you during the tour to tell you what he thinks you want to hear.

The goal of the interview is to get to know the real person, so you must let him or her do most of the talking.

An effective interview must be conducted in a quiet, dignified setting, free from distractions. Use comfortable seating with no desks or tables between the interviewer and the applicant. This allows the interviewer to observe body language in addition to the oral responses.

Some questions should require the interviewee to repeat in his or her own words what is on the resume. Be alert to any inconsistencies between the two.

Never use questions that can be answered by a "yes" or "no." Instead, ask broad, open-ended questions. Some examples are as follows:

Tell me about yourself.
What did you like best about your last job?
What did you dislike most about your last job?
Describe your favorite boss.
What are your strengths?
Why should we hire you?
Describe a typical day in your last job.
What are your shortcomings?

Make every effort to show no judgments. If the interviewer's reactions are obvious to the interviewee, the candidate will probably adjust responses to fit the interviewer's biases.

Some employers will not make telephone checks because they believe that no one will tell them anything. Sometimes this is true. But if a person has had three or four previous jobs, the chances are good that at least one of the former employers will help you. Here are some good questions for the reference check:

- How would you describe the candidate's attitude?
- How did the candidate compare against peers?
- Was the candidate involved in any significant failures?
- What was the best way you found to motivate this candidate?

- Does the candidate work well with other people?
- What were the candidate's strengths (or weaknesses)?
- Why did he or she leave?
- Would you rehire?

Always keep the feelings of the interviewee in mind. Job hunting is an anxiety- producing experience—one that often yields a lot of wounds and hurts. Always treat the candidate with kindness and consideration. Treat a candidate as you would a customer and let him know of your decision as quickly and as considerately as possible.[3]

Drug and Substance Abuse in the Warehouse

Over ten percent of the people in the workplace have some kind of substance abuse problem. Probably the highest use of an illegal substance is marijuana, estimated at over nine percent of people in the workplace. Certainly a higher number use alcohol, a legal substance which is sometimes abused. Abuse of prescription drugs is fairly common, and cocaine use is growing rapidly.

As a warehouse manager, perhaps your greatest risk is an employee who steals to raise the money to acquire the substances which are abused. Some substances, like alcohol, can be acquired cheaply. Others, such as narcotics, are quite expensive. The abuser who is a thief is probably addicted to one of the more expensive substances.

Workers around a warehouse are engaged in driving motor vehicles, usually warehouse vehicles but sometimes also tractor-trailers. A mood-altering substance can affect the ability to drive safely and can be a danger both to the driver and to others.

Lost time accidents resulting from substance abuse may be job related or not, but if they prevent regular attendance at work, they create productivity losses.

Effective customer service and dealing with other people depend on maintaining a reasonably even disposition, and substance abuse which causes a mood swing can noticeably affect job performance at

---

3   The last three subchapters were by Robert L. Prior, "Hiring Good People," Warehousing Forum, Vol. 2, No. 1, Ackerman Co.

MAINTAINING AND IMPROVING PERFORMANCE

those places in the warehouse where people must deal with outsiders. This would be especially true in the office and at shipping and receiving docks.

Warehouse receiving and order picking depend upon accuracy in reading and recording information. This accuracy is severely impeded by some mood-altering substances. Other substance abuse may encourage horseplay on the job, and this can be disruptive and quite dangerous in a warehouse.

There are physical symptoms and conditions which create reasonable suspicion of abuse. These include bloodshot or watery eyes, as well as pupils which are abnormally large or small. A runny nose or sores around the nose are sometimes signs of abuse, as well as bloodstains on shirt sleeves. Watch for the individual who wears sunglasses in places where they should not be needed. Excessive perspiration or sudden worsening of complexion can be symptoms of abuse, as well as sudden weight loss. Abusers who are taking injections will wear long sleeve shirts in hot weather. Slurred speech, unsteady movement, or tremors are other signs. In some cases, mood swings will cause unpredictable responses to ordinary requests.

Some companies use undercover investigation to detect substance abuse. Private detective agencies have for many years provided undercover agents, people who take a job as a regular employee but are actually engaged in reporting on activities within that company. An undercover agent is carefully trained to be a good observer, to escape detection, and to understand the rules of evidence and what evidence will be acceptable in legal proceedings. As in any service industry, there are also incompetent detective agencies which offer undercover agents. The reputation of the agency hired should be carefully investigated. Undercover investigation requires top management commitment and no leaks at a lower management level in order to protect the cover of the agent.

The issue of testing or screening is the one which has generated the most controversy in the media. In our culture, screening is portrayed as denial of traditional human rights. Yet, warehouse workers can cause death or injury if they do their jobs in a reckless manner. If you fail to screen people whose habits may cause them to work recklessly, you could be failing to exercise reasonable care for the healthy people who work in your warehouse as well as for the goods which are placed in your custody. Therefore, consider whether you are in more trouble for failing to test or screen for substance abuse than you would be if you did engage in testing.

Many companies have denied or avoided the issue of substance abuse by failing to publish a company policy on this matter. Your policy should relate to your company as a warehousing, distributing or manufacturing operation. You should relate substance abuse to health, safety, property protection, and quality control. You should set standards of conduct which are expected of every employee in the company. If you have a rehabilitation program, the policy should recognize it. Finally, you must state the penalties you are willing to invoke for violating the policy. Here is a sample policy statement which you should consider if you do not already have one:

> This company is committed to providing its employees with a safe workplace and an atmosphere which allows its people to protect merchandise placed in their care. Our employees are expected to be in suitable mental and physical condition while at work to allow them to perform their jobs effectively and safely. Whenever use or abuse of any mood-altering substance interferes with a safe workplace, appropriate action must be taken. The company has no desire to intrude into its employees' personal lives. However, both on-the-job and off-the-job involvement with any mood-altering substance (alcohol and drugs) can have an impact on the workplace and on our company's ability to achieve its objective of safety and security. Therefore, employees are expected to report to the workplace with no mood-altering substances in their bodies. While employees may make their own lifestyle choices, the company cannot accept the risk in the workplace which substance abuse may create. The possession, sale or use of mood-altering substances at the workplace, or coming to work under the influence of such substances shall be a violation of safe work practices. This violation may result in termination of employment.

MAINTAINING AND IMPROVING PERFORMANCE

Managers often prefer to ignore the issue of substance abuse. However, ignoring this problem does not make it go away, and in fact probably contributes to its growth. There are those who will say that these matters should be left to the personnel department, but as a warehouse manager, you should consider whether substance abuse is a matter which should be delegated to others or one which you must handle yourself.[4]

Probation

Most labor contracts allow a probationary period of one to three months before the new employee becomes a permanent member of the workforce. It is obviously preferable to complete all investigations before the new worker is ever put on the job. This will prevent the possibility of personal injury or damage caused by an individual unsuited for the job he or she is performing, as well as the expense and pain of dismissing an applicant already on the job. But no matter how much care is taken with pre-employment investigation, the risk of error always exists. Therefore, it is absolutely essential to observe the new employee closely during the probationary period. Pre-employment investigations cannot reveal attitude problems that will show up only in the actual work situation. By eliminating marginal employees during the probationary period, the warehouse operator has a second chance to ensure the highest quality in his work force.

Personnel Management

Regardless of whether labor is unionized or not (See Chapter 27), there should be little difference in methods of personnel management in the warehouse.

A labor contract is simply a written confirmation of wages, fringe benefits and working conditions. This information is important to everyone regardless of a labor agreement.

Precise communication of rules is an essential means of ensuring

---

4 From "Substance Abuse in the Warehouse," Warehousing Forum, Vol. 2, No. 6, Ackerman Co.

that people on the job are working together without a misunderstanding.

A formal grievance procedure is necessary whether or not there is a trade union. A manager who merely proclaims an "open door policy" cannot expect unhappy workers to march into the office with their complaints. Many workers fear that they cannot express themselves effectively. They also may be concerned about retaliation from a line supervisor if they go over his or her head with a problem. For these reasons, a formal grievance procedure is a necessary safety valve in any plant organization. Most importantly, the grievance procedure should provide an independent arbitrator who will settle problems that cannot be resolved internally.

The Labor Contract

A labor contract usually formalizes actions that good management would take whether or not a union existed. Tough negotiations are customary in formulating labor contracts, but contract administration is even more important than negotiation. A warehouse manager who tries to cut corners or seek contract loopholes will usually reap a harvest of ill-will.

Wage Reviews

Money may not be the most important motivator, yet a person who feels underpaid will not be motivated. Therefore, periodic wage reviews are needed in any job that does not have compensation formalized by the collective bargaining process. The employer should be certain that pay levels remain comparable with those for similar jobs in other companies. Wage and salary surveys will determine whether the company is consistent with patterns within the community and among businesses in a similar industry. When comparing wages with other firms, the full package of wages and benefits must be considered.

Job-Performance Appraisal

A properly-conducted job-performance appraisal increases mo-

MAINTAINING AND IMPROVING PERFORMANCE

tivation and allows constructive criticism in both directions. The interview may also provide a needed pat on the back for good performance. There are many forms and procedures for appraisal, but the core of the program is the interview between subordinate and immediate supervisor.

Like wage reviews, appraisal interviews should be handled on a regularly-scheduled basis. Such interviews should be objective, emphasizing actions and things rather than individuals. At the same time,

ASSOCIATE PERFORMANCE EVALUATION

ASSOCIATE: _____   POSITION _____
D.C. LOCATION
DATE OF LAST EVALUATION _____   REVIEW DATE _____
PREPARED BY _____   TITLE _____   DATE _____

FOR EACH "CHARACTERISTIC" BELOW, CHECK THE MOST APPROPRIATE RATING CATEGORY. USE THE "COMMENTS" SECTION FOR ANY SIGNIFICANT REMARKS DESCRIPTIVE OF THE ASSOCIATE.

CODE "A" FACTORS RELATE PRIMARILY TO WAREHOUSE AND CLERICAL POSITIONS, CODE "B" PRIMARILY TO SUPERVISORY POSITIONS, AND CODE "C" RELATES TO MANAGEMENT POSITIONS

BELOW EXPECTED FACTORS ARE THOSE WHICH REQUIRE IMPROVEMENT IN ORDER TO MEET EXPECTED REQUIREMENTS. EXPECTED FACTORS ARE THOSE WHICH ARE ADDITIONAL AREAS FOR IMPROVEMENT POTENTIAL. ABOVE EXPECTED FACTORS ARE THOSE WHICH ARE POSITIVE CHARACTERISTICS.

CODE	JOB PERFORMANCE FACTORS	ABOVE EXPECTED	EXPECTED	BELOW EXPECTED	COMMENTS	CODE	JOB PERFORMANCE FACTORS	ABOVE EXPECTED	EXPECTED	BELOW EXPECTED	COMMENTS
A	WORK PERFORMANCE: Quantity or Volume					BC	JUDGMENT:				
AB	DAMAGE MINIMIZATION AND CONTROL:					BC	DEALING WITH CUSTOMERS:				
AB	WORK PERFORMANCE: Quality or Accuracy					BC	COURTESY:				
ABC	DEPENDABILITY: Follow up reliability, Meeting work deadlines, Attendance & Tardiness					BC	SUPERVISING:				
						BC	DISCIPLINE: Firmness				
ABC	PERSONAL CHARACTERISTICS: Disposition, Poise, Appearance & Sincerity					BC	DELEGATING:				
						BC	SELF-IMPROVEMENT:				
						BC	FOLLOW UP:				
ABC	JOB KNOWLEDGE:					BC	DEVELOPING SUBORDINATES:				
ABC	SAFETY:					BC	LEADERSHIP:				
ABC	EQUIPMENT & FACILITIES: Care & Conservation					C	SALES & COMPANY IMAGE BUILDING:				
ABC	HOUSEKEEPING:					C	WRITTEN COMMUNICATIONS:				
ABC	EFFICIENT USE OF STORAGE SPACE:					C	SELECTING COMPETETENT SUBORDINATES:				
ABC	COMMUNICATION: With Peers, Subordinates & Superiors					C	EVALUATING SUBORDINATES:				
						C	TERMINATING INCOMPETENT SUBORDINATES:				
ABC	CONSISTENCY:										
ABC	SELF IMPROVEMENT: Acceptance of Criticism					C	SETTING FAST PACE:				
ABC	DRIVE & INITIATIVE:					C	MAKES DEMANDS ON SUBORDINATES:				
ABC	COOPERATION:					C	FOCUSING ATTENTION ON GETTING JOB DONE:				
ABC	INNOVATION:					ABC	OVERALL EVALUATION:				
ABC	STABILITY:					REMARKS:					
ABC	ORGANIZATION:										
BC	PLANNING:										
BC•	OVERTIME CONTROL:										

Figure 29-1

the interview should clearly point out needed improvements and provide a plan for accomplishing them. If the employee participates, the plan is more likely to be successful. Figure 29-1 shows a performance evaluation form.

Attitude Surveys

When warehouse workers are unhappy in their work, it is a sad fact that management is frequently among the last to know. One way to measure the state of worker morale is a periodic attitude survey.[5]

Regardless of the existence of an "open door" policy, it is not easy for management to learn when there is trouble in the warehouse. Even when line supervisors are highly competent, the difficulties will not always surface. The typical case however, involves a problem with one or more of the line supervisors and here the channel of communication to management is often not open.

The goal of an attitude survey is to measure the satisfaction (or otherwise) which people have with their jobs, and if there is dissatisfaction to find the cause of it, though it is often not possible to find the people who are unhappy. While some survey programs will code questionnaires to identify the respondents, this is usually considered to be unethical.

The survey could be done internally, but typically the results are more satisfactory if it is handled by a consultant or some other neutral third party. The surveyor's first task is to establish the trust of the workers being surveyed. He or she does this usually by first assuring the working group of the absolute anonymity of the questionnaire respondent. An equally important assurance on the part of the surveyor and management is that changes will probably be made as a result of the conclusion of the survey. Indeed, if surveys are run by a management which never makes any changes afterward, the result is usually counterproductive. This telegraphs the fact that management asks questions but apparently does not listen to the answers.

---

5   The author is indebted to W.J. Ransom for suggesting that this subchapter be written and for providing some of the material used in it.

It is an unfortunate fact that most surveys are originally taken when a warehouse is in trouble. Sometimes the stimulus for the survey is a union organizing campaign, or when there is high turnover or other clear signs of dissatisfaction. If this is done, it is of utmost importance to run the survey again, perhaps a year later, when conditions have returned to normal. Surveys are most valid if they are run on a periodic basis with a comparison of results, and if they are run at times when things seem to be going smoothly.

If workers are to have confidence in the survey, it is important to provide some feedback on the results. The degree of detail involved in this feedback is highly variable, but at a minimum there should be some indication that the results were received, and that certain changes in policies or conditions have been made from the information learned in the survey.

The design of such surveys is not simple, and the task of tabulating and measuring the results is equally complex.

One survey prepared by Center for Values Research, a management consulting firm in Dallas, Texas, is noteworthy because of its brevity—there are just 20 questions in it. The consultant offers a tabulation and evaluation service to the users, as well as the ability to administer the survey itself.

Attitude surveys are an important means of measuring morale in the warehouse, but to be successful the following conditions must apply:

1. Run the survey on a periodic basis, not just once when the warehouse is in trouble.

2. The survey must be administered by an individual who is perceived as a neutral party and who can gain the trust of the people being surveyed.

3. Management must be committed to making changes as a result of things learned in the survey.

4. A demonstration of such changes must be made, together with feedback to the people surveyed telling them that changes came as a result of their input.

Promotion From Within

When seeking individuals for higher positions, management should look inside the company first. In spite of the dangers of the Peter Principle mentioned earlier, upward movement is an important morale builder. If no opportunities for promotion exist, talented employees will soon leave the company.

Pride in the Company

A hallmark of dedicated employees is their pride in the company. The task of creating such pride is more difficult today than it was a generation ago. Nevertheless, company identification can be stimulated through athletic teams, service pins and, sometimes, uniforms. A company newsletter also can be a fine means of building pride in the organization.

Amazing things can happen when a warehouse is staffed with dedicated personnel. They will suggest ideas for improving methods that might never be discovered by top management. And an unbroken chain of dedicated warehouse workers will provide the best possible security against cargo theft.

PART VII

PRODUCTIVITY AND QUALITY CONTROL

30

MAKING WAREHOUSING MORE EFFICIENT

Warehousing costs have climbed sharply and capital expenditures allotted to warehousing have failed to keep pace with other aspects of business. There are at least three reasons for selecting the warehouse as a target for productivity improvement:[1]

1. A typical manufacturing concern has 30-to-35 percent of its total under-roof space devoted to the storage of raw materials or finished goods inventory. In most parts of the U.S., increases in construction costs of warehouse space have exceeded the rise in the cost of living. On the average, warehousing costs climbed almost 80 percent between mid-1975 and mid-1979. Higher space costs, along with higher interest costs, have focused attention on the effective use of available space.

2. For manufacturers, warehousing costs (for both raw materials and finished goods) constitute six to eight percent of the sales dollar. Any reduction in warehousing costs quickly flows through to the bottom line. Being labor-intensive, the warehousing function has been especially sensitive to the rapid escalation of labor costs. So worker productivity is a prime concern in attempts to reduce costs.

3. The warehouse readily lends itself to productivity measurement and analysis. Since volume throughput varies significantly in

---

[1] This chapter is from an article with the same name written by the author and by Bernard J. La Londe of The Ohio State University for *Harvard Business Review*, March-April 1980.

many warehouses, the consistent application of work measurement standards offers a fruitful area for productivity improvement.

Productivity can be defined in many different ways. Figure 30-1 shows the distinctions between productivity, utilization and performance, as shown in a study on productivity measurement.

Improvement of any warehouse operation usually involves many small things. Although technology will affect warehousing advances in methods, it will not change the fundamentals of the function. A new data processing system cannot reform sloppy information-handling procedures. A stacker crane will not clean up a dirty warehouse.

Only the imagination of the creative manager limits the number of ways to improve productivity in a warehouse. This chapter will focus on 10 ways to make warehousing more efficient. Six of them will be covered briefly, with four high-priority targets considered at greater length. These latter four are reduction of distances traveled,

Measurement	Definition
1. Productivity	Ratio of real output to real input.
2. Utilization	Ratio of capacity used to available output.
3. Performance	Ratio of actual output to standard output.
	Examples
	1. Cases picked per labor hour.
	2. Percentage of available cubic feet used.
	3. Actual cases picked *versus* standard output.

SOURCE: "Measuring Productivity in Physical Distribution," Council of Physical Logistics Management (Chicago, IL, 1978).

Figure 30-1

increasing the average size of units handled, raising the number of round trips on the floor and improvement of cube utilization.

1. Improve Forecasting Accuracy

Because warehousing is largely a hedge against future uncertainties, an accurate near-term forecast may eliminate 90 percent of the need. A good forecast will prevent the deployment of items in, say, Chicago that are needed only in Florida. Better forecasting also reduces two prime sources of waste: the cost of "reserving" space that the user thinks he may need and the "hoarding" of workers he may need when volume increases.

At Eastman Kodak, the physical distribution group is responsible for estimating and planning, which includes market forecasts and the scheduling that follows them. Unfortunately, very few companies involve warehousing and distribution personnel in such planning.

2. Free Labor Bottlenecks

It has been said that bottlenecks are always found at the top of the bottle, which suggests that management must take the responsibility for correcting the situation. A typical bottleneck occurs in unloading a floor-loaded box car. Two laborers may be assigned to the box car to palletize cases for a fork-lift truck driver who removes loaded pallets. If the laborers' speed exceeds the driver's, they are obliged to wait until he returns to remove a load. Changing the crew size may reduce the bottleneck.

Another approach is the "one man, one machine" technique, which gives each worker his or her own power equipment so there is no need to wait for anyone else. As you tour your plant warehouse, look for workers who are waiting and find out what has caused the bottleneck.

3. Reduce Item Handling

In a factory warehouse a product is typically handled 16 times. Many of these handlings can be eliminated. Each movement of a product is an opportunity to damage it. Each lifting of the product fatigues the package.

As you examine your warehousing operation ask why it is necessary to stage every inbound load on the dock before moving it to a storage bay. Ask the same thing about the staging of outbound loads. In trying to reduce the number of product handlings, look for opportunities to circumvent temporary storage altogether.

4. Improve the Packaging

From the warehouse operator's point of view, the perfect package is made of cast iron and filled with feathers; it is indestructible and can be stacked 50 feet high with no artificial support. Such perfection is non-existent, of course, but the warehouseman is justified in his or her concern about packages that won't permit use of the available space in the warehouse.

For the last few decades, the ceiling height of most new warehouses has exceeded the free-standing stacking capabilities of the packages going into them. The packaging engineer, however, wants to design a container that is light and cheap—and, therefore, barely strong enough to get the item to the consumer. So the manufacturer must consider the tradeoffs. The thoughtful product manager will ask whether it would pay to spend more money on packaging and achieve comparable savings by avoiding product damage in warehousing and distribution.

5. Smooth the Flow Variation

Most companies have rush seasons based on demand peaks or sometimes on the responses to sales incentives. Such peaks can waste storage space and add to labor costs.

There is a way to control such waste. Smoothing the flow usually involves cooperation with marketing personnel. Can we give customers an incentive to do their own warehousing during the off-season? If a sales contest is involved, do the goods really have to be shipped? Perhaps entering the order at the warehouse might permit it to be counted for incentive purposes, but the shipment could actually be made at a later time to smooth the work flow.

6. Set Improvement Targets

Nearly every productivity improvement program involves quantifiable targets in order to take advantage of the instinctive competitiveness of human nature. Because warehousing involves a great deal more random operation than manufacturing, the development of work standards is more difficult. Yet the use of productivity goals, based on predetermined engineered standards (perhaps developed through multiple regression analysis, see Chapter 31) is feasible in a warehouse operation.

In warehousing, as in athletics, workers perform better in a competitive atmosphere. As long as the measurement is not used for disciplinary purposes, workers will cooperate in providing the input needed to measure their performance against a standard. When they become interested in the statistics, they will improve their performance in order to compete for better scores.

High-Priority Goals

The four productivity improvement techniques described here in greater detail involve methods peculiar to warehousing. It is important to understand the difference between the workplace in warehousing and the one in manufacturing. A manufacturing area is characterized by a more or less fixed layout and a consistent flow of work and materials. Workers are normally stationed in certain areas and move little from these stations. The work flow is readily visible and measurable. By contrast, the warehouse work area is large, seldom precisely defined, and in a state of constant change. Aisles are often relocated and workers are usually mobile. The large work area and the mobility of the workers make close supervision of the work impossible.[2]

---

[2] See Joel C. Wolff, "Work Measurement and Productivity," Warehousing and Physical Distribution Productivity Report, Vol. 14, No. 6. Marketing Publications Inc., Silver Spring, MD.

7. Reduce Distances Traveled

In controlling the distances typically traveled in moving material between storage bays and shipping or receiving docks, the warehouse manager finds one of the easiest ways to cut costs.

First, examine the layout. One of the best tools in measuring travel distances is a rule called Pareto's Law, or the 80-20 rule. This law states that in most enterprises, 80 percent of the demand is satisfied with only 20 percent of the stockkeeping units. Clearly, if management knows which 20 percent of the items account for 80 percent of the activity, those items will be located closest to the shipping and receiving doors. Yet, in most warehouse operations, goods are stored by product family. A grocery product operation, for example, may have all canned goods in one section, all paper items in another, housewares in another, and so on. In many cases, storage characteristics require such a separation, but often they do not.

To change the layout to conform to demand, the planner must first determine the demand pattern. In doing so, you must recognize that it is like shooting at a moving target. In any dynamic business, a study to reveal the effects of Pareto's Law must monitor future changes in its specifications.

Take order-picking documents, for example. Are they designed for the convenience of sales people, or are they engineered for maximum efficiency in the warehouse? If the document lists items in the same order in which they are found in an order-pickline, the stockpicker can start at the top of the sheet and at the head of the aisle, and move down the aisle and down the document at the same time.

Reduction of distances traveled is a function of planning, data handling and materials handling. The planner can use Pareto's Law or some other layout procedure to design aisles and staging areas that reduce unnecessary movement. Then the data processing personnel can design a locator system that functions effectively with random locations and order-pick lists that conform to the physical layout.

In materials handling, inbound movements offer good opportunities for trip economy. In some warehouses, the responsibility for finding a location for an inbound load rests with the inbound materials handlers. In the absence of instructions, they will put the incoming load in the first empty slot. The result is a needless proliferation of stock locations for the same item.

Management also should look at the location and use of dock doors. Many warehouses have dock areas dedicated strictly to receiving, while other docks are used only for shipping. But perhaps time and money can be saved through use of the same locations for both shipping and receiving. If management has the flexibility to use any dock door for any purpose, it can reduce travel by assigning inbound loads to the door closest to the storage area to which these items are taken. Similar placement can be made of empty vehicles brought in for outbound loads.

Unlike our national highway network, the warehouse highway system can be changed without substantial cost. The layout of aisles should be continually fine-tuned with a goal of reducing travel distance.

8. Increase the Size of Units Handled

Some retailers in Latin America sell aspirin tablets in individual cellophane packages because their customers, needing only a single dose, cannot afford a full box. While the cost of handling and selling single aspirin tablets is high, the distributor is meeting a consumer need. Contrast that situation to the scene at a dock in Baltimore, where a sealed, 40-foot marine container carrying bulk whiskey from Scotland is being unloaded. Handling in large units greatly reduces the opportunities for breakage and pilferage, as well as the cost-per-ton of handling. As you examine your own warehouse, ask whether the average size of units handled could reasonably be increased. Obviously, this move often requires changes in marketing policy. Some grocery product companies offer an incentive to buy in pallet-load quantities, with the discount offered only when the order involves an exact multiple of the pallet load, say, 50, 100 or 150 cases.

9. Seek Round-Trip Opportunities

Most fork-lift travel in a warehouse involves carrying a payload in just one direction. Therefore, this means that the lift is traveling empty half of the time.

In contrast, most high-rise facilities equipped with stacker cranes include a computer control with a memory unit. This memory is used to maximize round-trip travel opportunities. When the crane is putting away an inbound truckload containing item F, the computer already has determined the best storage address for that item. Meanwhile, the computer remembers that item J, stored in a nearby slot is wanted for an outbound order being staged. Under instructions from the computer, the crane takes item F to its storage address, then moves to the address for item J and returns with this merchandise for the outbound order.

The opportunities for reducing one-way travel are not limited to use of computers or stacker cranes. However, to have more payloads going in two directions the warehouse operation must be carefully planned and must have excellent communication. Equipping each fork-lift truck with a radio helps. If the driver advises the dispatcher that item C is being moved to a given storage address, the dispatcher can instruct the driver to return with a supply of item N.

Admittedly, maximizing round-trip hauling adds some complications. For one thing, it requires more precise planning by supervisors. Moreover, it creates new opportunities for errors by the lift driver, now involved in both shipping and receiving operations. Also, the shipping and receiving docks require more staging area.

But the supervisors can take action to control the complications. They can get the time cushion they need to plan and control round-trip travel if they can negotiate reasonable lead times for both the shipping and receiving systems.

10. Improve Cube Utilization

The cheapest space in any warehouse is the space closest to the ceiling. If you were constructing a new 22-foot clear warehouse, you might develop a cost figure of $20 per square-foot, or $.91 per cubic-foot. If, however, you raise the clearance from 22 feet to 24 feet, the total cost of the building will go up only slightly, to about $20.30 per square-foot. This means a drop in the cost per cubic-foot ($20.30 divided by 24) from $.91 to $.85.

While packaging strength limits the practical pile height, management can do something about the packaging problem. The most obvious solution, mentioned earlier, is to improve the package. However, high-density storage racks, such as drive-in racks, permit higher stacking of products with weak packages.

As cube utilization improves, the amount of travel distance in the building can be reduced. But there is always a tradeoff in using the overhead space. Elevating the product to the roof takes time and money. The most economical routine is geared to the speed of turnover. The fastest-moving product should be stored not only close to the door but close to the floor as well.

There are some limits to cube utilization. To protect the automatic sprinkler systems, fire insurance underwriters impose a maximum on pile height, particularly of hazardous materials. They require a certain distance between the sprinkler heads and the storage pile, the distance depending on the capacities of the sprinkler system, as well as the pile height and the type of merchandise stored.

Furthermore, it is important not to overemphasize the vertical. We recall the plight of a grocery warehouse manager who had inherited from previous management a system that was the ultimate in narrow-aisle design. As he gloomily surveyed the cramped environment, he tried to make the best of the situation. "One thing about aisles," he said, "is they cost a little less each year. Aisles don't bargain for pay increases or increased benefits, and if we save a few aisles and add a lot of people, we'll never be ahead of the game."

Making use of the overhead space is usually the cheapest way to get more product into your building. But the key to improvement of cube utilization is the tradeoff between storage costs and handling costs.

Technology to Achieve Increased Efficiency

Future productivity improvement through improved technology will be concentrated in three areas.

The first area is greater use of the cube. This will range from high-mast, narrow-aisle lift-trucks to carousel systems that fill unused space up to the ceiling.

The second area is less human input through reduction of man-hours to accomplish throughput and elimination of decision making on the warehouse floor. Building design, layout and technology will trim the number of direct labor hours required. Rather than have a worker move through the warehouse for the correct item, the order picker will have the inventory brought to him by a carousel. Microprocessing equipment will be programmed with decisions to follow optimal operational patterns. Reduction of human error and elimination of paperwork will be substantial side benefits.

The third area is improved use of fuels. Because of the cost of energy and uncertainty as to its ready availability, dual-fuel vehicles and equipment will gain greater acceptance. In addition, control systems and equipment options that cut the use of motor fuel and make warehouse heating and cooling more efficient will get widespread attention.

Until recently many U.S. manufacturers treated warehousing as a necessary evil, rather than an integral part of the business system. Commitments of capital, technology and engineering emphasized manufacturing and marketing and gave little attention to warehousing. As a result, warehousing has not kept pace with other corporate functions. Today, however, this situation is changing.

31

MEASURING PRODUCTIVITY

Why measure work in the first place? For four reasons: First, work measurement helps schedule labor requirements provided there is enough lead time for day-to-day planning. Second, work measurement will identify and correct problems. This doesn't mean we punish the lazy worker, but it could mean that we improve teamwork, stop bottlenecks created by an inefficient plant or equipment, or simply create an awareness of productivity among those who are best able to improve the situation. And third, work measurement will identify the need for change. Every business is undergoing constant change with new products added and old ones deleted. A work measurement system will show when there is a significant change to our operating environment. Fourth, and perhaps most important, measurement should insure that we charge our customers correctly for work performed.[1]

What to Measure?

This is always a problem because there is no single measure that will truly reflect the variations in labor requirements when averages do not hold. For example, most warehouse managers can quote you averages but don't really know what they mean. Every day is above or below average in most warehouses. Furthermore, it is unreasonable to expect otherwise because the work content of warehousing is not evenly distributed on a day-to-day basis. Over a period of time labor

---

[1] From "Work Measurement and Productivity," an article by Joel C. Wolff, President, Drake Sheahan/Stewart Dougall Inc. Published in Vol. 14, No. 6 of Warehousing and Physical Distribution Productivity Report. Marketing Publications Inc., Silver Spring, MD.

requirements may average out. But averages maintained on a monthly basis do not show the erratic rise and fall of the various activities that occur on a day-to-day basis. Therefore, it is essential to keep daily records. If time and experience in your warehouse show that daily summaries are not required, then you can reduce the amount of time spent in data collection and processing.[2]

There are four approaches to work measurement in the warehouse: historical goal setting, predetermined engineered standards, conventional stopwatch study and work study using multiple regression analysis techniques.[3]

Historical Standards

Historical goal setting is simply the development of records from past experience, sometimes taking the form of crew averages. Using this technique, one can record the average unit's output per day at a given warehouse and develop an average rate. Some companies prefer not to develop such time studies, and they may look to an industry association to develop average data for companies within their line of business. The individual will then determine how well he is doing compared with others in his industry.

The comparison technique will develop average data at different warehouses, or between different employees in the same general area in one warehouse within the same company and note the comparison. Historical standards are the least effective work measurement approach because:

- They usually are not predicated on well-defined methods.
- The same work is not performed in the same manner each time.
- They do not objectively measure work content.
- They cannot specifically define differences in superficially similar tasks.

---

2 *Distribution/Warehouse Cost Digest,* Vol. 9, No. 24, Marketing Publications Inc., Silver Spring, MD.
3 The next five subchapters are from Joel C. Wolff, op. cit.

MEASURING PRODUCTIVITY

TABLE 19.
Operation: Travel Times for Manual Pallet Jack
Code: EEPJ

This element includes the times necessary to complete a round of travel. The travel time includes a start and stop and normal turns and provides for one way travel empty and one way loaded. Time also is included to get pallet jack, run in empty, raise blades, run out loaded, position pallet to drop, and to drop pallet. Times are for a round trip based on one way distance.

Distance	Constant Time	Travel Time Empty	Travel Time Loaded	Total Round Trip Time
10'	1518	102	153	1773
20'	1518	187	272	1977
30'	1518	272	391	2181
40'	1518	357	510	2385
50'	1518	442	629	1589
60'	1518	527	748	2793
70'	1518	612	867	2997
80'	1518	697	986	3201
90'	1518	782	1105	3405
100'	1518	867	1224	3609

Each additional 10 feet one way distance—204 TMU.

ANALYSIS OF TRAVEL TIME PALLET JACK (MANUAL)

Distance Feet	Paces to Move (Loaded)	Time in TMU Loaded	Paces To Move Empty	Time in TMU Empty
10	9	153	6	102
20	16	272	11	187
30	23	391	16	272
40	30	510	21	357
50	37	629	26	442
60	44	748	31	527
70	51	867	36	612
80	58	986	41	697
90	65	1105	46	782
100	72	1224	51	867

SOURCE: "Digest of Warehouse Cost Calculation and Handling Standards," Marketing Publications, Inc. Silver Spring, MD. 1977.

Figure 31-1

Predetermined Standards

Predetermined engineered standards are useful in design activities. They may not provide a measure of the range of performance variation that exists in some warehouses. Such time values have been published by the Department of Defense, the AWA and others. Figure 31-1 shows a table from a Department of Defense publication.

Observation Techniques

There are two work measurement methods that depend on detailed observation and time measurement. Both of these study the elements of the methods being used that less precise measurement techniques, such as sampling, usually do not do. One method employs multiple regression analysis. The other, conventional stopwatch technique, commonly is employed in manufacturing work measurement.

Stopwatch Standards

While the conventional method results in engineered standards, it does not easily identify the more subtle variables affecting work content and performance inherent in demand labor operations. It also does not provide a means of measuring the normal variations in warehousing functions and thus cannot define properly the warehouse working environment.

Finally, since demand labor operations do not deal with tasks that are repetitive (e.g., one order is not exactly like the one before or the one following), either a very large number of operations would have to be taken or averaging would have to be applied to task measurement. Averaging will introduce unknown and unmeasured assumptions about work content.

Multiple Regression Analysis

Multiple regression is a mathematical technique utilized to identify variables defining work content and explain how they affect the time to perform a task. Regression analysis combines measurement of actual working conditions with work content.

More than one variable may have an important impact on time

MEASURING PRODUCTIVITY

and output. That is, the time taken to select an order may depend not only on the number of cases, but also on the weight or number of line items.

One public warehousing company designed a scheduling and productivity management system using multiple regression analysis. In this situation, the variables considered are shown in Figure 31-2.

The program provided a formula that projected the time needed to complete a task using significant variables. The formulas were entered into a hand-held calculator used by the foreman.

A sample output from a programmable calculator is shown in Figure 31-3.

In this program the calculator prompts the foreman by asking how many cases are to be picked, which the foreman answers by entering the appropriate number. The foreman continues to answer questions posed by the calculator. The calculator, in turn, displays the number of expected hours needed for receiving, order picking, shipping and total.

FUNCTION \ NUMBER OF	ORDERS-B/L	CASES	LINE ITEMS	PALLETS	SLIP SHEETS	WEIGHT	TRUCKS
RECEIVING		X	X	X	X	X	
ORDER PICKING	X	X	X	X		X	
SHIPPING	X	X	X	X		X	X

SOURCE: Warehousing and Physical Distribution Productivity Report, Vol. 15, No. 8 © Marketing Publications, Inc., Silver Spring, MD.

Figure 31-2 Variables considered

Uses of Regression Analysis

The warehouse foreman can make many uses of this tool.[4] First, the total expected job-performance hours are an accurate prediction the foreman can use in planning manpower needs for the day, whether

---

[4] From an article by Norman E. Landes, President, Slidemasters Inc., Vol. 15, No. 8, Warehousing and Physical Distribution Productivity Report. Marketing Publications Inc., Silver Spring, MD.

```
      FORECAST
      XYZ CO.
        PICKING
        CASES?
        4000.
          LINES?
          350.
          B/L'S?
          125.
          24.03    HRS.

        SHIPPING
          BILLS?
          90.
            LBS?
        165000.
          7.71    HRS.

        RECEIVING
          SSHEETS?
          88.
          4.00    HRS.

          35.74    TOT.
```

SOURCE: Warehousing and Physical Distribution Productivity Report, Vol. 15, No. 8. © Publications, Inc., Silver Spring, MD.

Figure 31-3 Calculator output

overtime will be necessary, or whether part-timers should be called in. Another benefit is that individuals assigned work know when they should complete the assignments. In other words, regression analysis is a good short-interval scheduling tool.

Another use ties in with less-than-truckload (LTL) pickups. Typically, LTL pickups are made in the afternoon, creating congestion due to the number of carriers trying to be served in a short time. Due to the accuracy of the measurement system, we can now group orders by truck line, giving each truck line a pickup time out of congestion hours. Though there is no requirement that the trucker come in at that time, it is to his benefit to make a quicker turn-around by avoiding the congestion.

Quality Measurements

One company, for example, uses several tools for measuring productivity, including measures of warehouse quality. It utilizes a service-level reporting system that measures internal order cycle time—the days elapsed from receipt of order to date of shipment—and this keeps tabs on the time required to process an order through both the office and the warehouse. The order cycle time is based on the number of work days between date received and date shipped, using a calendar that has weekends excepted. With this system an order received Friday and shipped Monday will be reported as being shipped the next day, and the same applies to holidays.

The same firm uses an error-reporting system to measure the error rate per thousand line items. Errors by sales representatives, plants, carriers and other non-identifiable sources are grouped, with non-warehouse errors also being compiled as an error rate per thousand line items. Both warehouse and non-warehouse error rates are then combined to arrive at a total error frequency.[5]

---

5   From "Tools for Measuring Productivity" by Richard E. Nelson, Pfizer Inc., in Warehousing and Physical Distribution Productivity Report, Vol. 14, No. 8. Marketing Publications Inc., Silver Spring, MD.

Common Methods of Measuring Productivity

In general, there are three areas commonly measured in calculating warehouse productivity. These are the storage function, the handling operation and the overall warehousing program—a combination of storage, handling and sometimes clerical.[6]

One measure of storage productivity is to compare theoretical with actual cube utilization of the building. Another storage productivity measure is inventory turns, comparing the most recent completed period with previous periods.

In looking at handling productivity, consider both receiving and shipping. Receiving merchandise might be measured in lines of stock keeping units (SKUs) stocked per day. In shipping, lines of SKUs could also be used, though weight handled per manhour is considered more relevant. The most common measure of overall warehouse productivity is to measure the total warehouse cost as a percent of sales. In some cases, just the payroll cost in the warehouse is measured as a percent of sales.

Productivity is not the only thing which is measured in warehouse performance audits—sometimes accuracy is considered as well. When this is done, a meaningful measure is obtained by comparing the cost of error credits to total sales. Another good measurement is to calculate total-period damage costs as a percentage of sales. Both of these measurements of course become more meaningful as progress toward a goal is achieved.[7]

Guidelines for Measurement Systems

Many warehouse measurement systems have not been successful. Those that have succeeded generally follow certain principles, which are listed below:

- Keep it simple. The best systems are those every warehouse employee can understand. They not only know how the system

---

[6] From Vol. 19, No. 3, Warehousing and Physical Distribution Productivity Report. Marketing Publications Inc., Silver Spring, MD.
[7] Ibid.

MEASURING PRODUCTIVITY

works, but what the system is and why it is there. Systems designed with participation from warehouse employees have an excellent chance of success.
- Avoid making frequent changes in standards. With warehouse standards, accuracy may be less important than consistency. Warehouse work changes frequently, and constant changes in standards may make measurements meaningless. Furthermore, frequent standard changes may be the source of employee distrust.
- Create measurement systems to cope with predictable changes in the product line. Every warehouse inventory has certain predetermined variations in product mix that can be anticipated when establishing a standard.
- Never use performance measurement systems for worker discipline. Maintenance of such systems depends on worker cooperation. If they are seen as a way for management to spy on the workers for disciplinary purposes, the entire reporting system will break down.

32

SCHEDULING WAREHOUSE OPERATIONS

Eliminating Delay

Perhaps the biggest drain on productivity in most warehouses is the "hurry up and wait" syndrome popularly associated with the military. One example is the unloading of a truckload of floor-loaded bags of product at a warehouse. There are at least three alternative loading methods. First, two men may work in the truck to palletize the freight for a lift driver who removes the loaded pallets. Second, a single man may load the pallets, receiving assistance as needed from the fork-lift driver who is transporting the pallets. Third, a single man could work the truck alone, equipped with either a lift truck or a pallet jack to remove loaded pallets. The third method eliminates lost time because there is no need for anybody to be waiting for anyone else. With the first method it is likely that the two workers will sometimes be waiting for the lift driver. With the second method the man in the truck will have no work to do if the lift truck driver is delayed in returning.

Scheduling to greatly reduce waiting is possible once the manager knows the time required for each element of each job, as well as the likelihood of delay or interruption.

Using the multiple regression analysis described in the last chapter, the manager can determine when one task will be completed and another started. With this knowledge, truck appointment times can be set to reduce waiting time and congestion. If carrier drivers know they will receive prompt service, they are more inclined to cooperate with the appointment demanded by warehouse management.

Truck Dock Schedules

No warehouse manager with any significant volume of truck activity should run an unscheduled truck dock. Common carriers today are generally accustomed to working by appointment, both for receiving and shipping. Establishing appointments places an obligation on warehouse managers as well as trucking dispatchers. If the carrier arrives at the appointed hour and then is not promptly served, no one can expect the trucking company to show any consideration in making charges for detention. On the other hand, some warehouse managers use failure of a carrier to keep an appointment as a reason for not providing loading or unloading service.

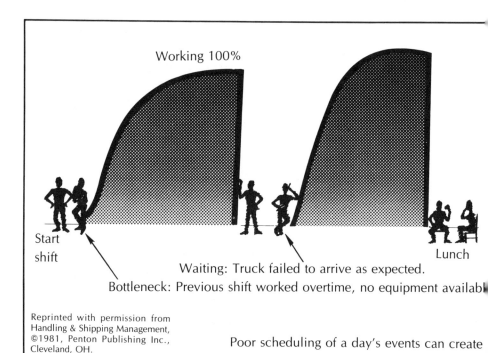

Figure 32-1. Uneven warehouse workflow.

Backlog

Another important value of scheduling is the ability to calculate current backlog. By using measurement techniques described in the previous chapter the warehouse manager can calculate the number of hours of work needed to perform all of the tasks assigned. When these are compared with hours remaining in the normal work day, the manager can determine whether some of the current work should be put off until the next day, scheduled for overtime, or possibly reassigned to another distribution center.

Scheduling for Peak Demand

In some businesses the scheduling problem of the warehouse is complicated by uneven demand. For example, in the retail grocery business there is a typical peak on Thursday, with abnormally low

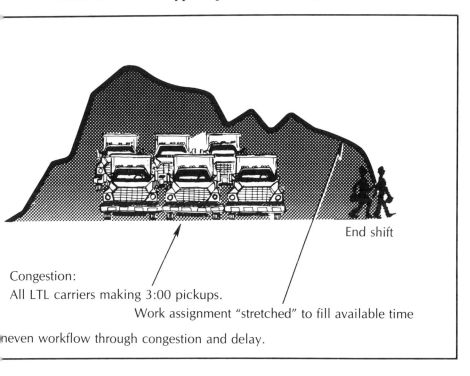

End shift

Congestion:
All LTL carriers making 3:00 pickups.

Work assignment "stretched" to fill available time

Uneven workflow through congestion and delay.

demand on the first three days of the week. As a result, the food chain warehouse will probably have too much work for Monday and Tuesday to supply the Thursday peak and not enough for Friday. Such unbalanced workloads can be partially balanced by delaying some work having minimal penalty, such as unloading of inbound box cars.

Real Time Information and Scheduling

"Real time" computer systems record and deal with events and conditions as they occur, or as they are recognized. In most warehouses, particularly those with fast movement, substantial savings can be made with the timely transmission of information to receiving docks. This will allow precise replenishment, as well as eliminating needless product movement.

If the receiving manager knows when to expect certain merchandise, plans can be made to speed up inbound merchandise. Handling requirements are reduced. For example, if the receiving department knows that stock number 123 is urgently needed to fill current orders, it expedites the unloading of an inbound load containing stock number 123. Then instead of putting that stock number in an assigned pick slot, today's outbound orders are picked directly from the merchandise staged at the receiving dock. Doing this will eliminate double handling of that particular stock number from receiving dock to pick line and back to shipping dock.[1]

Stock Locations and Scheduling

Inventory location substantially affects warehouse scheduling and location decisions for inbound freight should always be made by management, though in many instances they are not. If freight location is unsupervised, the same stock number may be scattered around the warehouse, thus hampering fast movement of the merchandise. Scheduling can be improved in most warehouses by better control of stock locations and by ensuring that the fastest-moving items are

---

[1] From Vol. 16, No. 10, Warehousing and Physical Distribution Productivity Report, Marketing Publications Inc., Silver Spring, MD.

stored in places that will minimize travel time to and from loading docks.

Aids to Scheduling

Warehouse scheduling can be improved by a variety of commonly-available scheduling tools. One of these is a pad or schedule board to assist the planner in tracking the schedule. Another is a record of work schedules during comparable earlier periods so that management can predict probable seasonal workload peaks. A building layout diagram is essential in scheduling; it should show stock locations, temporary pick lines or other arrangements that can be made to ease scheduling problems.

33

CUSTOMER SERVICE—
ITS ROLE IN WAREHOUSING

The concept of customer service evolved when management realized that growth in sales comes more easily from increasing sales to existing customers than from finding new ones. Customer service provides a way to sell more to present customers by improving service to them. Both customer service managers and warehouse managers obviously have the common goal of improving company profits, but frequently there is wide division of opinion on how this goal should be met.[1]

The core of the problem is that warehouse managers are typically rewarded for productivity; high throughput and low costs. Customer service managers, on the other hand, are rewarded for high rates of order fill and fast turnaround on orders—plus a variety of customer accommodations such as emergency orders, change orders and special packing or marking.

The two reward systems are often in head-on opposition to each other. The warehouse may "win" because it has more clout, but then Customer Service may suffer the long-term consequence of lost sales. Conversely, if Customer Service backed by Sales "wins" the confrontation, warehouse costs may go through the ceiling and seriously erode the profitability of the sales transaction itself.

---

1   This chapter is from an article by Warren Blanding, Chief Operating Executive of Marketing Publications, Incorporated. The article appeared in *Warehousing and Physical Distribution Productivity Report,* Vol. 18, No. 9 Marketing Publications, Inc., Silver Spring, MD.

In an ideal world—or perhaps a college classroom—this problem would be resolved swiftly and efficiently by setting more compatible goals for the two departments and rewarding the managers accordingly. In the real world it seldom happens that way, and so the following nine suggestions are given as practical-but-not-perfect ways in which acceptable changes in customer service policies and practices can generate sizable savings in the warehouse.

1. Is This Warehouse Necessary?

This question is often asked after the fact rather than before. Before making a major warehousing investment in order to satisfy perceived customer service requirements, make sure the perception is accurate.

Two case histories illustrate the importance of probing in depth before the commitment is made. A well-known office equipment manufacturer's marketing department determined that customers required a five-day turnaround, from placement of order to delivery. To do this would require setting up half a dozen more distribution centers. Somebody thought to question the source of this new customer service standard: "Who says customers need five-day turnaround?" It was the sales force, not customers, who wanted the five-day turnaround. The reason? Commissions. Customers, they found, would settle for 16 to 21 days—a time frame perfectly serviceable with the present network.

Case No. 2 was somewhat different: extremely poor customer service in a particular sales region, with customers threatening to change sources. The solution? A central warehouse serving that area exclusively. Great idea, but what to stock it with? A more objective analysis of the problem showed that it wasn't delivery but inventory. Orders were taking a long time to fill and deliver because the product needed to fill them had not been manufactured! The real problem that needed solving was in production planning and manufacturing, and in the clearly inadequate sales forecasting on which they were based.

2. Phone Orders and Toll-free Numbers

If you're processing orders at the warehouse site and a significant number come by mail, consider the potential advantage of taking more orders by phone. Telephone ordering provides the time and convenience that translates into improved warehouse productivity.

It's probably a fairly safe rule of thumb today that the typical time for an order that's mailed to transmit to your location will be four or five days. Switching to phone actually saves four or five days. If your service levels are satisfactory now, use those days to balance the workload and cut overtime.

Nothing says you have to give the entire benefit to the customer. There are some negatives, but most of them can be overcome. For example, what do you do when all the calls start bunching at peak periods and some customers can't get through? Set up schedules for customers in different areas to call during different time frames and tell them they'll get better service when they do. You'll get something better than 70% cooperation, which is all you need to level the workload.

Or, set up separate lines for emergency orders and stock orders; one firm actually has a person-less pushbutton order system for emergency orders and customers like it fine—it takes the pressure off representatives handling the larger and more complex stock orders. The system uses a touchtone telephone to code the order.

Should you go toll free? The debate goes on; eventually, 90%-plus of ordering could be via toll-free numbers. And while you're looking at phone orders, remember that in a person-to-person contact vs. mail there's always the opportunity to upgrade orders taken on the phone to more economic shipping units, offer substitutions and suggest taking advantage of consolidation schedules.

3. Consolidations

Consolidate orders as well as freight, but look for compromises that benefit the warehouse as well as the customer.

It's an imperfect world. in Company "A," Customer Service wants the warehouse to consolidate all orders received on a given day for a given customer and ship them together at the close of business; in Company "B," Customer Service wants every order to go through the system without delay—and above all, no consolidations!

To compound matters, in Company "A," the warehouse says it has no way of identifying and consolidating orders common to one customer received on a given day. At Company "B," by contrast, warehouse people are screaming because of the high percentage of small, "penalty" shipments that are going through the stream just because Customer Service won't tolerate consolidation.

In Company "A," the warehouse says why consolidate? What difference does it make? The customer pays the freight! In Company "B," Traffic says, "We pay the freight and we're paying through the nose."

In response to Company "A," let's point out that when the customer pays the freight and sees a dozen or more shipments coming in daily at $2 or $3 apiece, eventually that customer will get tired of paying $30 or $40 a day for freight that could have been shipped as one unit for $5 or $6. Writing a program that will "trap" such orders for staging in a dedicated space will probably pay off in more ways than one—including the inevitable change orders you get after the order has already been assembled. Also, don't forget the high cost of packing materials.

To the Company "B" traffic department we say: Yes, you are paying too much for transportation, but let's not go too far in the other direction: let's consolidate *some* shipments and let others go as small shipments on their own.

Here's what one company did. The warehouse offered a compromise. It said to Customer Service: "If you will give us the option to consolidate when we can, we will guarantee you that all orders received in the warehouse by 5 p.m. Thursday will be shipped Friday." To which Customer Service responded: "What do we tell our customers?" And the warehouse replied: "Regardless of what day they call the order in, tell them it will ship Friday. Which we guarantee it will. So a few customers will get orders earlier than they expected, and others may take a day or so longer than in the past. You also gain

CUSTOMER SERVICE

a few days in the meanwhile to look for stock at other locations, or suggest substitutions." The customer service department bought the compromise and the result is that the company is saving a few million dollars a year and customers are receiving equal or better service than in the past.

4. Free Shipping

Give your customers even a token incentive and chances are they will support programs where your warehouse is the primary beneficiary. Free shipping is another name for consolidation—but one that goes down easier with customers. It's also a way of sharing the benefits with customers and an incentive for next-to-100% participation in your program.

Here's how it works: the warehouse and customer service departments working together make an analysis of the workload by region. They determine that if all accounts in Region 1 place their orders on Monday, all in Region 2 on Tuesday and so forth throughout the week, they can balance the workload in both departments—not ideally, but significantly better than at present when orders come in randomly. The savings in labor as well as actual overtime will be substantial.

But how to get customers to participate? Give them "free shipping"—customers in Region 1 who place their orders on Monday will have them shipped free, Zone 2 on Tuesday. But—somebody objects—that's going to cost us money! Not really, if you look at the tradeoffs involved. It's a high-dollar density item, with freight charges only a fraction of product value; you could just about afford to ship it free to begin with.

By consolidating freight to the regions, the cost per-customer drops to an extremely small amount—which is recovered many times over by the savings in both the warehouse and the customer service department. It also imposes a discipline on customers and it eliminates about 50% of those "emergency" orders that have to be filled and cause expensive disruptions to normal operations.

5. Classify Orders by Size

Workload leveling can often be a sore point between Customer Service and the warehouse. It need not be . . . particularly when you can identify a clearcut benefit like reduced complaints for the customer service department.

Customer Service may want to handle all orders as they come through, regardless of size or destination. But when you provide an incentive to send small orders—less than palletload—for morning assembly and all larger orders for the afternoon, you can handle each type more efficiently and productively than when they are mixed.

What makes this more palatable to Customer Service is that the system virtually guarantees fewer complaints. The many small orders that go out in the morning include the hand-to-mouth buyers, and the one- or two-line orders, each one of which has the potential for generating a complaint for one reason or another. A certain percentage of these orders will be semi-emergency or need-now items. On the percentage alone, the relatively few large orders handled in the afternoon will generate fewer complaints.

There's an added factor, too, which can enhance the end performance of warehouse workers. The afternoon setup for larger orders is different from the morning setup, and perhaps less demanding; it injects variety and greater ease into the job at a time when fatigue is beginning to set in.

6. Backhaul

If pickup rules don't already exist, work with Customer Service to develop policies that are reasonable and acceptable to customers and permit you to schedule effectively. Then make sure that Customer Service reminds customers of those rules.

For many companies, backhaul is already a way of life. Yet every warehouse manager knows that it can be extremely disruptive when a customer's truck "drops in" looking for an order that may not even be assembled yet. You can't turn the driver away; his company is one of your best customers. But if you drop other work to assemble his order, a dozen or more other customers' orders are compromised.

CUSTOMER SERVICE

Here's where Customer Service can really help. They can develop policies (with the concurrence of Sales and Marketing) requiring specified lead times and pickup by appointment only. Then the customer's truck can be fit into the workstream as any common carrier pickup might be. Some companies go so far as to get customer pickup rules that say, if you don't pick up within X hours of the appointment, we will ship common carrier and bill you the charges.

How well do such policies work? Certainly not 100 percent. But if you could cut unscheduled customer pickups by 50 or 75% or more, wouldn't it be worth while? Certainly you're always going to have to make exceptions for some customers. A surprising number of customers will fall in line with your appointment procedure without protest—particularly when customer service personnel and literature remind them of the rules dealing specifically with pickup.

7. Accuracy in the Warehouse[2]

Discuss and plan with the customer service manager ways of improving the total system that will not only avoid errors but will also provide tangible savings and benefits for both parties as well as for customers.

Nobody likes errors, and warehouse managers devote a great deal of time and energy to avoiding order-picking errors in particular. System errors are avoidable and human errors are predictable. When the distinction isn't made, trouble arises. An order-entry error that originates in the customer service department but is passed along to the warehouse can be an expensive proposition if it isn't caught.

If the system isn't designed to allow for human error—for example, built-in parameters on cost and quantity, as well as a double-check of unusual orders at the warehouse level—mistakes will continue to occur. They'll continue to cost hard dollars to rectify plus substantial but less visible administrative cost in complaint-handling and problem resolution. To say nothing of long range but very real

---

2   See Chapter 38.

costs of loss in customer confidence and the threat to the company's status as preferred-source vendor.

Errors resulting from poor communications between the departments, rush orders (admittedly one of the most common sources of errors), or such things as customers trying to shortcut the customer service department and deal directly with warehouse operating personnel—many of these reflect system errors rather than human mistakes.

Some causes are basic: the design of the forms used to transmit orders or other data. We know one firm which uses a 14-part order form which invites errors or shortcuts. Specials and promotions and stock location systems represent another error potential.

While it's true that error-prone individuals should be removed from sensitive jobs, the system itself should at least allow for predictable (and acceptable) human error rates, and should itself be as error-free as possible within cost constraints. Don't overlook the fundamental fact that it's often less costly to let some mistakes get through and correct them later than it is to try to avoid all mistakes altogether. This is particularly true if you have a captive or closed loop system, in which product moves from company warehouse to company store.

8. Keep Communications Channels Open

Communicate with Customer Service and they will return the compliment, and you will both benefit. Just as expensive errors can often be avoided simply by querying orders that seem unusual or out-of-the-ordinary, so can potentially serious situations of other types often be avoided when there are good two-way communications between Customer Service and the warehouse.

For example, when physical inventory appears to be substantially below book (computer) inventory, advance notice will help the customer service department avoid embarrassing over-commitments and may even reach them in time to allocate or defer some orders. Or, if the customer service department anticipates unusual volumes of orders, or some other change from the norm, advance notice obviously helps the warehouse get ready. Orders that are delayed, shipments not picked up and other situations like absentee-caused bottlenecks that may bounce back unexpectedly on the customer service department are much easier to deal with when the department has advance notice.

It's obviously in the interests of warehouse efficiency to set up such communications, and it may be worth while to set up some informal meetings with line personnel of both departments to ensure that such communications actually take place.

9. Seasonality and Sales Contests

Join with your customer service department in discussing with Marketing some of the ways that seasonality and other peaks and valleys of activity can be offset by better spacing of special promotions and sales contests.

Seasonality, sales contests, end-of-month commissions, special promotions, deals and the like are the natural enemies of warehouse efficiency. Enough so that warehouse operating executives sometimes forget that the *primary* goal of the warehouse is to serve the marketing and profit goals of the company . . . and to do so as efficiently as possible consistent with those goals.

It's also true that the impact of some of these marketing events on warehouse efficiency can in fact be softened considerably without compromising those marketing goals. While neither the warehouse nor the customer service department has the authority to change the cutoffs, or the way points are awarded in sales contests, they certainly have a mutual interest in trying to get the rules changed to minimize the excessive peaks and valleys of activity that hinder warehouse operations.

Of course, you can't change the seasons of the year—or can you? At least you can take a crack at persuading marketing people to level out some of the valleys between the peaks of seasonal surges of activity by spacing out their promotions and sales contests. You have a good selling point, too: during special promotions, fast turnaround on orders and particularly good reorder capability, are critical to promotion success. Which means that both the warehouse and customer service departments can do a better job supporting promotions when they're not already busy with half a dozen other things.

It isn't easy to bring about fundamental change in marketing practices, but it's worth a try—even if it's something as basic as getting Sales to let you and Customer Service know a few weeks in advance when a major promotion is going to be cut loose, rather than letting you find out for yourselves after it begins. Just being able to plan and anticipate can make a significant difference in your costs as well as the levels of service customers actually get.

The Customers Pay the Bills

Rightfully, customers demand service because they pay the bills. The price they pay includes the costs of the service level you have chosen to provide. One of the first laws of business survival and growth is to give customers plus-service value for order-value received.

Philip B. Crosby, in *Quality Is Free*[3] said: "Quality is conformance to requirements . . . Requirements must be clearly stated so that they cannot be misunderstood." The same statement applies to customer service. Management sets the quality requirements for product and service levels. Therefore, customer service is a basic management responsibility that starts with management's strategic service plan. This plan sets the customer service level to market and distribute the business's products and services. Warehousing is a key service provider in the strategic service plan.

Delegating customer service is dangerous, because it is a primary management responsibility. Customer service activities, when delegated, remain a management responsibility. A customer service audit system will alert managers, and show any variance from the desired service levels.

In a competitive environment, customer service changes under the pressures of the customers' changing desires. Managers must

---

3   *Quality Is Free, The Art of Making Quality Certain* by Philip B. Crosby, McGraw-Hill Book Company, 1221 Avenue of the Americas, New York, NY 10020.

spend some part of every day attending to customer service needs or run the risk of losing customers.

Summary

Efficiency is not the only or even the primary goal of the warehouse. It's easy to forget that companies are run to make money, not save it. When we talk about saving money, we are presupposing that there is a way the job can be done less expensively . . . not that we are going to save money by shortcutting the job. That would be like stopping the clock to save time. Warehousing is an integral part of the firm's marketing strategy and marketing system, and what we have sought to do here is suggest ways in which it can fulfill its role in both respects with even greater efficiency than before.

34

IMPROVING ASSET UTILIZATION

Most productivity improvement programs concentrate on improving utilization of labor. Yet there are four other major cost areas in warehousing where making better use of assets can boost productivity. These four are space, energy, equipment and inventory.

Space Utilization

The warehouse space most frequently wasted is that closest to the ceiling. While there are limitations to the pile height for most commodities, it is not unusual to see cube space wasted because merchandise is not piled as effectively as it might be. Racks or stacking platforms will improve cube utilization as will good planning (See Chapter 15).

Reducing the size of working aisles also may improve utilization of cube. However, reducing aisles can involve a tradeoff between space and labor. Since better labor productivity is achieved in wider aisles, it is necessary to balance the cost of the space against the value of the time.

Energy Utilization

Perhaps no commodity has undergone such violent and unpredictable price changes within the past two decades as energy. With these sudden changes of pricing have also come great changes in our attitude toward energy as a resource. Buildings constructed when energy was cheap will typically show enormous heat losses through insulation which, by today's standards, is grossly inadequate. In contrast, buildings built more recently are governed by building codes which require substantial protection from heat loss. Changes in the price of energy have changed many procedures in construction. Extensive use of skylighting in older buildings was justified by a reduction in the price of illumination. More recently, some types of skylighting have been found to lose large amounts of heat energy which is more costly than the energy for lighting which they might save.

Warehousing practices of a few years ago were based on the assumption that energy was a cheap resource. Many storage buildings were designed to allow workers to be comfortable in shirt sleeves, even in mid-winter. More recently, warehouse operators have questioned some of the reasons for heating a building.

Essentially, there are three reasons to control warehouse temperatures:

1. Product environment: Many products are damaged by freezing, and others will deteriorate if the temperature is too cold. Other products must be kept frozen or cool in order to be preserved.

2. Fire protection: Most sprinkler systems are wet-pipe systems, which require protection from freezing in order to operate (See Chapter 21).

3. Labor productivity: Warehouse workers are likely to work more slowly or require warm-up breaks if they work in a totally unheated area in which temperatures fall substantially below freezing. On the other hand, a properly-dressed warehouse worker is likely to be nearly as productive in a 40°F building as in a 70°F building. The only exception might be precision warehousing jobs such as order picking or packing. Comfort levels in these special areas can be maintained with spot-heating units, such as infra-red fixtures.

With the emphasis in recent years on conserving energy, many practical means have been discovered to use it wisely. New sodium lighting fixtures save substantial costs in electricity. However, most sodium lighting systems require a high-voltage wiring which is expensive to relocate when aisles must be changed. Unless a lighting

IMPROVING ASSET UTILIZATION

system is specifically designed for ease of relocation, the cost of rearranging aisles can be prohibitive.

Any warehouse heating system designed before the price of energy started to climb is obsolescent today. Perhaps the greatest advance in warehouse heating is a realization that circulation is as important as heat. For this reason, heaters with fans are a great deal more effective than radiant heating devices. Radiant or infra-red heat is useful only for spot heating. Otherwise, warehouse heating can be enhanced by using fans which blow the heat from the ceiling back to the floor.

In addition to insulation, great progress has been made in better sealers for all kinds of warehouse openings. This includes the use of flexible vestibule-type doors which swing open when a lift truck or any other kind of transporting mechanism moves through the doorway opening. The same principle can be used to provide flexible doors for craneways or other openings used by more complex materials handling equipment.

Because of the size of most industrial warehousing operations, the energy saving opportunities through careful study of conservation

---

Energy Checklist

1. Are order pick-areas in the warmest part of the warehouse?
2. Are lighting levels high in order pick-areas?
3. Are picking vehicles equipped with headlights which could replace warehouse lights?
4. Are thermostats locked so that heat level is controlled only by management?
5. Are heat losses through doors and automatic dock boards controlled with weather stripping and flexible vestibule doors?
6. Is grounds lighting controlled by timers or daylight sensors?
7. Are all heat circulators functioning without blockage?
8. Are walls and ceilings painted white to maximize light reflection?
9. Are warehouse lights designed so that they can easily be relocated when aisles are changed?

Figure 34-1

can be impressive. For example, a ventilating system designed for a large warehouse could easily air condition the warehouse office.

Many warehouse operators purchase energy surveys of their buildings by professionals, but a simple checklist, such as that shown in Figure 34-1, will help.

Increasing use of semi-automated warehouse equipment is another means of saving energy costs. One warehouse which has standardized its stock-picking trucks has greatly reduced warehouse lighting. Each order picking truck is equipped with headlights which brilliantly illuminate the stock being picked. Because the truck has its own lights, lighting in the warehouse is kept to a minimum.

Saving energy in the warehouse is largely a matter of common sense. Remember that warehouse workers do not need to be in shirt sleeves to be effective. While personnel comfort is obviously desirable, in the warehouse a degree of comfort can be achieved through clothing rather than temperature control. Because most warehouse jobs are active, a reasonably-dressed warehouse worker can function effectively in temperatures which are somewhat more variable than those found in the office.

With reasonable vigilance and good equipment maintenance, any warehouse operator can achieve significant energy savings.

Equipment Utilization

Every lift truck should be equipped with an hour meter that can be used not only to show engine-running time, but also to indicate when to perform regular preventive maintenance.[1] Hour meters do not measure the amount of time a lift truck is actually in use. In-use time must include time when the engine may not be running, but the lift truck cannot be made available to another operation. To illustrate, a driver may be unloading a truck and at the same time must hand-

---

1   The next six sections are adapted from "A Program for Lift Truck Utilization/Maintenance" by Todd Benton and Albert D. Creel of Distribution Centers Inc., Columbus, OH. Published in Warehousing and Physical Distribution Productivity Report, Vol. 17, No. 12. Marketing Publications, Inc., Silver Spring, MD.

IMPROVING ASSET UTILIZATION

stack cartons or process paperwork. The engine may be turned off for that portion of the time, which stops the hour meter. Yet the truck cannot be used for another operation during this brief waiting period.

One warehouse operator uses a rule of thumb that lift trucks are actually in use for 75 percent of each shift. This means that an hour meter that shows 5.5 hours for a 7.5-hour-shift is actually at 100 percent of normal capacity.

The Equipment Utilization Ratio

There are three basic elements in calculating an equipment utilization ratio:

Calculation of Utilization Ratio At Location A

I. Data

Number of days worked	23 days
Number of active lift trucks	10 trucks
Total hour meter readings	1,199 hours

II. Base Calculation

Base = (# days) × (# trucks) × (5.5 hrs./truck)
= 23 × 10 × 5.5
= 1,265 hours

III. Utilization Ratio Calculation

$$\text{Ratio} = \frac{\text{Total hour meter readings}}{\text{Available base hours}}$$

$$= \frac{1,199 \text{ hours}}{1,265 \text{ hours}}$$

= 94.8%

SOURCE: Warehousing and Physical Distribution Productivity Report, Vol. 17, No. 12.

Figure 34-2

1. The number of days in the work month (less weekends and holidays).
2. The number of active lift trucks in the system.
3. The total hour meter readings from the active lift trucks.

The equipment utilization ratio is calculated by dividing the total hours from all hour meters by the available-hours base number. The base number is determined by multiplying the number of active lift trucks times the number of worked days in the month times 5.5 available hours per lift. Figure 34-2 is an example of how to calculate the ratio.

A monthly utilization report shows monthly equipment utilization figures for each operating location. Managers can see their own utilization, compare themselves with others and begin to inquire into the whys and hows.

Improving Performance

There are several uses for this program. At the corporate level, utilization ratios may be used to evaluate and plan equipment allocation among various operating locations. The combined utilization ratios also will provide an indication of whether there might be too much equipment at a particular location.

On the operating level, local managers should use this report to evaluate their own performance. When regularly confronted by a low-utilization ratio, they should ask themselves if they have a surplus of equipment. Because demand for equipment can fluctuate depending on the seasonality of distribution activity, managers should consider the alternative of short-term rentals to complement their base lift truck fleet during periods of high activity.

Although an equipment utilization ratio of 80 percent is certainly acceptable, managers can achieve ratios greater than 100 percent. This is accomplished by adding more shifts to the operation. The available hour base (*see Figure 34-3*) is determined by the number of lifts required to operate one shift. By spreading the workload over two or three shifts, less equipment would be needed to accomplish the same amount of work.

There are additional costs in multiple shifts, such as supervision, shift premiums and utilities. These extra costs must be weighed against the savings of decreasing the equipment asset base. Additional shifts are one way to absorb some periodic high activity without resorting to short-term equipment rental.

Lift Truck Rebuild Program

Reducing the investment required for a fleet of lift trucks is of major importance, but it's equally important to extend the life of that investment. A lift-truck rebuild program is an integral part of increased utilization planning.

Overall lift-truck utilization is calculated at 75 percent efficiency, or approximately 1,500 hours per year. From a controlled maintenance program you should expect to get about 10,000 hours'

Calculation of Utilization With Different Shifts

I. One Shift Operation:
Usage hours	1,199 hours
Available base (10 trucks)	1,265 hours
Utilization ratio	94.8%

II. Two Shift Operation:
Usage hours	1,199 hours
Available base (8 trucks)	1,012 hours
Utilization ratio	118.5%

III. Three Shift Operation:
Usage hours	1,199 hours
Available base (6 trucks)	759 hours
Utilization ratio	158.0%

SOURCE: Warehousing and Physical Distribution Productivity Report, Vol. 17, No. 12.

Figure 34-3

operation prior to requiring a major overhaul. Depending on usage, this would be six to 10 years.

If you rebuild in your own workshop, you can expect the average cost to be between 30-35 percent of a new truck. If you have the rebuilding done by one of the major truck dealers, you can expect the cost to be about half that of a new truck.

Key Points of the Rebuild Program

A good rebuild program requires the complete dismantling of the lift truck and inspection of every operating part.

The engines should be completely rebuilt, including the reboring of cylinders, or installation of a new block. It is normally more economical to have this job done by a contractor.

Transmissions and differentials must be inspected and worn parts replaced. Steering mechanisms, brakes and electrical portions (including the electrical harness) also should be inspected. Most chains, cylinders and hydraulics should be repaired and/or rebuilt as necessary.

Every part should be inspected and any one with discernible wear replaced. This includes seats, pedals, meters, tires, etc.

The truck should be completely repainted and equipped with new anti-skid strips, knobs, handles and decals. The vehicle should look, feel and run like a new machine.

In order to make sound decisions on rebuilding, the warehouse operator must have a meticulous maintenance program. Without accurate repair costs, it is difficult to determine whether it is time to rebuild, or whether the rebuild program will be economically feasible.

For example, a schedule might call for rebuilding at 10,000 hours. However, if one truck had a major failure at 8,000 hours, it would be sensible to schedule rebuilding at that point, rather than fix the failure and then rebuild just 2,000 hours later.

If hourly maintenance costs remain relatively low, a second rebuild may be justified.

IMPROVING ASSET UTILIZATION

The Value of the Program

A rebuild program has several advantages. For example, at a cost of a third to half that of a new lift truck, the rebuilt truck's service life is extended by 80 percent. Even with the extension, the resale or trade-in value of the truck will increase.

Materials handling equipment is a major budget item for any distribution facility. A recent study of the grocery products industry showed that equipment costs represented an average of 4 percent of total distribution center costs, nearly as much as the percentage spent for warehouse supervisors. Yet few organizations carefully control equipment costs.

When controls are lacking, equipment is purchased that is not really needed. Frequently, managers will "squirrel" spare equipment that is underutilized and should be sold or traded for newer models. In other cases excessive maintenance is expended on a piece of equipment not worth repairing.

Good controls based on meter utilization will show when there are too many lift trucks in a fleet. When you find excess equipment you might move some of the excess to another location where more equipment is needed.

Records of repairs expressed as a cost per-operating-hour will show which units are worth repairing, as well as which ones have been unusually good or bad from a maintenance standpoint. Judgment is an important part of the rebuild program, since there may be cases where an older truck has been made obsolete by technological advances.

With a combination of fleet utilization records and judicious use of a rebuild program, you can bring down the costs of a materials handling fleet as a percentage of total operating costs. And you can do it with no sacrifice in warehouse productivity.

Inventory Performance

Not every warehouse manager is responsible for inventory performance and very few public warehouse operators have any input at all into inventory planning.[2] Some private warehouse operators are not held responsible for this phase of the operation either. Yet failure to achieve satisfactory inventory performance has caused the closing of many a warehouse, both public and private; so every warehouse operator should understand the essentials of effective inventory performance.

After personnel, inventory is the most important asset in most companies. Inventory is volatile, dynamic and complex. It is affected by both external and internal forces. External forces that affect inventory include general business conditions, inflation, customers, market conditions, technology, vendors, seasonality and competition. Internal forces include management awareness, company policy, engineering development, inventory control techniques, forecasts, availability of capital and storage capability at multiple levels (raw materials, work-in-process, finished goods).

Realistically, a company will never have a completely accurate inventory. It is the degree of imbalance that is critical, since there will always be some excess inventory or shortages. This can be managed, however, since inventory problems do respond to an organized, focused effort.

The Management Factor

The most influential factor affecting inventory and service level—more important than the computer system or forecasting methodology—is senior management's interest or awareness. Observe the effect on inventories when a president or senior officer isn't happy with present inventory levels (whether they are high or low). Frequently, the source of bungled decisions is senior management's sud-

---

2 The remainder of the chapter is from an article by John A. Tetz, Coopers & Lybrand, Columbus, OH. Published in Warehousing and Physical Distribution Productivity Report, Vol. 17, No. 11. Marketing Publications Inc., Silver Spring, MD.

den realization that inventory levels are too high, or service levels too low. Management, simply by expressing concern, can have a significant impact on inventory levels and their profit contribution to the business.

Internal forces are more significant than external forces in determining inventory and service levels. Decisions to reduce inventory, when not part of a planned program, generally result in shortages of fast-moving items while the slow-movers sit in the warehouse.

Why Measure?

Why should you measure inventory performance? Why should you circulate the results of your measurements when, after all, they could be embarrassing? Measurement highlights many tough issues—many that some managers would prefer not to face. Yet measuring inventory levels is necessary to evaluate and monitor the effectiveness of your inventory management system. It identifies where you can plan improvements, track progress and highlight success. It provides a forum for communicating and working with other managers on a topic of critical importance to the profitability of the business.

By measuring inventory performance you can determine the most desirable inventory levels for the top 20 percent of the high sales/usage items, develop programs to use (or scrap) obsolete items, and organize the warehouse so the fast-movers are located nearest the shipping doors.

The performance measurement system can be used as a base from which to measure improvement in inventory turns. Too often, inventory management and control policies or systems are implemented without the means to measure their effect.

Danger Signals

Here are some of the red flags to watch for:
- Your inventory turns are less than industry experience. Is the majority of your inventory turning markedly less except for a few items, or are a few items responsible for the majority of your turns?

- Sales volume shows significant growth, with little increase in profitability. Could the cause be pricing structure, product mix, product cost, or, perhaps, excessive inventories?
- Inventory is growing at a faster rate than sales. Should production be slowed or purchases delayed? Perhaps it is time to consider reducing the labor force—a layoff may be in order.
- You are hanging on to surplus or obsolete inventory with the hope of selling it one day. Do the costs associated with warehouse space, handling and administrative time exceed the potential that can be realized from selling the inventory at a reduced price?
- Your inventory and on-order records are not 95-to-98 percent accurate. Inaccurate records generally result in additional safety (or fudge) stocks, with a corresponding increase in expenses for inventory carrying cost.
- Your deliveries are consistently too late or your lead times are too long. In a manufacturing company late deliveries or a longer-than-necessary lead time most often means excessive work-in-process inventories. In a distribution company, you may be holding or storing items while waiting for an order to be shipped complete, or you may be taking longer than necessary to process, pack and ship.
- There are material shortages, or production is delayed due to lack of material. Perhaps order reviews should be more frequent, or the inventory management methodology revised.
- There is a reluctance to mark down stock. Often, inventory carrying costs for an item over weeks or months will exceed the markdown of a special promotion.
- Inventory is growing as a percentage of assets. Instead of investing in productivity improvements, is capital going to unnecessary inventory?
- Cost systems are inadequate. Perhaps your inventory is overvalued or undervalued (particularly if scrap is not properly accounted for).

IMPROVING ASSET UTILIZATION

- Sales forecasts are used to manage inventory. Are the forecasts consistently too optimistic, resulting in excess stock?
- Sales or usage is not analyzed by customer, product line or profitability. If not, do you really know the inventory or service level—or what is desirable?
- Senior management does not review inventory performance. It *should* review inventory performance, obsolescence and service level periodically. Is senior management attentive, interested and aware of inventory performance?
- Your warehouse or storeroom is running out of space. This can be a sign of serious trouble, particularly if sales (adjusted for inflation) are not increasing significantly, or lead times are lengthening.
- Your service level is increasing significantly with no apparent reason, or backlogs are declining. Excessive inventories may be just over the horizon.
- Your company has had a large inventory adjustment or write-off recently. This can indicate recordkeeping (receipt/issue control) that frequently results in simultaneous excessive inventory and stock-outs.

If any of these red flags come up in your company, you should take a hard look at your inventory performance on an item-by-item basis.

Inventory Turnover and Distribution Analysis

Every company with inventory can and should apply straightforward, tested methods to measure its inventory performance. Two basic reports are the key: a turnover analysis and a distribution analysis. Both should include a one-page age summary by major inventory category. Properly prepared and used, these analyses can provide an objective and quantitative diagnosis of your inventory.

It is important to measure performance on an item-by-item basis as customers buy individual items, not groups or product lines. Then, summarize your individual measurements by category to help identify particular problem categories or situations that are not obvious from either the detail or the total.

Often, managers judge inventory turns by dividing annual sales or usage dollars by the present or average inventory level, thus failing to recognize the fluctuations—which can be significant—or status of the individual items making up the totals.

For example, take the case of the firm whose inventory turned an average of five times a year. This was acceptable to management and in line with other firms in the industry. Yet an item analysis showed that 15 percent of the items were moving 25 times annually, while 85 percent of the items were turning only twice a year. On an item-by-item basis, the firm was in trouble. The average turnover of five turns simply did not reflect the true condition of the inventory.

Always use a common denominator in performance measurement. Dollars are generally accepted, although for some companies pounds or gallons are more relevant.

A turnover analysis shows the number of months' supply on hand, calculated at the current or forecasted rate of usage. Inventory balances are divided by the average usage per month, and arrayed in descending order. Stated another way, turnover is computed as the ratio of annual usage divided by the inventory.

Identifying low-stock or no-stock items in the inventory with demand during the year helps you spot a potential service-level problem. Conversely, you can age the inventory to determine the value that is potentially excess by identifying that portion of inventory that has had little or no turnover during the past year.

Case History

Figure 34-4 is a turnover analysis report, using actual data from a small company that manufactures products sold primarily through distributors. Thirteen-hundred finished goods products, with sales or inventory, have a manufacturing lead time of two to six weeks.

The company has received occasional complaints from customers on product availability. The firm has had a computer for six years, used primarily for financial reporting. Production schedul-

IMPROVING ASSET UTILIZATION

Figure 34-4. Turnover Analysis (Abbreviated format)
(Snapshot of 10 items out of 1,300 in the full report)

Item/Part No.	Cumulative % Items	Inventory Units	Inventory $	# Months of Coverage**	$ Coverage More Than 12 months	$ Coverage More Than 24 months	Comments
1001	1 (13th item)*	0	0	0	-	-	
4050	10	0	0	0	-	-	Probable service
4123	33 (#430)	0	0	0	-	-	level/lost sales!
9701	40	3	136	.7	-	-	
5945	50	391	586	1.9	-	-	
6173	60	4	20	4.0	-	-	4 to 6 weeks
1711	66 (#858)	50	2,000	12.0	400	-	mfg. lead time!
2105	80	30	891	24.0	446	-	At 30% per year!
1796	90	617	23,175	57.4	18,337	13,500	4.7 yr. supply
8731	100	28	1,498	999.9	1,498	1,498	All stock inactive!
Totals			$1,262,000		$402,811	$243,877	
					32%	19% of total inventory	

*of 1,300
**At a forecasted rate of usage

Figure 34-5. Turnover Analysis Summary (Abbreviated format)

Inventory Category Code	# of items	% of items	$ Coverage more than 12 months	$ Coverage more than 24 months
01	332	26	38,782	9,094
02	201	15	55,341	30,247
03	640	49	288,439	189,060
04	38	3	4,514	4,232
05	89	7	15,735	11,244
	1,300	100	$402,811	$243,877

SOURCE: Warehousing and Physical Distribution Productivity Report, Vol. 17, No. 11, © Marketing Publications, Inc., Silver Spring, MD.

ing was done by plant management. A tour of its crowded warehouse indicated that some stock was inactive.

The red flags clearly indicated a potential problem. A turnover and distribution analysis was run to quantify the problem and to aid in developing solutions.

The charts shown here are "snapshots" of the actual reports. Data are shown on only 10 out of the 1,300 line items that were actually analyzed. The item/part number has been disguised to maintain confidentiality, but the balance of the data is real.

In this illustration only those columns critical to a turnover analysis are shown. Other data pertinent to an item-by-item analysis within a company (such as description, unit of measure, category codes, new item identification and on-order status) were included on the actual report but are omitted from the illustration.

The full report shows that 33 percent of the total number of items (430 out of 1,300) are stock out. This is an indicator of a potential customer service level problem. Sixty percent of the items have on-hand inventory equal to or greater than a four months' supply with 33 percent of these items having a greater than 12 months' supply. Many of these are surplus or potentially obsolete and should be reviewed for discontinuance and disposition.

The turnover analysis report summarized by stocking category code is shown in Figure 34-5. Note that 77 percent of the excess inventory is in Category 03, which has only 49 percent of the part numbers. This summary exposes areas that require more or less emphasis.

A distribution analysis (or ABC analysis) is illustrated in Figure 34-6. This company's inventory situation follows Pareto's Law (the 80-20 rule) very closely. The careful control of a few items will provide a high degree of overall control.[3]

---

3   Vilfredo Pareto (1842-1923) was an Italian philosopher who determined that income distribution patterns were basically the same in different countries. That is, a small percentage of the population had the majority of the income. Translated more broadly, a relatively small number of items account for a large proportion of the activity. This phenomenon is one of the most applicable and effective, yet least used, of the basic principles of inventory management.

The items in the distribution analysis are arrayed in descending value of annual sales (or usage) dollars at inventory cost. In Figure 34-6, through item number 2004 (the 13th item on the total report of 1,300 items), the forecasted annual costed sales are $58,289, with a cumulative sales usage of 19.9 percent (of the total sales). These can be considered the "A" items; therefore, they need to be tightly controlled.

Figure 34-6 also shows that 20 percent (through Item 4004) or 260 of the items account for 80.8 percent of the total sales, but only 55.6 percent of the inventory.

What was learned from the turnover analysis and the distribution analysis? At first management was not concerned with the total number of 4.8 turns a year. However, when the inventory was detailed on an item-by-item basis, an alarm was raised.

Figure 34-4 shows that 33 percent of the items (below the third line) have an inventory greater than a 12-month supply. Another 33 percent of the items (above the second line) with demand during the most recent 12 months have no inventory. Clearly, this inventory is out of control. The company was losing sales to competition because of lack of inventory, yet its inventory investment was excessive, and there was a disproportionate amount of slow-moving or inactive items.

What was done? After an in-depth management analysis of the situation, the company developed and implemented an action plan that included:

- A weekly senior management review of schedules and inventories of the top 10 percent (A items) of the sales items.
- Hiring a production planner to schedule production based on needs *versus* ease of manufacture.
- A rework disposition program for inactive items.
- A product line review that resulted in dropping many slow-moving, unprofitable products.
- Installation of a forecasting system to help the company better anticipate its requirements.

Distribution Analysis
(Abbreviated form — snapshot of 10 items out of 1300 in the full report)

Item Code	Cumulative % Items	Descending Order Annual Sales/Usage $	Cumulative % Usage $	Inventory Units	Cumulative $	Cumulative % Inventory
2004	1 (13th item)*	58,289	19.9	0	0	9.8
6041	10	10,692	65.0	0	0	44.4
4004	20 (260th item)	4,981	80.8	19	830	55.6
5321	30	2,841	88.8	0	0	65.6
4013	40	1,824	93.7	45	1,026	72.5
4348	50	1,008	96.6	13	222	77.8
5632	56	775	97.7	8	304	79.3
5721	70	342	100.0	46	2,097	87.3
7051	91	1	100.0	12	3	93.5
8017	100	0	100.0	102	2,703	100.0
* of 1300	TOTALS	$6,057,922			$1,262,300	
				Average turnover rate: 4.8 times		

SOURCE: Warehousing and Physical Distribution Productivity Report. Vol 17, No. 11. © Marketing Publications, Inc., Silver Spring, MD.

Figure 34-6. Abbreviated distribution analysis

- Rearrangement of warehouse stocking locations to make it easier to pick high-volume items.
- An ongoing inventory performance measurement system to provide the base from which to measure improvement.

An Action Plan—And Potential Benefits

To implement a comprehensive inventory performance measurement system within the company, you must have a specific action plan. The major steps in the plan include: designing a turnover and distribution analysis, with appropriate summaries; reviewing requirements with the data processing staff; establishing targets or goals for service and inventory levels; analyzing the reports generated to identify opportunities for improvement; summarizing the analysis and recommendation for a management presentation; formulating a strategy for the best approach; and making specific corrective action recommendations.

Corrective action could include developing an obsolete/surplus utilization (or disposition) program; improving the inventory management or forecasting system; emphasizing the need for better sales forecasts; improving inventory recordkeeping; designing and implementing an improved cost system; or looking at warehousing operations or facilities. Specific corrective action recommendations can include anything that improves your inventory or service level.

What benefits can you realize as the result of an aggressive inventory performance program? Improved cash flow with less inventory; recovered cash from the sale of surplus or obsolete items; deferred manufacturing cost; recovery of space—perhaps avoiding the cost of additional storage or racking.

This results in improved warehouse productivity from reorganizing the warehouse with high activity items stored more effectively; better product quality by allowing optional engineering change sooner, and increased sales with improved product availability. All of these potential benefits are directly related to profitability.

Summary

Of the four assets considered—space, energy, equipment, and inventory—the last represents the greatest potential saving, although warehouse managers may have a greater degree of control over space, energy and equipment. However, a good manager should exert influence, if not control, over every asset that could affect warehousing performance.

35

'JUST-IN-TIME'

The last two decades have seen the popularizing of an old distribution strategy referred to as "just-in-time" or alternatively as "kan ban." *Kan ban* means signboard in Japanese, and it was so named because suppliers to a Japanese auto manufacturer were instructed to place inbound materials beneath posted signs which indicated the appropriate part number. Like many popular business developments, just-in-time or JIT is a product of both fact and myth.

- The first myth is that it is a Japanese idea. It is something that the Japanese learned from others. In fact, JIT is a variant of the postponement strategy (See Chapter 18) which has been practiced in business for many decades.

- A second myth is that JIT is a production strategy. In fact, JIT was used by wholesale and retail distributors at least as early as it was in production. Its use in merchandising is probably more widespread than in manufacturing, and that merchandising use nearly always involves warehousing.

- A third fable is that JIT requires that a supplier's plant be close to the user. This in fact is the case in Japan, a compact country where suppliers and customers are pushed close together by geography and culture. In the United States and other countries accustomed to larger distances, use of air and other premium transportation systems will permit JIT to work effectively with a supply line which stretches over hundreds or even thousands of miles.

JIT in Manufacturing

A Japanese auto builder some years ago ordered that tires of a certain size be delivered to a specific spot in the plant at 8:00 a.m., with another delivery at 11:00 a.m., another at 2:00 p.m., and another at 5:00 p.m. Instead of having a tire warehouse adjacent to the assembly plant, the manufacturer relied upon his supplier to have the right tires in a spot which was convenient to the assembly process. Also, instead of the normal rehandling and quality checking, the auto assembler relied upon his supplier to have the right product in perfect condition at the right place at a precise time.

The previous system of storage, rehandling and checking had generated excessive costs, and the Japanese auto maker demonstrated that the *kan ban* concept was one way in which such costs could be eliminated.

Because the term *kan ban* was associated with one Japanese company (Toyota) the phrase "just-in-time" was adopted as a more universal method of describing a timed delivery program.

The practice spread to other industries. One U.S. computer manufacturer now uses air freight transportation to provide reliable timed delivery to its manufacturing lines. Parts to supply the assembly lines are classified as A, B or C. Many of the "A" parts are supplied just-in-time; the remainder are warehoused at the assembly plants. Not all of the "A" parts come by air, particularly those moving only a short distance. Quality charts are maintained to measure every carrier under contract. A delivery record of 98% or lower is not acceptable. The manufacturer provides routing instructions to all suppliers to maintain reliability. Quality charts are updated monthly to be certain that delivery reliability is under control.

JIT in Service Support

Sometimes JIT is used to eliminate parts and service centers in many locations around the country. One health-care supplier, for example, produces an appliance used in hospitals. If it breaks down, a premium air service is used to fly in a new unit and to return the one which has malfunctioned. Then the repair of the malfunctioning unit can be done at one central location rather than in scattered repair depots. Furthermore, the service is usually better than it would be in field repair stations.

Similarly, a computer service company maintains detailed records on the equipment used by each customer. If a computer breaks down, the service company isolates the problem through telephone consultation. In about one third of the cases, a technician is able to solve the problem by providing advice over the telephone. When this cannot be done, the failing component is replaced rather than repaired, again using air freight to bring in the replacement unit and return the one which has malfunctioned.

JIT As a Merchandising Strategy

Both wholesalers and retailers have successfully used JIT to provide improved service with less inventory. The pinnacle of success for a merchant is the ability to sell something to his customer before the bill from the supplier is due, thus creating an almost infinite return on invested capital. Merchandisers for years have recognized that one way to do this is to delay taking delivery from suppliers to the latest possible moment. One chain grocery company gives grades to its suppliers based upon timing of delivery, and early delivery is considered almost as harmful as late delivery. If goods are sent in too soon, they increase the space which is needed to store inventory, they slow the turnover, and they create the probability that the goods will have to be purchased before they are re-sold.

Because of this emphasis on turnover and reducing capital investment, merchants constantly seek ways to push the inventory investment back to the supplier. One way to do this is to delay receipt of goods until they are about to be sold.

One auto parts retailer has a goal of providing service which will always save the sale and keep the customer from going to a competitor. Yet the typical store of 5,000 square feet cannot carry the 30,000 stockkeeping units which are maintained in the product line. So the retailer keeps a master warehouse in its headquarters city to hold all

PRACTICAL HANDBOOK OF WAREHOUSING

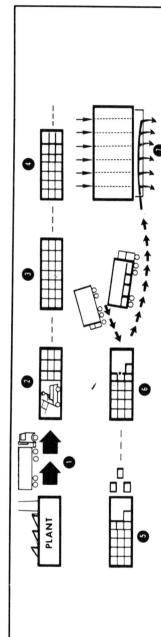

1. Product moves in bulk shipment to a public warehouse in the center of the market area.
2. Product is put into storage at the manufacturer's expense, until it is sold by the producer to wholesalers or retailers in the area.
3. Full pallet loads of product are used for order selection.
4. The warehouseman provides continued communication to the processor about the number of cases which remain in stock. Sometimes product is transferred to the new owner without being moved from one storage location to another.
5. Local buyers place orders on the warehouse, and these are picked and either shipped or held for pick-up. Records are adjusted to show withdrawals.
6. The buyer's truck arrives at the warehouse to trade empty pallets for full pallet loads of selected merchandise.
7. At the buyer's grocery warehouse, containers with store orders are merged with the merchandise which was just received from the public warehouse. Because of the speed of this merger, frozen foods can be treated the same way as dry groceries.

Figure 35-1

Diagram courtesy of Customer Service Institute

'JUST-IN-TIME'

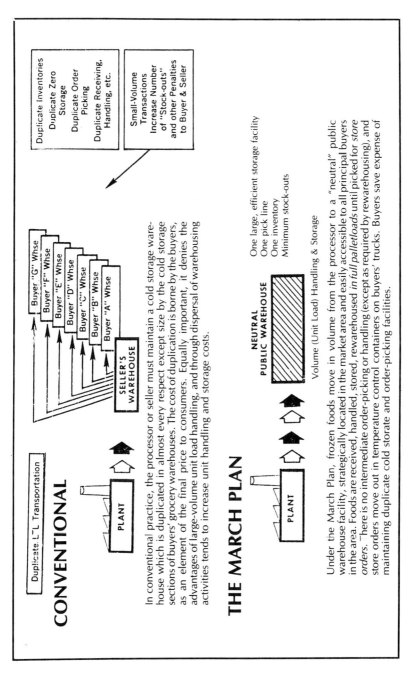

Figure 35-2. The March Plan eliminates duplicate warehousing.

of the items which are sold. If a customer stops by a store and cannot find the part wanted, store managers look at three options:
- The first is to see if another nearby store has the part in stock. If so, the customer is asked to go to the other store where the part is held in reserve for him.
- The second option is to ship from the express parts warehouse overnight by air, with a promise to the customer that the part will be available on the next day.
- The third option is to resupply the item overnight by air, directly from the supplier to the store. This "save the sale" policy has contributed to rapid growth of the company as a chain retailer.

JIT in Warehousing

The late Howard Way, a writer and consultant in warehousing, was employed by March Cold Storage Company of Indianapolis. There he developed the March Plan, which used a public warehouse to hold frozen foods at a location convenient to the local distributor. Rather than receive small quantities of perishable frozen merchandise by common carrier, the local distributor was able to pick up his frozen foods at the last moment, thus eliminating rehandling with its potential for damage and spoilage. Unlike other shipments, the frozen food order was handled as a pickup rather than as a delivery. A refinement of the plan suggested that a number of different suppliers keep their inventory at the same public warehouse, and the local distributor was able to eliminate a quantity of small deliveries by stopping at one warehouse and picking up a variety of merchandise. Figures 35-1 and 35-2 show schematic drawings developed to describe the March Plan.

While early writings about JIT suggested that this was a strategy to eliminate warehousing, it is probable that JIT has created more new warehouses. The difference is that these newer JIT centers are flow-through terminals as much as they are warehouses. JIT is a method of increasing the velocity of inventory turns, but that increased ve-

'JUST-IN-TIME'

locity is not accomplished by eliminating warehousing, but simply by changing the way in which the warehousing function is performed.[1]

---

[1] This chapter is from Vol. 3, No. 5 of Warehousing Forum, The Ackerman Company.

36

IDENTIFYING AND CONTROLLING COSTS

The true costs of operating a warehouse are actually a combination of fixed and variable outlays. In the cost examples to be shown here, costs are shown as totally variable and this reflection is accurate if volume is known. However, if the volume is subject to substantial change, the assumed costs for space, labor and handling equipment must be examined.

Three Cost Centers

Public warehouse operators generally recognize three distinct cost centers.

The first is storage, and storage costs are usually defined as those costs associated with "goods at rest," or those costs which would occur whether or not any product in the warehouse were ever moved. Handling costs are those costs associated with the movement of material in or out of the warehouse. Clerical costs are those associated with the record keeping and communication activity necessary to administer the warehouse.

The allocation of these costs represents a matter of judgment, and there is room for honest debate about whether certain costs should be applied to storage or to handling. One operator assigns all fixed costs to the storage function, recognizing that these costs cannot be readily reduced when activity decreases. Such fixed costs would include all building rent and long-term equipment leases, and some utilities and insurance. Depreciation on owned equipment and real estate is also a fixed cost allocated to the storage function. The base supervisory cost for those supervisors who would be needed at all times is a fixed cost charged to storage. The same for executive salaries and all general and administrative costs. Taxes are also regarded as fixed. Using this approach, one typical public warehouseman's budget spreads approximately $1.5 million of costs over 650,000 square feet, for a cost per-square-foot of $2.31 per year or $.192 per month.

Materials handling costs are typically divided between two major areas—cost of labor and cost of handling equipment. In one warehouse operation, a base labor rate of $9.00 per hour is built up to a total cost per manhour of $18.18. Payroll taxes are $1.20, and fringe benefits are $1.45. An allocation for supervision adds another $2.00, and allocation of general and administrative expenses adds another $1.80. This provides a subtotal of $15.45, which is divided by 85% productivity, since there is a lost-time factor of 15% in the operation.

A calculation of lift truck costs includes only operating expenses, since the fixed costs of the trucks have already been applied to storage. In this example, the result is a cost per fork-truck-hour of $4.64.

Once a budgeted cost is developed for labor per-hour and another cost for fork lift truck operations, the two figures can be combined when both worker and lift are necessary.

Space Utilization

No warehouse can be fully utilized, so any conversion of overall costs into unit costs requires allowances for space which cannot be utilized. Loading dock and utility areas will account for 20% of lost space in a typical warehouse.

Work aisles and honeycombing must also be accounted for in lost space analysis. Honeycombing results from the fact that all of the usable space in a warehouse cannot be completely full. Removing cases or pallets of one item leaves an empty space. This empty space cannot be filled with another item without blocking the product stored behind. Other space must be left for sanitation and cleaning. As a result, in a typical warehouse, net utilization is only 55-65% of the gross square footage of the building. But there can be significant variations in space utilization in different warehouse operations, and it is important for you to analyze the actual space utilization in your own building.

The Effect of Product Turns

Storage costs are typically calculated as a cost per-month. However, within one month the same spot in a warehouse could be occupied by more than one unit of a very fast-turning inventory. Therefore, the throughput inventory must be related to average inventory to calculate the number of turns involved. An inventory which turns fast is less costly to store than one which turns slowly. Throughput is generally defined as the total of units received and units shipped divided by two. The average inventory is defined as the total of beginning-inventory plus ending-inventory divided by two. The relationship of throughput to average inventory indicates the speed of product turn.

Other Performance Specifications

Warehousing costs will be influenced by many factors which affect the ability to store and handle a product. One is stacking height.

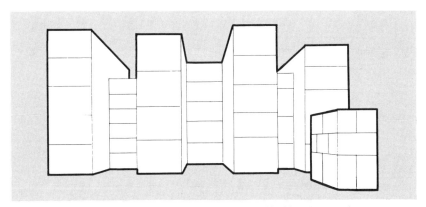

Lost storage space in front of partial stacks is called honeycombing. It exists in all warehouses to some degree.

Stacking height is limited by the clearance available in the building, and the strength of the product or the package being stored. If you have a 20-foot building and you are storing product which can be stacked only five feet high, you will need to buy pallet racks to make

---

Warehousing Performance Specifications

1. Describe the product to be stored, including weight, dimensions, stack height, type of packaging, number of units in average inventory and turns per year.
2. What is the number of stockkeeping units (SKUs)?
3. Is lot number or serial number control needed?
4. Carrier mode used for inbound material:
    truck .... _____%
    rail .... _____%
5. What percent of material is shipped to the warehouse in a unitized form?
    unitized .... _____%
    loose .... _____%
6. What is the average receipt at the warehouse?
    No. of cases .... _____
    No. of SKUs .... _____
    Does carrier unload? .... _____
7. Describe the average outbound shipment:
    _____ % via truck
    _____ % via rail
    _____ % unitized
    _____ % loose
    _____ % in order pick quantities
8. Describe order pattern:
    how many orders per month? .... _____
    average order size .... _____
    average SKUs per order .... _____
9. Is first-in first-out (FIFO) required? _____
10. Is there a pallet exchange program? _____

Figure 36-1

IDENTIFYING AND CONTROLLING COSTS

best use of the cubic space available and the cost of these or other necessary storage equipment should be included in your calculations. Handling rates are based on the actual cost of doing the materials handling job, and that cost can be affected substantially by size and

---

11. How are orders received?
    _____ % mail
    _____ % wire
    _____ % telephone
    _____ % other (describe) _____
12. Describe cost format needed (cost per case, cost per hundredweight, through-put cost)
13. Percent of outbound cases which require marks or stencils? _____%
14. How frequently are physical inventories required? _____
15. How is damage disposed of?
    Reject to carrier .... _____
    Repackage at warehouse .... _____
16. How often are warehouse reports required?
    end of month .... _____
    daily .... _____
    with each transaction .... _____
17. How much is shipped via mail or parcel services? _____
18. How much of inventory is hazardous? _____
    Describe:

19. Provide average:
    weight per case .... _____
    weight per unit of pallet ..... _____
20. Average value of product per case _____
21. What is past experience for storage and handling productivity?
    receiving cases or units per hour .. _____
    shipping cases or units per hour .. _____
    order picking cases per hour ..... _____
    repackaging cases per hour ...... _____
    storage cases per square foot ..... _____

weight of products, packaging, average order size, and method of transportation used.

A public warehouse operator typically asks the prospective customer to provide information about the handling and storage characteristics of his products through a questionnaire. This questionnaire allows the operator to simulate various costs which will be involved. Figure 36-1 shows a survey which you might use to establish performance specifications of the product which you now have in your warehouse.

Remember that it is not easy to collect all of this information, and many warehouse users can't or won't take the time to gather all of it. The closer you can come to getting all of these answers, the more accurately you can identify unit warehousing costs.

Measuring Productivity

The last question in Figure 36-1 asks about number of units or cases that can be handled. What if you don't know? There are published and available sources for warehouse productivity, based on the experience of many operators. The American Warehouseman's Association has software designed for personal computers which will display typical materials handling times. The same information is available in a written collection of calculations. The federal government has published warehouse productivity figures developed by the U.S. Department of Agriculture and by the Department of Defense. The AWA calculations are based on "Master Standard Data," by Crossan and Nance, published by McGraw-Hill. The government figures, the AWA standards, and other measured standards show a reasonable correlation. Most importantly, you should be able to measure your own experience simply by tracking the total units handled through your warehouse and relating that volume to the total time spent in materials handling (See Chapter 31). As your records are developed over time, these figures can be refined to reflect the true results of your own operation.

IDENTIFYING AND CONTROLLING COSTS

A Cost Calculation Example

Figure 36-2 shows a computation in which the cost of storage is developed. Lines 1 & 2 establish the average inventory and monthly throughput, and they demonstrate that the throughput will not be in excess of the average inventory. Line 3 shows that 50 of the 115 stockkeeping units are fast movers, but in overall volume, the customer has told us that 80% of the volume is in fast-moving items. We know that three-high stacking is involved, and after allowing for a honeycomb factor, the equivalent of 1,250 storage spots are needed, as shown in line 5. The fast movers go in deep row spots, with the slow movers in shorter rows, and the 80-20 rule is used to allocate the number of deep rows and short rows needed. If a 12-foot aisle is used, we have simulated a storage arrangement with short rows two-deep on one side of the aisle and 6 deep rows on the other side. Each 45-foot warehouse bay allows for only 33 feet of storage after 12 feet of aisle is deducted. Since each pallet takes 4 feet of space, there is room for 8 pallet rows in each bay ($8 \times 4 = 32 + 12 = 44'$).

Line 12 shows that 167 facings will be needed at 6-deep to hold the 1,000 pallet spots needed. And lines 13 & 14 extend the mathematics to show the square footage needed for deep spacings. Lines 15, 16 & 17 go through the same math to show the square footage needed for short-row facings, and the total of these is then divided by an 80% efficiency factor to allow for dock and staging area losses. Line 21 shows the calculation which multiplies square-feet-used by cost to develop a total dollar cost for storage. This is then divided by tonnage throughput expressed as billing units. Note that product received is reduced by half, since if goods were flowing in at a steady rate throughout the month, on average the inventory is only there for a half-month. The resulting calculation is converted to a cost per-hundred-weight.

Figure 36-3 develops a computation to arrive at a handling rate of $.303 per hundred-weight. The customer has supplied the figures given in lines 1 & 2, and from this we have developed productivity standards shown in line 4—14 lift-truck runs per hour. Line 5 shows the extension of the total time required, and when this is multiplied

Storage Computation

Volume/Criteria
1. average invoice   6 million lbs   300,000 cs 3000 pallets
2. monthly thruput   4 million lbs   200,000 cs 2000 pallets
3. 50 SKU = fast movers   65 SKU = slow movers

Layout
4. 3000 pallets ÷ 3 high stacking = 1000 spots
5. 1000 spots ÷ 80% honeycomb factor = 1250 storage spots
6. 1250 storage spots × 80% fast movers = 1000 deep row spots
7. 1000 deep spots = 3000 pallets with 50 SKUs
8. 1000 spots ÷ 50 SKU = 20 spots/SKU
9. 250 spots ÷ 65 SKU = 3.8 spots/SKU
10. 45' − 12' aisle = 33' storage
11. 33' ÷ 4/pallet = 8 pallet rows per bay

Pick Facings

2 deep   12' aisle   6 deep

45'

Space Calculations
12. deep facings = 167
13. (6 plts × 4' deep) × 4' width + (4' × $\frac{12'\ \text{aisle}}{2}$) = 120 sq.ft./facing
14. 120 sq.ft./facing × 167 facings = 20,040 sq.ft.
15. short row facings = 125
16. (2 plts × 4' deep) × 4' width + (4' × $\frac{12'\ \text{aisle}}{2}$) = 56 sq.ft./facing
17. 56 sq.ft./facing × 125 facings = 7000 sq.ft.
18. 20,040 sq.ft.
19. 7,000 sq.ft.   includes storage & aisle space
20. 27,040 sq.ft. ÷ 80% eff. = 33,800 sq. ft.

Rate Calculations
21. 33,800 sq. ft. × $.195/sq. ft./mo. = $6,591.00
    beginning inventory = 60,000 cwt.
    product rec.   = 40,000 cwt. × .5 months = 20,000 cwt.
    total billing units   = 80,000 cwt.
22. $6,591 ÷ 80,000 cwt. = $.082/cwt. cost

Figure 36-2

IDENTIFYING AND CONTROLLING COSTS

Handling Computation

Criteria
1. 40,000 lbs/tl   20 pallets/tl   100 tl/mo
2. 1-5 SKU/tl   avg. order 15,000 lbs
                      12 SKU/order   6 full pallets/order

Inbound Calculation: (based on 1 tl)
3. Work to be performed — 20 pallets to storage
4. Productivity — 14 runs/280 cwt. per hour
5. Time required 20 pits ÷ 14 runs/hour = 1.43 hrs.
6. Calculation   1.43 hrs. × $22.82/hr = $32.63 per trailer load
   $32.63 ÷ 400 cwt = $.082 cwt.

Pick Calculation: (based on 1 order)
7. Work to be performed 12,000 lbs. pallets to stage
                        3,000 lbs. cases to stage
8. Productivity — 280 cwt/hr (full pallets)
                  40 cwt/hr (case pick)
9. Time required — 120cwt ÷ 280 cwt/hr = .43 hrs
                    30 cwt ÷ 40 cwt/hr = .75 hrs
                                          1.18
10. Calculation — 1.18 × $22.82 = $26.93
    $26.93 ÷ 150 cwt = .180/cwt

Stage & Out
11. Work to be performed load 8 pallets/150 cwt
            check order
            pallet exchange
            dock board
            paperwork
12. Productivity 30 pallets/hr
13. Time required 8 pallets ÷ 30 pallets/hr = .27 hrs.
14. Calculation .27 hrs × $22.82 = $6.16
    $6.16 ÷ 150 cwt = $.041 cwt

Rate Calculation
15. Inbound cost .... $.082
    Picking cost .... $.180
    Loading cost .... $.041
    Warehouse handling rate = ... $.303/cwt.

Figure 36-3

by the dollars per-hour to support a man and a lift truck, a per-hundred-weight calculation for inbound work is developed. Lines 7-10 show the calculations involved in picking one order. The customer's specifications have shown that part of the order will be full pallets and another portion of it will be case-pick. Line 9 shows the calculations used to develop known productivity for full pallets and case-pick into the total hours required to fill the order. This hourly commitment is multiplied by a cost per-manhour to develop a per-hundred-weight cost for order picking. Lines 11-14 develop a calculation for the cost of staging and shipping the picked order. Line 12 shows a productivity benchmark of 30 pallets per-hour which is taken from the warehouse operator's time standards.

Line 15 shows that the total handling rate is developed by adding the three elements shown above.

Figure 36-4 shows a clerical computation to develop a cost of $1.43 per processed document.

The Throughput Rate Concept

You may wish to express cost as a total cost of warehousing per-unit. If so, storage, handling, and clerical costs will all be applied to each unit which moves through the warehouse. In order to do this, the operator must know the average turns per-year for the product, and with this knowledge the three costs are bundled together into a total cost per-unit. Any throughput rate will become inaccurate if there is a change in the conditions on which it was calculated. For products which turn at a constant rate, a throughput rate is a simple way to express warehousing costs.

Adjustments

After unit costs are established, they should be constantly checked by comparing them with current warehouse budgets. When the unit rate no longer produces the return which is necessary to cover actual costs, or when it produces a return greatly in excess of that required, some adjustment is clearly necessary.

You should understand those factors which frequently cause the

IDENTIFYING AND CONTROLLING COSTS

need for rate adjustment. The most common is a change in product turn. When merchandise turns faster or slower than calculated, storage revenues will change accordingly.

A major change in average order size (either inbound or outbound) can substantially affect costs of handling. A change in the number of stockkeeping units controlled can affect both handling and storage costs.

The development of a unit cost for warehousing services is based upon the cost of providing those services. To develop accurate per-unit warehouse costs, we must know the gross warehousing costs presently incurred. We then track the number of units which moved

---

Clerical Computation

Criteria
Cost/hr = $11.00
No batch control
Normal reports

4,000,000 lbs ship/mo
Customer terminal

Work required
Process 100 warehouse receipts
Process 266 orders
Customer service contract

Productivity
6 warehouse receipts/hr
10 bills of lading/hour
Misc. telephone calls, reports, tracers, etc.

Time Required
100 warehouse releases ÷ 6 wr/hr = 16.6
266 bills of lading ÷ 10 bl/hr = 26.6
Misc. = 10%   43.2 × 110% = 47.5 hrs/mo   43.2
  Postage
  Telephone

Rate calculation
47.5 hrs. × $11.00/hr. = $523 cost
Rec'd cwt   523 ÷ 40,000 cwt   = .013/cwt
Bill of lading   523 ÷ 255 bl's   = 1.97/bl
Document   523 ÷ 366 doc   = 1.43/doc

Figure 36-4

through the warehouse. If unit costs are to remain accurate, they must be constantly compared with a detailed warehouse budget, and adjustments are made as variances occur. In this way, the accuracy of your warehouse unit rates is constantly checked and adjusted for changing conditions.[1]

Summary

The calculations in Figures 36-1, 2, 3, 4 show the logic used by a warehouse operator to determine total costs. The public warehouseman will add a margin for profit to these figures to develop a selling price for his services.

By basing warehousing rates on cost measurement, both public warehouse operators and customers can ascertain whether the prices they negotiated are reasonable. Both recognize that a non-compensatory rate will not serve either party for the long term. The private warehouse operator, on the other hand, needs some benchmarks to determine the economic justification of his operation. He may also use these benchmarks for development of incentive goals for both workers and management. By using cost measurement rather than rules of thumb, many warehouse operators have learned a great deal about improving control of warehousing costs.

---

[1] Most of this chapter is from "How Much Does Warehousing Cost?" by Robert E. Ness. This article was published in Vol. 2, No. 3 of Warehousing Forum. Ackerman Company, Columbus, OH.

37

MANAGEMENT PRODUCTIVITY

If managers are not productive, it is unlikely that productivity on the warehouse floor can be improved. In a company where senior managers look down on junior managers, or in a warehouse where supervisors show contempt for workers, even the best external techniques for improving productivity are likely to fail. Proper management is a pre-requisite for any productivity improvement program.

Defining Management

Management has been defined as getting things done through others. A critical difference between worker and manager is that the manager doesn't pick up his tools and do the job—he or she causes that job to be done by *others*. A primary requisite to success in management then, is the ability to identify and hire good "others," the hourly people who will get things done as the manager directs. Some managers are promoted from the hourly workforce into management, and must make the transition from being an "other" to performing as a manager. Everyone finds this transition difficult, and some never quite accomplish it. Signs of failure to manage can be seen in the manager who repeatedly leaves the desk and does a job which should have been delegated to an hourly worker. Another sign of failure is the inability to find and keep good people. Still another sign is a manager's failure to recognize the difference between having a skill at doing a job and having the ability to lead others in performing that same job.

Why Warehousing Is Different

Warehousing creates its own peculiar problems for managers. In most warehouses, a high percentage of supervisors and managers are ex "others," former fork lift drivers or checkers. Therefore the problem of transition from an "other" to a manager may be found more frequently in warehousing than in most businesses.

Entry level work in warehousing has an undeservedly low image. Some managers believe warehousing is a job for a strong back and a weak mind, and that pay scales can be lower than those for manufacturing. It is a false image. A warehouse employee often needs greater skills than those of the more closely-supervised assembly line worker. The very nature of warehousing works against close supervision, so a warehouse worker must be a self-starter, a person who will produce high-quality work at a highly productive level without someone looking over his shoulder. The problem is management's failure to build community respect for warehousing as an occupational specialty. As we enter a time of growing labor shortages, the problem and opportunity will become even more apparent.

The Effective Manager

Have you taken an inventory of your own strengths and weaknesses as a manager?

A first step in managing yourself is getting to know yourself. It isn't easy for most of us to engage in critical self-appraisal. One way to learn objectively about strengths and weaknesses is by working with a consulting industrial psychologist. The psychologist usually provides no real surprises, but the interview will give you an objective appraisal of your own strengths and shortcomings as a manager.

If your organization uses an industrial psychologist you should welcome the chance of an interview. You'll find out a few things you may not have been aware of before. This can give you a chance to compensate for shortcomings by working on changing your behavior or by hiring others in your organization whose skills complement your own limitations in those areas.

If you have no access to an industrial psychologist, run through

MANAGEMENT PRODUCTIVITY

the self-appraisal checklist shown in Figure 37-1. And, of course, you probably already have a pretty good idea of the things you do well and where you may fall down. The next step is to figure out what to do about it.

Measurement Tools for the Manager

Nothing can be managed well without being measured. In business, the ultimate measurement for every organization is the balance sheet with the profit and loss statement. The expression "the bottom

12 Factors to Rate Yourself as a Manager

	Excellent	Average	Needs Development
Administration			
Job knowledge			
Planning			
Innovation			
Communication initiative			
Responsibility			
Team player			
Sales person			
Decision maker			
Leader			
Selector and developer of personnel			

Figure 37-1

line" is overworked, but its meaning is still important; the last line on a financial statement, the profit or loss of your organization, is still the vital statistic.

Depending on your rank in the managerial organization, and the size of your company, your ability to affect your company's bottom line may be great—or it may be nearly insignificant. Yet there are many things you can do to measure managerial progress.

The first of these is productivity. In the warehouse, you can measure productivity in many different ways. Whatever you measure, your measurements must be simple, readily understood and relevant to the tasks performed. This should include percentage of damage, frequency and severity of customer complaints, frequency and quantity of overage or shortage, or frequency and quantity of customer returns. All can be valid measurements of effectiveness.

Always remember that what you measure is important, but it is equally important that the measurements be reasonable and consistent. And if no action is taken, the measurements are meaningless.

Consider the following example of how measurement affects managerial skills in warehousing. As a warehouse supervisor, you are giving out assignments at the beginning of the day. You have given a two-man crew the task of unloading a box car of merchandise. Work standards available to you indicate that two men will take four hours to complete the job.

When you give the assignment, you describe the job to be done and indicate to the lead man that you will expect to see him at lunch time ready for a second assignment. In making the work assignment this way, you communicate your expectations of the time you expect the work to take. You also communicate your knowledge of your operation. If the job can really be done in three hours, for example, your communication of a four-hour expectation can be seen as an invitation to slow down. The opposite is true if the job would normally take five hours.

Workers will typically accept and conform to standards which are perceived as fair and soundly arrived at. Above all, use of standards such as these shows that you as a manager have measured and

MANAGEMENT PRODUCTIVITY

will continue to measure the output of your work crew, that you have communicated your expectations and they are fair and reachable.

The Critical Tasks of a Manager

Since management involves getting things done through others, how do you find those other people who make your job a success? Some managers have little experience or training in attracting and selecting the people they need for their team. A first step is to set the specifications for the kind of people you need to work in your warehouse. If the work is strenuous, you need people in good health, free of ailments which could be aggravated by warehouse work, such as back injury. Substance abuse is another health problem which must be recognized. You need people who are honest and unlikely to fall prey to temptations to steal. The right people are those who will be content with the working situation in your warehouse and who are not anxious to change the existing situation. Some feel that it is desirable to find workers who have past experience in the kind of work done in your warehouse, though others feel that training of an inexperienced person will quickly produce a quality worker who does not need to "unlearn" bad habits from a previous job. The desires or specifications of the people you are seeking should be placed in priority. Some may be absolutely essential, and others could be subordinated if the essential talents were relatively strong. As you set priorities, remember that it is easier to teach new skills than it is to change attitudes. (See Chapters 27, 28, 29).

The interview process is a key step in selection. Entire books have been written about interviewing, and courses are available to enhance this skill. A successful interview is one in which the applicant does 80 to 90 percent of the talking. The best questions asked are nondirective ones which typically cannot be answered in few words. The goal of the interview is to learn as much as possible about the attitudes as well as the experience of the applicant. Not all successful managers are good interviewers, but they have delegated someone to do this job who is good at it. The checking of references is equally critical to successful selection. Finally, creative use of the probationary period for new employees is essential to be sure that any mistakes made in the initial selection process are corrected during probation (See Chapter 29).

The development of people begins with employee "indoctrination" that starts with the first day on the job. The impressions of your company which are gained on that first day of work are usually the strongest impressions the individual gains of your company. Every worker will be motivated and oriented by somebody—if you have no program, the job will no doubt be done by a leader of the work crew. If you want the image of the company to be a good one in the minds of its employees, then it is necessary for management to have a well-designed program.

New employees, hourly workers or managers, are "socialized" through several processes. The new worker develops skills and learns the tasks involved in the job. At the same time, that new person learns what behavior is acceptable or unacceptable within the company. Often these things are learned from fellow workers rather than from management. If the socialization of the new worker comes from other workers, the new employee is likely to be exposed to conflicting demands. The supervisor may provide one set of demands, the union steward another, and the informal leader of the work group still another.

Training is a critical part of the development process. Sometimes the training given is not relevant. In other cases training is given only to new employees and is not available to older workers. If this happens, an older employee who is given a new task may fail at the new job because training was not provided. This is one valid reason why many workers resist the chance to change jobs.

Part of the development task is to provide up-to-date procedures manuals. Such a manual is effective only if kept up-to-date.

Appraisal is perhaps the most critical part of a manager's task in developing people. Many managers don't know how well they are doing, because their performance is kept in the dark. Many managers are "too busy" to provide appraisals, which usually means that they are not comfortable in giving them. Others lack a format or training in conducting appraisal interviews. One format is shown as Figure 37-1. A successful appraisal emphasizes performance. It provides both positive and negative feedback, recognizing that nobody is perfect and everyone has a few good traits. The appraisal interview should include discussion of a step-by-step approach for achieving improvement in future performance.

Coaching and counseling is another aspect of developing the people who report to you. Much of this coaching and counseling is providing feedback, and an equally important part is the role of being a good listener.

The final result of your efforts in developing people is the promotion to greater responsibilities and the awarding of pay increases for those who have met or surpassed management's expectations. After you have communicated your expectations on job performance, you should reward those who meet or exceed them.

The Manager's Hardest Job

Promptly firing incompetent subordinates is probably the most emotionally-demanding task required of managers. Many hesitate to fire unsatisfactory employees because they like them, or feel sorry for them, or are afraid to fire them, or shy away from the unpleasant task. Firing an employee is often considered a personally hostile act. Whatever the reason, many managers have a difficult time terminating an unsatisfactory worker.

No matter what the circumstances, firing an employee is never pleasant. The following steps are suggested to make it as easy as possible for both you and the employee.

1. Never surprise the employee. Be deliberate and take the necessary time. Never do it in haste.

2. Never fire an employee in anger. You must establish that the employee cannot be salvaged, that there is no chance for improvement, that all the responsibilities have been clearly understood.

3. The employee has to know in advance that the company is dissatisfied because the expected standards of performance have not

How Effective A Manager Are You?

1. How productive are you? Do you spend your time putting out fires or do you concentrate on managerial tasks?
2. Are your work habits ones you would want your employes to adopt?
3. Do you allow employes room to grow?
4. Do you have any employes you should have fired, but haven't?
5. Do you have a formal training program for new employes?
6. Do you have a continuing education program or does formal training stop at the entry level?
7. Are long-time employes continually challenged, given new tasks, rotated in other jobs?
8. Do you have written procedures? When was the last time they were updated?
9. Are your supervisors trained to manage others? Do they teach employes to think, rather than simply tell them what to do?
10. Do your supervisors regularly recognize performance of workers?
11. Does your company make an effort to recognize superior performance?
12. Do your fellow workers recognize good performance?
13. Is there a performance rating system? Is it fair?
14. Do you have a work measurement standard? Is it understandable? Can you influence the standard? Is it easy to follow? Does it measure in positive terms?
15. Do your managers/supervisors look down their noses at the workers?
16. Do managers/supervisors further their own education?
17. How often do you ask, "What are my firm's distinctive traits? Where are we ahead of—behind—the competition? Where will our corporation be in five years?" Answering these questions—planning—is what management is all about.
18. Do you tackle the tough decisions first?
19. Are you happy to tell outsiders good things about your company?
20. Can you easily defend your company if it is criticized?
21. Do you frequently produce creative new ideas for improving the job?
22. Are you willing to do everything that is necessary to get a job done?
23. Do you challenge others if a job is not being done correctly?

SOURCE: Vol. 17, No. 2, Warehousing and Physical Distribution Productivity Report. © Marketing Publications, Inc., Silver Spring, MD.

Figure 37-2

been met. Make it clear that termination will follow if there is no improvement.

4. Conduct the termination interview on neutral territory. Allow the employee to leave without an audience. Be sure to have ready all monies due the employee and give them to him or her at the end of the interview.

5. Don't drag it out or make a lengthy recitation of the employee's mistakes. Just do what has to be done.

6. Later, try to find out what went wrong in the hiring process that eventually led to this individual's dismissal.

The Buck Stops Here

Another problem for the warehouse manager is the intensity and the nature of the business pressure which builds up at the warehouse level. The shipping dock is the place where "the buck stops" in the corporate shell game which begins with salesmen's promises and ends with the question of whether or not the order has been delivered—which must be answered with either a *yes* or a *no*. If the answer is no, senior management often displays scant interest in the reasons why, and this creates a unique pressure seldom found elsewhere in the corporate scene. With this pressure come risks of failure which are peculiar to warehousing. A late shipment is a failure, but so is a shipping error due to shipping the wrong merchandise, shipping the wrong quantities, or even shipping too early. Failure of a common carrier is often blamed on the warehouse, even when warehouse management had no authority to select that carrier. Damage, whether caused by the carrier or warehousing people is a cause of failure, particularly if the damage is not discovered promptly. Theft, from whatever cause, is generally regarded as something which good management should have been able to control. Even casualty losses from fire or windstorm are sometimes considered as events which might have been prevented with better management. It is neither accurate nor fair to blame these failures on warehouse management, but that's the way it happens in many companies. Managers in other businesses face their risks of failure, but warehouse managers have more risks to cope with.

How Good Are Your Warehouse Supervisors?

One of the hallmarks of any good supervisor is skill and effectiveness in delegating. Those who do not know how to delegate are doomed to failure.

In many cases, supervisors receive little training in delegating the job to be done. And too many supervisors show this lack of training. Here are some points you should caution your supervisors about:

1. Be careful to whom you delegate a task. Particularly if your supervisors are new to delegating, caution them to pick carefully their first choice for the job at hand. It's essential that the first efforts be met with success. Not everyone is right for every job. And not everyone wants added responsibility. So it's important that the right person be chosen for the job.

2. Be sure instructions are given clearly. The effective supervisor gives instructions clearly and explicitly and then checks to be sure those instructions were received and understood. Poor initial communications often cause things to go wrong. It's particularly important that both the supervisor and worker understand precisely *what* is to be done, *how* it is to be done, and *when*.

3. Ask someone to do a task, rather than order that it be done. One of the responsibilities of a supervisor is to motivate workers. Ordering someone to do something is apt to have the opposite effect. Unfortunately, many improperly-trained supervisors don't realize the power of asking rather than ordering. Emphasize this point to supervisors in their training sessions.

4. Remember that someone may do a job other than the way you would do it—but, of course, this doesn't necessarily mean it has not been done correctly. An effective supervisor allows for different approaches and methods—and, in fact, may be amazed by the freshness and perception of an employee's approach, even if it is different.

5. Follow up. The supervisor needs to realize that delegating a task doesn't mean delegating responsibility for the success of the task. The old Army saying is true; you can delegate authority but you can't delegate responsibility. So follow-up is crucial. The supervisor should check the employee's progress at specified intervals—and also

on a spotcheck basis. Thus the supervisor is aware of the potential for a problem situation in time to take steps to prevent it.

6. Give credit where it's due. When a task is completed, the supervisor should be sure to give credit to the worker. Similarly, when one of your supervisors delegates a job successfully, be sure to recognize appropriately both the supervisor and the individual who actually did the job.

The Supervisor Promoted from Within

In warehousing it is not uncommon for supervisors to be promoted from the workforce. One of the most difficult tasks of supervision is to forge a new constructive relationship with people who were once peers. Supervision requires a certain detachment and freedom from "cronyism." It requires a certain measure of discipline, and the ability to instill pride in performance. It demands a responsibility to management, and a responsibility to the workforce. And the worker who is promoted into supervision must make the transition without being a crony and at the same time without making enemies.

This is a delicate balancing act, and requires sensitivity and tact on the part of the worker promoted from the ranks. It's important that warehouse management prepare the new supervisor for such problems, and suggest ways to handle them.

The Importance of Clarifying Expectations

Do the workers in your warehouse know what their supervisors expect of them? Expectations are best met when they are clearly stated—when both worker and supervisor know without question what is expected. Chances are good that your supervisor can tell the workers exactly what management expectations are. But it might be an interesting exercise to find out if workers have a clear understanding of management expectations—and if their understanding agrees with their supervisor's. Not knowing what management expects can be a major cause of worker dissatisfaction.

An atmosphere of positive achievement, with praise given when it has been earned, is a big help. Positive discipline—constructive criticism - involves keeping track of results, as well as acknowledging success. Workers in any field of activity including warehousing will tolerate strict discipline as long as it is accompanied by positive feedback (See Chapter 28).

In one large operation, a straightforward system of recognition is used. When a supervisor sees a worker doing a superior job, the supervisor gives the worker a card with a comment about the work. Records are kept of the number of cards awarded, and a worker who receives a certain number of cards is awarded a small prize, such as a ticket to a sports event or a dinner out.

While incentive programs have strong supporters, they also have vocal critics. Whatever approach you choose to use, the basic underpinning is your supervisors. Do they understand the value and importance of positive feedback? Do they recognize progress or the opportunity for a better job? Or are they negative-news carriers who recognize only failure? In some cases, there's a widespread feeling that it is impossible to please a particular supervisor. Yet, when counseled, the supervisor honestly believes that he or she is fair, and does use positive feedback properly. So part of management's responsibility is to make supervisors aware of the value and proper use of positive reinforcement.

Putting It All Together

Effective management is much more than symbolic behavior. It is based on meaningful, fair measurement. This means more than simply establishing productivity standards. It means getting management commitment—throughout the department, throughout the company—to a productivity improvement program. It means evaluating the standards, giving workers feedback and giving recognition when a job is well done.

Without effective management, all the productivity improvement programs in the world can go only so far. With effective management, fancy programs and gimmicks aren't necessary to motivate workers. Figure 37-2 is a questionnaire you can use to help determine how well you and your managers, peers or superiors are doing. It is ultimately management style that will set the tone needed for productivity improvement in the warehouse.

38

REDUCING ERRORS

One of the best ways to improve productivity is to get every job done right the first time.[1] The cost of errors in warehousing is high; the cost of an order picking error, for example, has been estimated as ranging from $10.00 to $30.00 for each occurrence. These costs are derived by considering the writeoffs of undiscovered warehouse shortages caused by picking mistakes, the cost of handling returns, cost of supplemental shipment to replace shortages, cost of reversing errors caused by substitution of the wrong product, correspondence to handle credits and adjustments and, finally, the almost unmeasurable cost of customer dissatisfaction or loss of confidence.

In many warehouse operations, errors are the worst problem which management faces. Errors not only mean lost customers—in a few situations, an error can mean the loss of a life. Regardless of the consequences, as long as you have human beings in the warehouse, there will be some errors. Still, it is management's job to reduce them to the greatest possible extent. In most warehouse operations, a decrease in errors may be the best way to increase overall productivity.

The Value of Order Checking

There is little evidence that checking orders substantially reduces order-filling errors. No amount of double checking or triple

---

[1] Much of this chapter is based on an article written for *Warehousing and Physical Distribution Productivity Report* by the late W.B. "Bud" Semco, President, Semco, Sweet & Mayers, Inc., Los Angeles. ©Marketing Publications, Inc., Silver Spring, MD. Also from Vol. 3, No. 3 of Warehousing Forum ©The Ackerman Company.

checking eliminates the errors; only a few of them will be caught. How few is evidenced by the fact that customer complaints about order errors continue despite all the checking.

The very fact that there is a checking department (or an order checker) seems to constitute management recognition that mistakes are inevitable. Thus, the checker becomes responsible for the accuracy of order-filling, rather than that being the responsibility of the order-filler himself. The fact that double-checking frequently turns up additional errors not caught by the original checker should signal management that order checking is futile and pointless—but apparently it does not do so.

Illustrating the futility of order checking, in one company five full-time order checkers were required in a particular broken case order filling department. Even with these five order checkers, customer-reported errors averaged 25 per day. When the order checking department was eliminated, management saved five man-years of labor at a cost of (an increase in) only 10 additional reported order-filling errors per day. Since no effort was made to upgrade the caliber of work or worker, it seemed that 10 additional order errors per day was a small price to pay for a saving of five annual salaries.

Order Picking Errors

Good order filling includes not only error-free picking, but also reaching a standard picking speed. High work standards exist in any number of warehouses where management has been motivated to take some action. But it takes a concerted effort on the part of management and the worker to achieve even that minimal concentration needed for reasonably error-free order filling. If management refuses to accept substandard work performance for standard pay, an acceptable rate of order filling accuracy is more likely to be reached.

Between 10 and 12 percent of all errors are not due to poor performance in the warehouse, but occur because the order taker fails to take down completely what the customer wanted or had ordered. The remaining 90 percent of order errors are equally divided among wrong items, wrong quantities and items completely missing from the order.

Order Taking Errors

A misunderstanding when the order is taken by telephone is a common source of errors. Consider the procedure used by a fast food company which makes home deliveries. This firm asks for the telephone number whenever a phone order for its product is taken. Then the operator phones back to the customer to repeat the order. This second phone call prevents a high percentage of errors.

For error-free order filling, it is important that the order sheet submitted to the warehouse for filling is correct or the customer is going to be dissatisfied—even if the warehouse is right. Clerical order-taking errors are easy to eliminate if the order taker takes the time to get it right.

Error-free order picking has four principal requirements:
1. A good item location system.
2. Clear item identification.
3. A clear description of quantity required.
4. A good order-picking document.

A Good Item Location System

This is one in which a number is assigned to each order-picking location and only to that location. The code system must be clear enough so that any order picker can quickly find any picking location in the warehouse.

A six-digit locator system will accommodate the needs of all but the very largest of warehouses. The first two digits indicate the aisle or rack-row number. The next two digits refer to the shelf or rack-"section" (ranging from three feet in width to eight or 10 feet). A section is defined as the space between two adjacent uprights, so the system will work not only for shelving, but pallet racks as well. The fifth digit indicates the elevation of the shelf or the rack beam. The last, or sixth, digit indicates the item location on the shelf, from location one up to nine counting from left to right from the nearest shelf or rack upright on the left.

Some warehouses have more than 100 aisles and in others the number of small parts on a shelf is too large for a single digit "last-number" system. Accordingly, a larger system may have to be considered. In any event the shorter the location number the better.

In the six digit system, the digits are generally three, two-digit numbers as follows: 12-34-56. This indicates to the order filler that he should first go to aisle 12, section 34. This system normally has even numbers on the right-hand side and odd numbers on the left side of the aisle, as the order filler enters, so section 34 would be found on the right side of the aisle. The order filler now counts five levels of shelving up and six items over from the left-hand side of the shelf (the upright support) and comes up with item location 12-34-56.

Clear Item Identification

Items are normally identified first by trade name, then by size, color or retail value, as for example, XYZ mouthwash 16 oz., or ABC toothpaste, giant size. Some industries require more precise item identification: "Hex head bolt, ¾×2½ cad. plated" becomes "1-6-¾×2½ C."

Wherever a product can be concisely listed by its name or name and specifications, errors are reduced. Avoid further complicating item identification by the use of in-house numbers or computer number or manufacturer's numbers, unless no other practical means of item identification is possible. Confusion is created by identification systems that include information that does not appear anywhere else but on the order document. Internal or "house numbers" that have been programmed into the computer for precise identification tend to create confusion unless they also appear on the product.

Most important in item identification is that the product on the shelf, the description of the product on the order-picking document, and the description on the shelf label are all identical. If the product on the order says XYZ toothpaste, the label on the package says XYZ dental cream, and the shelf label says XYZ TP, the order filler will not know whether the location was wrong, the order was wrong, or the item on the shelf was wrong.

One of the causes of item identification confusion is the use of manufacturer's catalogs—particularly where the catalog sheet has not been prepared by the advertising or sales department. Make sure the item description on the catalog sheet is the same as the item identification on the package that is going to be picked off a shelf.

The problem of item identification can be eliminated when the computer program used to produce shelf labels also produces the order-filling document. When the order filler arrives at a numbered location and checks the shelf label against the product description on the order document, they must correspond or else he's in trouble.

Clear Description of Quantity Required

Industries such as wholesale hardware, wholesale drugs, auto parts and toys have a practice of shipping merchandise in shelf packs, overwraps, tray packs and even reshippable containers within master cases. This creates confusion with respect to quantity, and many order-filling errors result.

An example is a brand of hacksaw blade that comes 10 to a package and 10 envelopes to a carton. The outside cover is marked "10 units saw blades." We must assume that the manufacturer of this product did not deliberately design the package to induce the order filler to ship 10 times the quantity required. The proper way to describe the quantity required would be "one envelope containing 10 hacksaw blades." A conscientious order filler, finding the cartons and hacksaw blades on the shelf and recognizing that the order did not call for a carton but an envelope, would either open the carton to determine the contents, or would check with his supervisor.

An opposite example is wholesale drugs, where ampules are prepackaged in lots of 25 units. While a customer's order for 25 ampules should not be a difficult problem to an order filler, one wholesale drug house described the product as XYZ ampules 25×1 cc, and in the quantity column "one each." Hopefully the order filler would know that "one each" meant one carton of 25, but the order did not say so. One package plainly marked in red letters, "do not break this carton," was broken open and contained 24 ampules. Evidently an unhappy customer had received one instead of 25, with the remaining 24 on the shelf constituting an unsalable package that must go to the return goods room for a lot of handling to get the credit. This could probably have been avoided by indicating clearly in the item description that XYZ ampules come 25 per-box and, in the quantity column, indicate "one box of 25 ampules."

'Good' Order Picking Document

The order-picking document should be specifically designed for and used only by the order filler. Frequently, however, it is designed for so many other purposes that its original function (error-free order picking) is forgotten, with the order itself buried in other important information of little interest to the order filler. The best possible order-filling form is one that has only essential information on it, such as customer identification and an order number or date so orders can be filled sequentially.

Order Identification

It is easier to control order numbers if they are reduced to the fewest possible number of digits. The order number system has to be workable within the framework of the particular warehouse. The reuse of sequential numbers is best limited to a 10- or 15-day cycle. The order number is best listed on the top of a page on the right-hand side.

Eliminate all unnecessary information on the order-filling document. Such items as "bill to" or "ship to" are not needed by the order filler. Likewise, credit terms, sales data, cost at retail or cost at wholesale have nothing to do with order-picking efficiency or accuracy.

Order entry systems should be designed to convey information of value to the order filler, rather than to the accounting or sales departments. The order filler's task is to proceed to the proper location, verify that this location is correct and the wanted product is stored, then select the right number of pieces of that item. Superfluous, nonessential information tends only to be confusing.

The Order-Filling Form

In many warehousing operations the order-filling document is

REDUCING ERRORS

one part of a multi-part form. Sometimes it is possible to remove the unnecessary information from the document by blocking it out on the order-filling copy only.

The body of the order-filling document should have no more than four columns of information. The information should be supplied to the order filler in proper sequence, as follows:

First column—Product Location
Second column—Product Identification
Third column—Check Marking
Fourth column—Quantity Ordered

The sequence of the columns must appear as indicated above, since the information needed by the order filler is in that order. If the picker can get to the right location, there is an excellent chance the correct product will be selected. A good order filler checks the order-picking document against the shelf label and against the merchandise on the shelf to verify that the correct product is selected. The order filler then counts the desired number of units or pieces of the product ordered. The order-filling document should be prepared in the logical sequence of these actions.

If the quantity filled is not the same as the quantity ordered, the place to make changes is in the checking column—not in an additional column. (*See Figure 38-1.*)

Importance of Uniform Procedures—Frequently, warehouse management allows each worker to use his own method of indicating a line has been filled. One order filler will use a checkmark, another will circle the quantity ordered, still another might circle the quantity when it was *not* filled. All order fillers should use the same means of identifying what work has been completed, what changes in quantity were made, or what items were omitted.

Two other important factors in the preparation of a good order-picking document include:

Using Item Location Sequence—The order should list line items in order of location sequencing. Without this, order filling may be accurate, but the rate of order picking will be slower.

Clarity in Spacing or Ruling—Single-spaced order picking documents should be avoided—double-spacing is an absolute minimum. It is even better to rule the pages horizontally and print the picking information within the rules.

Illustrated in Figure 38-1 is an example of an order-filling document for error-free broken case (repack) order filling. There is no superfluous information on the form. It was designed specifically for use by the order filler and contains only information needed by that person. The "heading" is limited to basic requirements only, including an identifying order number, the date and the customer's name and address. The customer's identification can be enhanced by the addition of a four-digit matrix-print number (Figure 38-2) indicating the route and stop number.

ORDER NUMBER 123456
DATE September 27, 19--

Smith's Hardware Store
100 S. Main Street
Glendale, CA 91436

LOCATION	ITEM		QUANTITY
11-08-34	Dixon Lumber Crayon #510		3 box 12 ea
14-02-52	#350062 Hose Coupling ½"	0	1 box 10 ea
16-11-44	Am. Tack & H. Nail ¾" #29BN34 2 oz	(10)	12 box
24-04-23	Collins Latex White #14 Gallon		2 case

The first and fourth items were picked completely. Item number two was out of stock and check marked "0." The third item was filled incompletely and was marked "10" in the check column, indicating the number of boxes picked.

SOURCE: Warehousing and Physical Distribution Productivity Report, Vol. 16, No. 19. ©Marketing Publications, Inc., Silver Spring, MD.

Figure 38-1. Order filling document for error-free broken case order filling.

The check ( ) column is placed to the left of the quantity column so that any changes in quantity or any omissions appear next to it.

Other Factors to Consider

Several other factors can help to reduce order-filling errors. These include:

1. Sufficient illumination—When illumination in the order-filling aisle drops below 20-foot candles at 36-inch elevation, there are more errors of the wrong-item-pick type. This applies to both full-case picking and repack picking. When it comes to order picking broken case or repack quantities, a minimum of 40 to 50 foot candles of illumination at 36-inch elevation is recommended.

2. Shelf locations—Merchandise not easily accessible tends to be "overlooked" or omitted from the order. Frequently, the items that cannot be found are located on the upper and/or lower shelves beyond easy reach. This includes shelves that exceed six feet in height, or less than 14 or 16 inches off the floor. The order filler may have to get down on hands and knees. So it's not surprising that he may "forget" to pick items located too high or too low. The answer to this sort

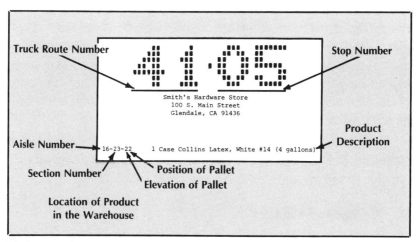

SOURCE: Warehousing and Physical Distribution Productivity Report, Vol. 16, No. 19. ©Marketing Publications, Inc., Silver Spring, MD.

Figure 38-2. Label for error-free case filling.

of problem is no shelf picking below a 14-16 inch shelf, or above a shelf 5 feet 8 inches high.

3. Item placement—Naturally, the owner of a trademark wants to make that trademark familiar to the public. But what works well in attracting the buyer works against the order filler. An order filler, looking for a particular color of hair dye may find five or six dozen packages on a shelf, all with similar design, color and decoration. The only difference may be an inconspicuous secondary identification. So the door is open for mistakes, by both the order filler and the stockman who replenishes the shelves. One way to prevent this type of order-filling error is to separate trademarks, so that no trademarked product constitutes a full shelf of merchandise.

4. Document clarity—If the order filler's document is one part of a multi-part form, the warehouse copy should be the first or second copy of it and be clean and legible. In too many instances the order filler gets the copy on the bottom of the stack and sometimes has to guess what's written on it.

Multi-Line Order Filling

Few order fillers are accurate enough to pick and check multiple lines. This involves selecting three or four items from various shelves, placing the merchandise in the tote box or shipping case and checking each of these items. A better procedure is for each order picker to select an item from the shelf, return to the order line where the merchandise is placed in the tote box and the picked document is marked, indicating that the pick has been completed.

While multiple line order filling is not necessarily ruled out, it should be limited only to those order pickers who have proved they are able to carry out the more complicated procedure.

Personnel Factors

The way you deal with people can have a great deal to do with error reduction in your warehouse.

Tell your people what you expect. Your expectations may be expressed in a written procedure, or meetings and other sessions dedicated to reduction of errors.

The first and best place to communicate your expectations is with the training of new people. Your people should have a chance to review written procedures and discuss them in training meetings. When new order pickers are retained, there should be supervised dry runs to teach them to do the job right the first time.

Order filling training is too frequently neglected. Many order fillers graduate from the shipping room to stockroom, and from stockman to order filler. Without actual training, a worker is likely to start with bad habits. And an order filler who is trained by a worker on the order line may pick up that worker's bad habits.

Identification With Work

The worker who is identified by his work usually does it better. One method of doing this is for each order filler to sign each page of an order document as it is completed. If an entire page isn't completed, the picker initials those lines that he or she handled.

Yet another means of encouraging worker identification is to have a picking ticket placed in every order filled. This ticket should include the picker's name, instead of a number (which is anonymous). A particularly effective picker's ticket reads: "This order was picked by Thomas E. Jones. If you find any discrepancy between the packing ticket and your order, please call 123/456/7890. Ask for the Claims Department, and give them my name when reporting the error."

A further refinement is to put the order filler's name and photograph on the picker's ticket. This personalization has resulted in a great improvement in order accuracy.

Posting Error Rates

Using peer pressure to encourage order pickers to be more careful sometimes works by listing each worker's error rate in declining sequence on the warehouse bulletin board. The individual making the most errors has his name at the top of the list—those making fewer errors follow. Always be sure that workers are graded on errors as a percentage of total lines picked. The fastest order picker may appear to be the worst error maker, when this may not be the case. If you post an error-rate list, be sure also, however, to provide recognition and praise for those workers who made no errors.

Using Case Labels

Most warehouses use the same document for case picking as for broken case or repack picking. But using a pressure-sensitive label for case picking will reduce picking errors.

The case label contains this information:
1. Location of the product to be picked.
2. Product name, description, size and number of units per case.
3. Customer's name and address.
4. Matrix print number identifying the customer (if possible).
5. Delivery particulars (such as, "do not deliver after 5:30 p.m.," etc.)

Case picking is usually done from a pallet in a rack, a pallet stacked on the floor, or from a multi-case inventory. With a good location system, the order filler can go to the right location, verify that the item on the case label and the case on the pallet in that location are the same, and be reasonably sure that the proper item is being picked. The worker then peels the label and applies it to the case.

Because of the high-dollar value of case picking as compared with most broken case picking, the cost of a quick check can be justified. If the case labels have been computer-printed, the only checking necessary is to verify that the product description on the label and the product description on the case are identical. If they don't match, the case must be held back until appropriate decisions are made on its disposal.

'Picking Rhythm'

"Picking rhythm" describes a method by which the order filler performs the job while simultaneously checking each step of the order-filling procedure.

Order picking rhythm is somewhat like learning how to drive a car. When we learned to drive, we had to concentrate on each separate step. We had to start the car, learn to shift gears by depressing a clutch while applying pressure to the accelerator, and moving the gear shift lever. However, it wasn't long before we were able to drive automatically, with the individual steps performed almost as one.

This same approach applies to learning order-picking rhythm. The order filler goes to the proper location, checks the order document against the locator sign on the aisle, section and shelf. The next step is to compare the product on the shelf with the shelf label and the order document. If everything tallies, the order filler checks the order for the required number of pieces, selects them, and makes the final check of verifying that the number of pieces picked is the number originally required.

It will not take long before the order filler can perform each of these steps without thinking, just as he or she learned to drive a car.

Summary

Order errors can be reduced, if not to absolute zero, at least to a few hundredths of a percentage point. This is done, first, by eliminating the clerical errors that occur before the warehouse ever gets the order. Then warehouse order-picking errors are controlled by improving the stock locator system, clarifying item identification, avoiding confusion regarding quantity of items to be picked and using a clear, legible picking document.

Improving several environmental factors also will cut errors. These include installing good lighting, providing convenient shelf locations, effective item placement and restricting multi-line order filling. Finally, it's up to management to hire competent order fillers, then train them thoroughly.

Part VIII

HANDLING OF MATERIALS

39

RECEIVING AND SHIPPING

In many organizations, particularly manufacturing companies, there is a distinct division of responsibility between receiving and shipping. Goods being received are controlled by the purchasing manager, while goods to be shipped are controlled by the traffic manager. Unfortunately, however, in too many cases there is poor communication between the purchasing and traffic departments. As a result, there is a tendency for receiving department personnel to have a life full of surprises—such as trucks showing up without notice to the warehouse manager. When such surprises provide substantial unloading volume, they can be seriously disruptive to the warehouse work schedule.

Warehousing tradition says that receiving docks should be at one end of the building and shipping docks at the other. Where a manufacturing process is involved, such separation may be quite logical, since the receiving dock takes raw materials that move to the head of the production line, and the shipping dock takes finished goods from the other end of the assembly line at the end of the building.

Regardless of warehousing tradition, however, it makes little sense to separate receiving from shipping except in manufacturing. If receiving and shipping docks are adjacent, dock space can be used flexibly to provide additional doors for either, depending on changes in demand pattern. Furthermore, goods destined for immediate shipment can be left on the dock for turnaround, rather than being hauled across the building from one dock to another. Workers, too, can be rotated between receiving and shipping.[1]

One furniture manufacturer makes an important case to its dealers about the benefits of receiving and shipping over the same dock. Yet very few dealers follow this manufacturer's advice, simply because old warehousing traditions seldom fade away.

Receiving and shipping in the warehouse bring the greatest opportunity for error, as well as for pilferage and theft. Because accidental or deliberate discrepancies will occur, it is particularly important to create a carefully-established routine for both functions.

Receiving

A typical receiving routine could be described as follows:

- A receiving clerk provides a specific time for the inbound receipt. Avoid any unannounced arrivals.
- When the vehicle arrives, assign the driver a specific dock door.
- Be certain that the inbound trailer is properly secured at the dock by firmly bracing its wheels.
- Record and check the seal on the inbound vehicle to be certain that the load has not been tampered with before arrival.
- Unload the freight and carefully tally the contents.
- Reconcile the receiving tally with the advance notice, a purchase order, or some other independently-prepared documentation.
- Record all exceptions, including overages, shortages or damages.
- Photograph any unusual conditions found as soon as they are found.
- Make any needed quality control checks without delay.
- Assign a warehouse stock location for the inbound load.

---

[1] From an interview with Burr Hupp, Executive Director of Warehousing Education and Research Council, in *Traffic Management,* Cahners Publishing Co., November, 1982.

RECEIVING AND SHIPPING

- Move freight from the receiving dock to the assigned location.

Verification of count on receipt is of critical importance. There are more thefts and fraud at the receiving area than at the shipping area, simply because it is easier to lose control of counting large quantities while receiving than while shipping. Here are several effective techniques for minimizing receiving losses.:

1. Require receiving employees to work in pairs and check each other's work. Emphasize team work rather than competition.

2. If inbound goods are to be palletized, calculate the number of pallets needed to hold the entire load and assign that number to the receiver. Filling a specified number of pallets can serve as a double check of the count. Another way to double check the count is to use pre-counted labels, stickers or tags to apply to each case and be exhausted when all merchandise is received.

3. Provide a separate receiving tally rather than trying to use marginal space on a delivery receipt or bill of lading. *(See Figure 39-1)*

4. Require an independent verification based on blind receipt; do not tell receiving clerk the count expected.

5. Always count twice, before signing once.[2]

Shipping

The ideal shipping routine is roughly the reverse of that specified for receiving. If an order destined for shipping is pulled from inventory by one employee and checked and loaded by another, the likelihood of discovering errors is greatly increased. Insisting on a shipping schedule will do a great deal to prevent unexpected peaks in workload or unpleasant surprises. A specified count of shipping labels or tags is a good way to create a double check of quantities in shipping. The shipping dock is also an excellent place to discover mistakes made by order pickers. *(See Chapter 38)*

---

2 From a technical paper by Leon Cohan of T. Marzetti Company, Columbus, Ohio and published by Warehousing education and Research Council.

DISTRIBUTION CENTERS, INC.
RECEIVING TALLY

DATE Jan. 2, 1990 LOT NUMBER

ACCOUNT Franistan Company

SHIPPER

ORIGIN Chicago

CAR NO. OR FREIGHT BILL NO. 8004

ROUTE Truckers' Dispatch

SEALS ABC 2102 + 3

TOTAL WEIGHT 36050

QTY. REC'D.	DESCRIPTION	UNITS PER PALLET	TALLY RECORD	LOCATION
1400	Green Widget 801	140	||||| |||||	C-10
2700	Blue Widget 850	270	||||| |||||	C-12

DAMAGE INFORMATION

QTY.	PRODUCT DESCRIPTION	DAMAGE DESCRIPTION	CAUSE OF DAMAGE	RESULT OF RECONDITIONING: GOOD STOCK	REJECTED	MAN HOURS
3	Blue 850	Creased cases	Load Shift		3	.5

UNLOADER J Jefferson LIFT OPERATOR A. Burr CHECKER B. Arnold

FORM # 1-001-0368

This is the receiving tally, the first document prepared at the warehouse.

Figure 39-1

RECEIVING AND SHIPPING

Dealing with Shippers Load and Count (SL&C)

As shipping and receiving agent for his customers (in the case of a public warehouse) or in his own right (in the case of a private warehouse), the warehouse manager should know exactly what a shippers load and count (SL&C) notation on the bill of lading (B/L) means. Such a notation can make it quite difficult to collect on in-transit loss and damage claims. On the other hand, a knowledgeable warehouse manager can take certain precautions which will either mitigate or remove entirely the effects of an SL&C notation.

An SL&C notation means that the freight was loaded and counted without the customary assistance of the carrier's truck driver. The notation is most commonly used when, for shipper's convenience, the carrier drops or spots a vehicle at the shipper's and returns and picks it up (without inspecting or counting its contents) after it has been loaded and sealed.

There is a certain truck driver mythology surrounding SL&C. Drivers often scrawl "SL&C" on the bill of lading in the belief that this will have the magical effect of relieving the carrier of liability for in-transit loss and damage. Regrettably, this myth has wide acceptance among uninformed shippers and many untrained warehousemen.

The fact of the matter is that SL&C has absolutely no effect on carrier liability for in-transit loss and damage. Even if a load is traveling under a valid SL&C, the carrier's liability for in-transit loss and damage remains the same as it would have been without such a notation. What does change, however, is that the shipper or claimant's burden of proof is substantially increased.

The U.S. Supreme Court has ruled that although a carrier is not an absolute insurer, it is "liable if the shipper makes a prima facie case and the carrier does not meet its burden to show both its freedom from negligence and that the loss was due to one of the causes excepted by the common law rule." Exceptions have been expanded to include damage caused by (1) act of God, (2) the public enemy, (3) inherent vice or nature of the product, (4) the public authority, and (5) shipper's act of default. Making a prima facie case requires showing (1) the carrier received the goods in apparent good order (producing a carrier-executed bill of lading with a statement to that effect), (2) that the goods arrived in damaged condition, and (3) the amount of such damages.

Once a shipper has proved a prima facie case, the burden of proof shifts to and remains with the carrier. However, in an SL&C movement the shipper will obviously be unable to produce a carrier-executed B/L stating the goods were received in apparent good order. Thus he will be unable to establish a prima facie case and, therefore be unable to shift the burden of proof onto the carrier. The point to remember, however, is that the only difference is the burden of proof. Should the shipper be able to establish, through a preponderance of evidence, that the shipment was in good order and complete as described when loaded, the normal laws of in-transit loss and damage would then apply, that is, the claim would be much the same as in a non-SL&C shipment.

Here are some of the specific ways a warehouse manager can support shipper/claimant attempts to collect on claims for in-transit loss and damage in SL&C movements.

1. Seals. In the case of shortages in an SL&C shipment the carrier must show that the load was sealed at all times except while the shipper/consignee (or their agents) were supervising the (un)loading. Correct handling of a sealed shipment means that the receiver physically removes the seal. It doesn't mean that he stands on the dock and watches while the driver cuts the seal. The tally/warehouse receipt should show "seal intact" only if the receiver has removed the seal himself; otherwise, the load should be considered as received with "seals not intact." If the seal is not intact upon arrival, the carrier cannot claim SL&C protection against shortage claims. If the seal is intact and removed by the warehouseman, it must be saved complete. Should a shortage be uncovered, the seal should be returned to the shipper for verification that it is indeed the same one applied at time of shipment.

2. Improper loading. If damages are caused by improper loading (in the case of an SL&C shipment), the carrier may not be held

RECEIVING AND SHIPPING

liable. The carrier must prove, however, that improper loading caused the damage. If, for example, an act of carrier negligence contributed to the damage, the shipper may still be able to recover. Additionally, if the damage was caused by other than improper loading, the SL&C protection should be irrelevant. Therefore, when damage is discovered while unloading an SL&C shipment it is extremely important that the extent, nature, and circumstances be accurately and fully documented. Pictures should be taken, statements collected from the unloader and his foreman, and if possible, the driver. In other words, it ought not to be assumed that as it is an SL&C shipment, there is no point collecting documentation for a possible claim.

3. Loading In those cases where, for the shipper's benefit, loading is done under SL&C there are certain precautions that may be taken to preclude a carrier's subsequent SL&C defense. Among them are:

—*Loading correctly.* The best way to avoid losing an in-transit loss and damage claim is not to have the claim in the first place. However, if a trailer/car must be loaded for later pickup by the carrier, i.e., as an SL&C load, extra attention should be given to loading it correctly. Additionally, that extra attention should be fully documented, e.g., a loading diagram may be prepared showing what, where and how the goods were loaded. This can be supplemented with pictures and a statement by the actual loaders and the foreman who inspects the load before it's sealed. In the event damage is discovered at unloading, the carrier must prove—not merely allege—improper loading. Documentary evidence such as noted here might outweigh the carrier's claim of improper loading and thus remove his ability to invoke SL&C protection.

—*Having the load inspected by the carrier.* If the load is sealed before being tendered to the carrier, the carrier has no obligation to break the seal and inspect the load—that is, unless the shipper (or his agent) specifically instructs the carrier to do so. Therefore, if possible, a notation should be placed on the bill of lading instructing the carrier, at pick-up time, to break the seal and inspect the trailer for proper loading. This should be sufficient to defeat the carrier's later claim that damage was due to shipper's improper loading.

—Verifying SL&C notations. When receiving do not proceed as if the shipment were SL&C just because it is indicated on the bill of lading. There's always the possibility the driver at the other end added the notation after pick-up. If there is damage or shortage when the doors are opened, treat the task of documentation the same as if the load were not SL&C. Who or what caused the damage is a matter that should be decided later on the basis of the evidence provided when the claim is filed. The warehouse manager's job is to ensure the claimant has the most complete and accurate documentation package possible.

4. Not allowing the carrier to mark the bill of lading as SL&C. If the carrier is present during loading but for whatever reason chooses not to observe the loading, do not allow him to later mark the bill of lading SL&C. DOT Safety Regulations require drivers to ensure, before moving a loaded vehicle, that the cargo is properly and safely stowed. The driver should be given an opportunity to check the load; if he refuses, both the refusal and the fact that an opportunity to inspect was available should be documented on the bill of lading. A note also should be made on warehouse records showing what the driver was doing while the trailer was being loaded.

5. Reading the bill of lading correctly. There are certain published commodity rates that apply in shipments where shipper loads and/or consignee unloads. These are not necessarily the same as SL&C. The carrier may not have been precluded from counting and inspecting the load just because the shipper did the actual loading. Notations on the bill that the shipper performed the loading should not, therefore, be automatically construed as SL&C.

It should be noted that all of the procedures described above apply primarily for shipments moving under common carrier rules. When your shipments are moving under a contract carrier arrangement, the rules might be quite different. Since deregulation, an increasing number of shipments are moving on something other than common carriers. When this happens, the shipper must carefully study the rules under which the freight was moved.

Shippers load and count makes the claimant's task more difficult, but it does not make it impossible. Avoidance of abuses of SL&C and careful documentation of shipments made in this manner can still make claims possible.[3]

Controlling Damage Claims

Claims of damage due to improper loading are often difficult to defend. A Polaroid camera to record the condition of the doorway of a box car or the end of a loaded trailer can provide pictures to use when the consignee or customer claims the load was not properly stowed.

These photographs, to be valid in a dispute, must be referenced. Therefore, it is wise to make a written notation on the reverse side of the photograph which shows the date the picture was taken and all details surrounding the shipment. If the carrier driver signs the photograph and acknowledges its authenticity, it's even stronger proof.

Careful inspection of the trailer or box car presented for the shipment will eliminate the possibility of loading a defective vehicle.

Perhaps the most important thing to remember about shipping is that it represents the last chance to avoid a warehouse error before goods reach the customer. In addition to watching for errors in picking, the warehouse manager should check on accidental shipment of damaged goods, errors in count, or shipping of goods not in first-class condition. One warehouse has this sign on the shipping dock: "The next inspector is our customer."

---

3   From a technical paper by Leon Cohan of T. Marzetti Company, Columbus, Ohio and published by Warehousing Education and Research Council, December, 1984.

40

SPECIALIZED WAREHOUSING

Most warehouses are designed to store bulk merchandise for volume shipments at ambient temperatures. These warehouses have normal fire risks and typically handle full loads of freight in-bound and smaller shipments for the out-bound move.

But there are many kinds of highly specialized warehouses. In this chapter we will look at just three which are growing in importance and complexity.

The first is temperature-controlled warehousing, which has grown substantially in the past several decades with the growing popularity of frozen foods and fresh produce requiring temperature control. Even the use of temperature-controlled warehousing for non-food products has grown with the increased distribution of chemicals which require refrigeration.

Second is warehousing for hazardous products, which may not have grown in popularity, but has certainly grown in complexity. As public awareness of the dangers of hazardous chemicals has increased, those warehouse operations which must safely store such products have experienced growing regulation and risk-management problems.

Finally, a new growth area in specialized warehousing is fulfillment, which is the handling of mail order and express shipments moving directly to the consumer.

Let us examine these three special warehouses in detail.

Temperature Controlled Warehousing: The Essential Differences

There are at least four kinds of cold warehouses. *The standard freezer* operates at 0° to -10°F; *ice cream storage freezers* operate at -20° to -25°F; *blast freezers* combine extreme cold with rapid air circulation to quickly freeze freshly-packed products; and *chilled warehouses* hold product at 35° to 45°F. When you are designing a cold storage building, consider potential future uses and the cost of conversion. For example, if you are building a large chilled warehouse design the building to allow for future conversion of part of the space to a freezer warehouse. The lower the temperature one wants to maintain, the higher the cost of the building.[1]

"Temperature-controlled storage" is usually cold storage, and its prime use throughout history has been to preserve foods. Before the development of refrigeration machinery in the 1890's, products were frozen by blowing air over salt and ice. Many of the cold storage plants were branches of ice plants. The quality of this cold storage was questionable, and some state laws required retailers to warn their customers of goods that had been in cold storage. Early cold storage warehouses were used primarily for dairy products, meat and poultry, and they were filled on a seasonal basis. In the early decades of this century, most temperature-controlled products were chilled rather than frozen. Consumer-sized packages of frozen foods were introduced in 1929 and did not become popular until the 1940's. By 1970, freezer space was over 75% of the total public refrigerated arehouse space in the United States. "Cooler" space is used primarily for fresh fruits and vegetables, dairy products and eggs. There is increasing use of chilled storage for non-food products such as plastics, film, seeds and adhesives.[2]

One significant difference between temperature-controlled and

---

[1] From *Temperature Controlled Warehousing: The Essential Differences,* by Tom Ryan and Joel Weber, United Refrigeration Services, Inc. This article appeared in Warehousing Forum, Volume 3, Number 11 © Ackerman Company.

[2] From *Operational Training Guide,* ©International Association of Refrigerated Warehouses, Bethesda, MD.

dry warehousing is the cost of the facility. A freezer warehouse will typically cost two to three times as much as a dry storage warehouse of similar size. The cost difference in utilities is even more extreme, with electricity costing as much as five times the amount per square-foot. So the successful temperature-controlled warehouse operator depends on excellent conservation and building maintenance to control these energy costs.

The freezer room should be monitored by temperature gauges, one at eye level showing the temperature as you walk into the room, and others of the same type in different corners of the room. In addition to this, a temperature recorder with a weekly disk should be used as a permanent record. A dry sprinkler system is mandatory. This cuts down exposure to liability for temperature fluctuations and fire. A 24-hour monitoring system that gives a warning of temperature fluctuations of three degrees or more should be installed. Plastic curtains with the hydraulic doors help in controlling energy costs.[3]

In a dry warehouse, walls may be thin steel panel, protected by a modest amount of insulation. In contrast, the walls of a freezer are an important part of the insulation system. A freezer may have 6-inch foam insulation panels, clad by sheet steel both inside and out. The floor too is an important insulator in a frozen warehouse. A typical specification would be six inches of concrete poured on top of six inches of foam insulation. Below the insulation, to protect against heaving of the earth beneath the floor, heat is provided by piping warm ethylene glycol through a layer of sand. The insulation layer thus serves a dual purpose—keeping the cold in and the heat below out.

The roof of a dry warehouse may consist of nothing more than a thin steel deck with a small amount of insulation. Most temperature-controlled warehouse roofs will start with a steel deck, but the structure must be designed to hold refrigeration equipment which either is mounted on top of the roof or hangs below the ceiling of the warehouse. Above the deck is a layer of ¾-inch fiberboard. Above

---

3   From Jesse Westburgh, Wales Industries, Inc., Columbus, Ohio.

that is an additional 10-12 inches of foam insulation, and then another layer of fiberboard. The weather seal is a single-ply rubber roof protected by a layer of stone ballast.

The operation of lift trucks is affected by cold, especially the lower temperatures of a frozen-products warehouse. Because of tight insulation and recirculated air, internal combustion engines simply cannot be used. Electric trucks need modification to operate at zero degrees. Heaters are required for the electric contact points as well as heavier-duty batteries to allow for a full shift use without recharge. The harsh conditions also produce additional wear and tear on the equipment. In this environment, thorough preventive maintenance is even more critical than in dry storage warehousing.

The batteries on electric equipment will last longer if they are charged every four to six hours rather than the eight hours recommended in normal temperatures. The warehouse which has both dry and cold space should have a plan to rotate fork trucks so that each truck spends only part of the time in the freezer. Stand-up lift trucks are somewhat easier than the sit-down types for people working at below-zero temperatures. Some trucks are specifically designed for this environment, with an enclosed cab to allow the workers some relief from the cold temperatures.

Workers should have a break-room where they can relax with coffee or soft drinks. A ten-minute break each hour will improve productivity for people working in the freezer. Rotating workers from all areas of the warehouse so that each works for some of the time in the freezer, is another useful tip.[4]

Most warehouse operators provide protective clothing for employees. This includes insulated boots, gloves and freezer suits. Working in a freezer results in a greater fatigue factor, since a significant portion of body energy is spent in keeping warm. In an ambient-temperature warehouse, a work crew is capable of handling an overtime or emergency assignment of well over 12 hours without significant loss of productivity or accuracy. Fatigue takes its toll in a

---

4   Ibid.

SPECIALIZED WAREHOUSING

much shorter time in a frozen warehouse. Most operators see a greater amount of sick leave among workers in a frozen environment.

The task of supervision is more difficult in a temperature-control than in a dry-storage warehouse. A supervisor in a dry warehouse can watch the loading dock and gain a good idea of what is going on throughout the warehouse. In a temperature-controlled warehouse, the dock is separated from storage areas by walls and doors to preserve the temperature in the storage rooms. A supervisor on the dock cannot see what is happening in the storage rooms. Effective supervision in the cold storage area requires additional foremen in the cold rooms. Supervisors have even greater risk of health problems, because they may move in and out of cold rooms more frequently than the workers, and because they are less physically active.

The nature of cold storage forces changes in the way work is scheduled. Because the cost of space is very high, staging of outbound orders may not be practical. Temperature-control warehouse operators tend to select most outbound orders as close as possible to the time of shipment in order to minimize space committed to staging. Product quality considerations may prevent staging outside the freezer area. When freight cannot be pre-selected, the operator cannot use staging as a scheduling buffer and when staging is difficult, performance of motor carriers is particularly critical. The cold storage operator must run a scheduled truck-dock.

Housekeeping is another function that is affected by the harsh environment in a frozen warehouse. Spills which are not cleaned up quickly may freeze, cause accidents and stain the floor. The method for scrubbing a freezer floor is far more expensive than in dry warehouses, because it requires the use of a non-freezing solution. The lowest-cost solutions are banned by FDA rules. Sometimes the only way to clean a floor in a frozen environment is to scrape it. The cleaning equipment, like lift trucks, is restricted to electric power. The cold causes similar maintenance problems and shorter battery life.

The temperature-controlled warehouse requires an extra measure of precision because the consequences of failure are serious. Failure to maintain temperature control of the product can have very costly consequences for the owner of the merchandise. Therefore the warehouse operator must not only protect the product, but provide ample proof that such protection was always provided.

As the use of chilled and frozen products seems to be growing faster than the economy as a whole, it is likely that temperature-controlled warehousing will be a part of your future even if it has not been part of your past.

Hazardous Materials Warehousing

Hazardous materials can be grouped generally into the following categories: flammables; explosives; corrosives; poisons; radioactive materials; and oxidizers. In addition, as the EPA's classification of materials continues to be expanded, those materials having "Reportable Quantities" are likely to increase.

For the warehouse operator who has never handled hazardous products, life suddenly becomes more complicated. The relationship changes between the warehouse operator and other concerned parties, forcing the operator to deal with agencies and individuals who would not be involved with other kinds of warehousing. Storage of hazardous materials creates increased responsibilities and liabilities.[5]

From an administrative standpoint, the distinguishing feature of hazardous material distribution is the regulation to which it is subject. There are several authorities who either have regulatory power over hazardous materials or with whom you should consult:

> *OSHA*—The Occupational Safety and Health Administration has regulatory power over warehousing hazardous materials with regard to protecting your employees from exposure to these products. OSHA's concerns are satisfied by taking the hazardous material information provided by the manufacturers on their container labels and Material Safety Data Sheets (MSDS's) and conveying that information to your employees through an approved communication program.

---

[5] From *Hazardous Materials Storage: The Essential Differences* by Lake Polan III of Allied Warehousing Services, Inc., Volume 4, Number 4 © Warehousing Forum, Ackerman Company.

EPA—The Environmental Protection Agency regulates hazardous materials through its Superfund Amendments and Reauthorization Act (SARA) Title III Program dealing with Emergency Planning and Community Right to Know. Emergency planning is carried out through state and local Emergency Planning Councils, both of whom must be supplied with either the previously mentioned MSDS's, or an appropriate listing covering any hazardous materials you store. The local Emergency Planning Council could conceivably be represented by your local fire department. Any release of a hazardous substance must also be reported to both state and local councils.

DOT—The Department of Transportation regulates the movement of all materials including hazardous materials. It has established 20 different classifications of hazardous materials, and it requires, among other things, that all materials be labeled properly. All trucks leaving your facility with hazardous materials must be appropriately placarded, and you must also properly annotate your bills of lading when hazardous materials are shipped. For the most part, warehouses accept the classifications provided by the manufacturer, but this is not always the case. For LTL outbound shipments, special care must be taken. When combining products with different classifications, the quantities involved in the shipment may require a special "dangerous" placard to be placed on the truck.

IMO, IATA, & ICAO—The International Maritime Organization, International Air Transport Association & International Civil Aviation Organization control overseas shipments of hazardous products. They too require product labeling, container placarding and regulatory descriptions on bills of lading, and their regulations can be even more stringent than those of the DOT.

While all the agencies listed above regulate hazardous materials, they have little to say about how the warehouse operator actually handles or stores product in his own facility. Therefore, the most important source of procedural advice is the supplier, the product manufacturer, or the product owner of goods in a third-party warehouse. These are the individuals who should know the most about the product involved, and should provide the warehouse operator with comprehensive documented standards for safe storage and handling. The operator also has a responsibility to acquire a maximum amount of knowledge and experience with the hazardous product. The possibility arises that other product owners who share the use of the warehouse will be concerned about storage and handling practices which increase the risk to their products—storing plutonium near baby food, for example.

If you lease your space, your property owner will also take an interest in hazardous materials storage. While your lease will probably not prohibit this business, you will need to address such issues as indemnification, the adequacy of your sprinkler system, increased insurance costs and your ability to return the building in usable condition at the end of the lease. "Reasonable wear and tear" does not contemplate a building left contaminated with methyl-ethyl-badstuff.

Most manufacturers, even if they self-insure, may use some or all the guidelines for storage developed by the National Fire Protection Association. NFPA membership represents both the manufacturing and insurance industries, with standards sensitive to the needs of the latter. Your insurance carrier conceivably uses NFPA standards as well, with the noted exception of the Factory Mutual Companies, and some others who have developed their own standards. You will need to establish a close working relationship with both your insurance agent and carrier(s) to help you evaluate the costs and risks of hazardous material storage, to determine that you can insure those risks and the price of the coverage.

State and Local Building Inspectors

If you plan new construction or are remodeling your facility to equip it for hazardous material storage, you may also need the approval of your local building inspectors to insure that you conform to the requirements of a local or municipal building code. These codes may or may not conform to NFPA guidelines. Local codes can be

SPECIALIZED WAREHOUSING

more restrictive than state codes, but if they are less restrictive then state codes take precedence.

Because a number of different authorities are involved, it is inevitable that there will be inconsistency or even conflicting regulation or instruction. Therefore, a key to functioning with these different agencies is to determine the authority having jurisdiction:

> Where public safety is primary, the "authority having jurisdiction" may be a federal, state, local or other regional department or individual such as a fire chief, fire marshall, chief of a fire prevention bureau, labor department, health department, building official, electrical inspector or others having statutory authority. For insurance purposes, an insurance inspector, department, rating bureau or other insurance company representative may be the "authority having jurisdiction." In many circumstances the property owner or his designated agent assumes the role of the "authority having jurisdiction."[6]

It is not unusual for the warehouse operator to find that the person responsible for regulation knows less about the product than the operator does. In such situations, it is wise to influence selection of the authority having jurisdiction, since this can have a dramatic effect on cost. Sometimes the warehouse operator becomes the authority by default, simply because no other authority is willing to make a decision.

Unfortunately, it is not practical to provide a simple "how to" which encompasses all the product variables and provides the storer with instructions on safe procedures. So where does the operator go to learn the best way?

If the storer is a third-party warehouseman, he will probably seek information from other third-party warehouse companies handling similar products. Storers of hazardous products typically indicate that they have this experience in the popular warehouse directories published by Chilton's Distribution and the American Public Warehouse Register. If you are a private warehouseman, you still might share information with third-party warehouse operators and with other op-

---

6 *NFPA 30 Flammable and Combustible Liquids Code,* 1987 edition, p. 30-5.

erators of private warehouses. Even storers of competitive products are willing to share information about safe warehousing practices, simply because everyone is anxious to learn from others in this field.

Consultants who specialize in environmental safety are able to assist both producers and storers of hazardous chemicals. The names of reputable consultants in the field can be obtained from your local or state fire marshall or emergency planning office. *Hazmat World*, published by Tower Bornes Publishing Company, has a number of advertisements for experts in this field.

Finally, NFPA is the leading authority on fire safety, and the cost of membership in this association is nominal.

Because regulation in the field is changing rapidly, there is a danger that the information you have is now out of date. To be sure that your warehouse is in compliance with current standards, it is wise to use the services of an agency specializing in interpretation of the federal register.

Of all the information sources available, the best is the manufacturer of the product who should have the most information available about the safe keeping of that product.

Becoming a competent warehouser of hazardous materials requires a serious commitment in organization, manpower, and capital resources. The warehouse operator is exposed to increased risks, uncertain requirements, greater costs, and a potential regulatory nightmare. Successful hazardous materials warehousing requires an increased discipline throughout the organization, since the consequences of failure can be most severe.

For storers of hazardous materials, the first concern is safety, the second is service, and the third is cost. As the requirements become more stringent, only the most competent warehouse operators will remain in this line of business.

Fulfillment Warehousing

Six special features make product fulfillment warehousing different from public and private warehousing. These are the following:

1. The warehouse operator has direct interface with consumers.

SPECIALIZED WAREHOUSING

2. Information requirements are instantaneous.
3. Order size is much smaller than typical warehouse orders.
4. The order-taking function at warehouse level is much more precise, particularly because it involves contact with the consumer.
5. Customer service requirements are different and typically more demanding.
6. The transportation function is more complex.[7]

Storage and materials handling functions are quite different. Fulfillment warehousing involves more than simple storage. Because of fast turns and low volumes, gravity-flow rack and high-security areas are almost always needed.

The handling function will also differ from that of the more conventional warehouse. While there may be some LTL shipments, there will be a much higher concentration of parcel service and mail movements and handling will include the metering of those shipments.

Thus the materials handling equipment investment will have more emphasis on scales and meters to control outbound movements.

Paper flow for a fulfillment warehouse is more complex than most other warehouses. A large number of orders are received by telephone. A significant amount of time is spent in handling customer returns. Because a fulfillment operator is dealing directly with individual consumers, the customer service function is particularly critical.

Some users want the fulfillment center to create invoices or even dunning notices. Accounts-receivable aging reports are another service which can be a by-product of the handling of invoices.

Nearly every fulfillment center must handle the major credit cards easily. Credit card authorization systems through either a local bank or an outside service become a necessity.

Many fulfillment centers handle banking for their customers, and a smooth relationship with the center's bank is a necessity to handle these accounts.

---

[7] From presentations by Jeffrey A. Coopersmith of Directel, Columbus, Ohio and by James E. Dockter of Fulfillment Unlimited, Atlanta, Georgia. From Volume 2, Number 2 © Warehousing Forum, Ackerman Company.

The best of fulfillment centers offer a 24-hour turnaround on orders. A few take two or three days to process and ship, but fast turnaround is becoming a standard.

Because of this fast turnaround requirement, labor flexibility is needed to deal with seasonality and variance in work load. The best fulfillment centers maintain a pool of part-time workers who are available if a second shift must be added or if extra people are needed quickly.

A fulfillment center can be a major headache in terms of claims and theft. Product shipped can be misdirected by a dishonest person who is running a postal machine. Operating the postal machine is a vulnerable position, and care should be taken in selecting the individual who takes on this job. Many fulfillment centers negotiate the inventory variance to be allowed in advance, based on the customer's own experience with errors. Some users will allow a predetermined formula for shrinkage which is in line with their internal experience.

Typically, the fulfillment center must absorb the dollar consequences of all of its mistakes—the public warehouseman's usual limitations do not apply in this service. The fulfillment center normally needs more equipment for communications than for shipping. This includes the ability to handle credit cards and toll-free phone lines. Compared to the conventional warehouse, the order volume is extremely high. The ability to do a great many things with a computer is far more critical in the fulfillment center than in the conventional warehouse. Finally, the number of stock keeping units controlled is usually relatively high in a fulfillment center.

As companies constantly seek new ways to promote their products, fulfillment is a warehousing service which is destined to increase in popularity.

Summary

The three specialized warehouse operations outlined here have all grown in number as well as complexity. Changes in our food consumption patterns promise to stimulate continued growth of cold storage. Public concern about hazardous materials will force those who handle such commodities to be better equipped and prepared. Finally, the continued popularity of credit cards and marketing by television, telephone and catalog will create growing need for fulfillment centers.

41

ORDER PICKING[1]

Accurate order picking is typically the most important warehouse operating responsibility. Actual costs of an incorrectly picked order are estimated at ten to thirty dollars per bad pick, not counting the customer dissatisfaction.

In many warehouses, order picking is the largest single expense category in the operation. Good order picking demands high levels of management in planning, supervising, checking, and dealing with personnel. The order picking operation is not easily or economically automated. Even with all of our automation advances, order picking often remains a manual operation because brain, eye, and hand coordination have not yet been equaled by any machine.

Because of its high labor content, order picking presents the greatest opportunity for error (Chapter 38 deals with reducing errors). A good order-picking document is the first step for accurate and efficient picking.

Seldom do warehouses receive goods in the same quantities or packaging required for shipping. Shipments from the warehouse must be orders assembled from stock, since economic shipping loads to the warehouse are not often the same quantities which customers actually purchase.

Quality in Order Picking

Quality here means to perform to standards and nothing more. Warehouse management must design, implement (through training),

---

1 This chapter was revised by William J. Ransom, Ransom & Associates.

and then insist upon standards of order picking that are error-free. Those managers who *know* there are going to be errors because there always *have been* errors will *always* have errors.

On the other hand, error-free order picking is done by those who believe in zero error and plan for it to happen. Any order picking error could be a customer lost, never to be regained.

Order-Picking Forms

A good order-picking form is one that has only essential information on it, i.e., customer identification, order number or date, location of items to be picked, item description, and a specific quantity to be picked.

Newer order-picking systems based on computer assistance should have the built-in elements of a good manual system. When you implement automation, choose a system that will accomplish your objectives. Consider these principles of automation: define your needs in detail, get the software that meets those needs, and then look for hardware.

The use of color coding in the order picking operation will reduce errors. Color coding should be used in any operation where it can be applied.

To reduce errors, be sure that the same terminology is used for the same items throughout the system in the warehouse.

Inventory Relationship to Order Picking

The accounting department with its dollar inventory record-keeping has no relationship to the inventory record-keeping of the warehouse operation. From an accounting point of view, the inventory may be treated as first-in-first-out (FIFO) or last-in-first-out (LIFO).

In good warehousing practice, regardless of the accounting system, inventory is treated on the FIFO basis. Since nothing improves with age, it is always best to move the oldest stock first.

Order Picking Systems

When setting up your order picking system, plan on generating data that will allow you to measure performance once the system is operating. Some common performance ratios used for order picking are the following:

1. $Orders\ per\ hour = \dfrac{Total\ number\ of\ orders\ picked}{Total\ labor\ hours\ used}$

If orders arc uniform, this is a valid measure, but if the number of items or quantities varies greatly from order to order, the following ratio will be more indicative of performance:

2. $Lines\ per\ hour = \dfrac{Total\ number\ of\ lines\ picked}{Total\ labor\ hours\ used}$

This ratio is a more accurate measure of the work performed by the picker as each line represents a task. Lines picked per hour is probably the most commonly used measurement ratio.

The order-picking system consists of pick slots (locations) where the product is available for selection in the quantity called for on the picking document. The location of the pick slots depends on the system, but they must always provide the necessary picking identification and be physically conducive to low fatigue and error-free picking. In other words, put the most popular items between waist and eye level.

Order picking is seldom done by only one method. Variances in package size and configuration, picking quantities, stocking quantities, and inventory requirements often necessitate more than one system. Most order-picking operations are hybrids of three order-picking methods, listed here in order of system complexity:

1. *Unit-load* picking is done when a pallet load of product is pulled from stock. An example of unit-load picking is a major appliance warehouse.

2. *Case-lot* picking is the selection of full cases of a product. However, the order is less than a full-pallet unit-load. Case-lot picking is best done by staging a unit load in a pick line and pulling case quantities until the unit load is depleted.

3. *Broken-case* picking is done when less than full cases are

called for by the customer's order. This kind of picking may be done from shelving or flow rack, depending on the size and volume of orders.

Fixed Versus Floating Slots

Where do you locate the stock? The shortest travel route in order-assembly sequence will yield the lowest picking cost.

Fixed locations are assigned to each product and the product is always located in the same place. Simplicity and elimination of errors are the greatest advantages of this system. The disadvantage is the waste of space that occurs when a slot is reserved for unstored product. A fixed-slot system can be combined with a preprinted order form to produce a simple order-picking system that uses the stock numbers as locations. This system is ideal where the order quantities are small.

Floating slot systems use the next available location in the warehouse for storage instead of reserving an assigned area. This random location system requires that a precise locator system be instantly available for fast moving items.

Floating slots are likely to increase picking travel and thereby erode order-picking efficiency. Travel will be reduced if random storage is arranged by zones of activity. Keeping fast movers together in short-travel locations will help control travel.

Order Picking Methods

Order picking can be manual, power-assisted, automatic, or it may be a combination of these methods.

A manual system uses two- or four-wheel hand trucks or carts pushed through the pick line and hand-loaded.

A powered system uses unguided or guided vehicles to transport and/or elevate the warehouse worker through the pick line. Pallets, carts or other containers are manually loaded by the order picker.

An automatic system uses the computer to guide the picker to the pick location, elevate him to the proper pick-height, instruct him as to the pick location, and indicate the proper pick quantity. Automated picking may be any combination of these, accomplished through computer control.

Discrete Versus Batch Picking

Discrete (single order) picking requires the picker to assemble the total order before moving on to another one—in other words a complete pass through the order-picking area for each order to be picked. Discrete picking has these advantages:

—Maintains single order integrity.
—Simplifies the picker's job.
—Avoids rehandling or repacking.
—Provides fast customer-order service.
—Allows for direct error checking and establishes direct error responsibility.
—Is highly efficient when the number of stock keeping units (SKUs) per order is small.

Discrete picking has these disadvantages.

—Requires full order-picking route travel for all orders.
—Doesn't allow for speed-picking of large quantities of an individual item.
—Requires the highest number of picking personnel for a given number of orders.

Batch picking is selecting of the total quantity of each item for a group of orders. In a break-out area, batches are re-sorted into the quantities for each order. Batch picking has these advantages:

—Reduces travel to pick the total quantities of a group of orders. Picking travel time can be reduced as much as fifty percent.
—Minimizes picking time for quantities of an item.
—Permits volume picking from large quantity or bulk storage, thus reducing the need for constant restocking of the pick lines.
—Provides a second check of the quantity picked by comparing the batch picked against the individual quantities in each order.
—Improves supervision by concentrating the final order-assembly in a smaller area.

Batch picking has these disadvantages:

—A second pick, or a distribution of the picked quantity is required to fill individual order requirements.

—Space is required for the distribution and order-assembly operation. Additional equipment may be required depending on the size of the batch-pick area.

—Individual orders are open until the entire batch of orders is complete.

—Counting is done twice and differences in count will require reconciliation time.

A variation of batch picking is to have the picker first pull the total item quantity and then place the proper quantity in separate containers for each order. The order picker is now doing discrete picking in a batch-picking mode. When combining these two methods of picking, higher skill is needed since the potential for error is greatly increased. A pre-pack check may be desirable if a pattern of errors develops. This picking method should be done only with well-trained and experienced personnel.

Zone System of Picking

Zone picking is accomplished by arranging pick lines in zones that handle similar types of items. For example, a "family" series like carburetor parts may be located in one zone.

The arrangement of the zones can provide for an order-assembling system in which each zone is used to build each order. Three zone arrangements may be used:

1. *Serial* zones are arranged in sequential order. The order picker must always go from zone A to zone B, then to zone C, etc.

2. *Parallel* zones are an arrangement of independent pick lines, perhaps with one on each side of the aisle. The picker need not take them in any particular sequence.

3. *Serial/parallel* zones are those where a number of serial zones are arranged in a parallel configuration.

Zones have the advantage of bringing together similar products in family groups. Where an order may be pulled from one family, this system arrangement expedites the picking process. Zones are very adaptable to use of flow rack systems.

Designing Your Order Picking System

In designing your system, consider how restocking will be accomplished. A minimum of one day's picking requirements is usually kept in the pick area. When restocking volume begins to equal the picking volume, it may be time to pick from the bulk supply area.

Scheduling of order picking depends on the necessary lead time to produce the picking documents. Order-picking systems work best when documentation is done on the day preceding the scheduled pick-date.

Consider "reverse engineering" your system. Take the finished orders that are to be shipped and start walking from the shipping point back through the flow process. As you move "upstream," see if the step just completed or the one before it could be eliminated. Equipment selection is easier to determine when approached from the output need rather than the input possibilities.

The following outline will help you develop your order picking operation:

1. Define your objectives by listing the customer's expectations and requirements.

2. Define your objectives in terms of:
   A. Cubic displacement of loads received, stored, picked and shipped.
   B. Volume of packages stored, picked and shipped.
   C. Flow rate of receipts, stored and picked, and shipped material in each classification.

3. Find out the state-of-the-art by requesting literature, specifications, and installed project reports from equipment suppliers.

4. Establish a set of operating specifications that say what you want to accomplish, not how you want to accomplish it. Operating specifications should encourage ideas and alternatives.

5. Ask two or three suppliers to review and bid on your objectives, giving them access to all of your information.

6. Evaluate the system's total economics for its life cycle. Be sure to place a fair value on flexibility and resale value, since you can seldom be sure that your product or needs will not change.

7. Purchase at the value price, not at the best price. All equipment needs service and supplier support. If you deal with the lowest bidder, it is wise to add something for the risk you run.

8. Implementation programming for a new order picking system should be a part of the entire process and should start with the beginning of the project. Planning implementation with the entire warehouse staff involved will insure a successful operating system.

Modeling Order Picking

Modeling is the process of simulating operations of a proposed system. Modeling can be done by computer programs or simply by flow-charting with operational standards. The purpose of modeling is to prove the usability and through-put of the system before going to the expense of building it.

In automated operations, modeling of the system is necessary before a final structure is determined. Automated systems are flexible in the design stage, but they can be altered only at great cost when they have been committed to bricks, mortar and iron. For this reason, modeling the system in detail is a small insurance cost for a successful system.

Safety

Because order picking requires people interfacing with machines, there is a significant safety hazard. The order picker often has both hands full and may be handling awkward items, so good footing is essential. Hand and body braces as well as safety belts can add security. Safety devices such as automatic brakes should be checked on a daily or weekly basis as a precaution.

Hardware

Effective order picking involves a combination of both storage and materials handling equipment. The order-pick function essentially has two options: either move the picker to the stock or move the stock to the picker. The most common means of doing the latter is through a carousel or a computer-controlled stacker crane. To some extent, a gravity-flow rack also moves stock to the picker, since it allows merchandise to slide from the back to the face of the rack.

When the picker must move to the stock, the storage layout should be arranged so that this movement is minimized. In a high-cube warehouse this is sometimes done by establishing pick mezzanines that allow some order pickers to stand on an elevated platform and pick from racks at a higher level.

(Chapters 42 and 43 provide more detail regarding equipment.)

Summary

Order picking efficiency is dependent on planned storage of material to be picked. This planning minimizes the distance traveled and the pick time involved. Labor content for order picking is usually the highest of all the jobs in the warehouse. Therefore, picking offers the greatest opportunity for cost reduction through improved layout, better methods and faster equipment.

Nothing is more important than improving picking accuracy. The importance that accurate order picking has on customer relations makes it a concern of senior management.

Accurate order picking is recognized not as a cost, but as a valuable sales asset.

42

EQUIPMENT: STORAGE AND LIVE STORAGE

A warehouse is more than a storage building. In order to handle cargo, it needs equipment. And it is the selection and use of that equipment that may spell the difference between profit and loss.

The operator may choose from a vast array of machinery and hardware designed to improve the efficiency of both handling and storage. The equipment choice is usually governed by the following criteria:

- Degree of flexibility desired for different uses.
- Nature of the warehouse building.
- Nature of the handling job—bulk, unit load, individual package, or broken package distribution.
- Volume to be handled by the warehouse.
- Reliability.
- Total system cost.

Defining the Job

It is easy to overlook the lowest-cost space in any warehouse—that which is close to the roof. Most contemporary buildings are high enough to allow at least a 20-foot stack height, and some allow substantially more. Fire regulations typically require that high stacks be at least 18 inches below sprinkler heads, but with this knowledge you can and should calculate the highest feasible stack height in the building. Then determine whether or not it is being used.

Saving space usually also saves time, since storage in a more compact area allows picking travel to be reduced. The prime justification for storage equipment is to increase cube utilization. Such equipment ranges from the simplest pallet rack or shelving to a rack-supported building designed for a stacker crane.

"Live storage" equipment not only increases the use of cube but

Pallet racks require more square feet per pallet than bulk storage areas. However, racks allow access to individual pallets.

EQUIPMENT: STORAGE AND LIVE STORAGE

also moves material when a movement is needed. One example is gravity flow rack, which both provides storage for merchandise and allows it to move from the rear of the rack to the picking face. More about this later.

Improving Storage with Racks

Because the storage rack is relatively simple, it is easy to overlook ways in which storage capacity can be greatly increased by using a rack. The most common storage rack found in warehouses is the three-high rack system. Installed throughout a warehouse of 100,800 square feet, this rack will allow up to 6,930 pallet positions as shown in Figure 42-1. Yet, in many warehouses, there is enough cube to permit this system to go higher, so the rack can be replaced with a higher one or rack extensions fitted as shown in Figure 42-2.

Once the rack is extended, it also is possible to bridge it over with cross aisles and further increase storage capacity. If this is done,

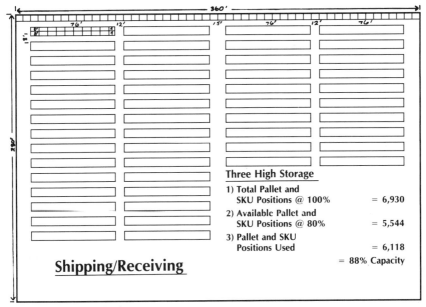

SOURCE: David L. Schaefer.

Figure 42-1

PRACTICAL HANDBOOK OF WAREHOUSING

the storage capacity shown in Figure 42-2 is increased 39 percent, from 6,930 pallet positions to 9,600.

Reducing Number of Aisles

One way to reduce the number of aisles is a rack system designed for two-deep storage. The only disadvantage is the possibility that overcrowding and insufficient volume may cause one item to be blocked behind another. As shown in Figure 42-3, this will increase

Drive-in rack allows high density storage of pallet loads which cannot be high-stacked on themselves because of crushing. Business forms are being stored.

EQUIPMENT: STORAGE AND LIVE STORAGE

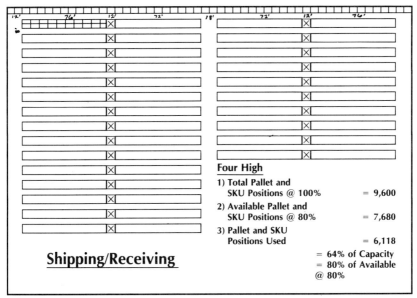

Figure 42-2

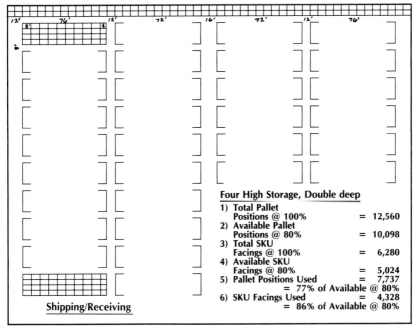

Figure 42-3

the total pallet positions to 12,560, nearly double the amount shown in Figure 42-1. However, because the double-deep rack denies access to the inside pallets, the total number of SKU facings available is less than the plan shown in Figure 42-2. Therefore, the double-deep system would be used only when storage capacities must be increased without increasing the number of SKU facings.[1]

Another means of improving space use is movable storage racks. These are roller-mounted racks which can be shifted sideways to create an aisle whenever needed. This equipment is costly, but it is justified by the space it saves. It may be impractical for a very fast-moving warehouse, because access to some facings must be blocked while the rack is rolled to create an aisle on the opposite facing. Mobile racks can be moved to close and open different aisles. A warehouse which is busy enough to require all aisles to be open at once cannot use the mobile rack.

Other Types of Pallet Rack

Greater storage density can be achieved by using a "drive-in" rack in which each load is supported by a flange that grips the edge of the pallet. While the drive-in rack achieves maximum density, it may do so at a sacrifice in handling efficiency as the driver guides his lift truck through the narrow alley between rows.

Tier-rack is a self-supporting framework that covers the unit load and permits free-standing high stacks supported by the rack structure rather than the merchandise. Tier-rack is frequently used for stacking auto tires or other products having no packaging or structural strength.

As mentioned earlier some storage racks use gravity to move stored loads from the back of the row to the face of the rack, permitting retrieval from the aisle. This is known as live storage.

Stacker-crane installations have been designed to heights of over 100 feet. A stacker-crane can be adapted to either bulk or small-lot

---

1 From "Improving Existing Warehouse Space Utilization," by David L. Schaefer. Published by W.E.R.C., 1981.

EQUIPMENT: STORAGE AND LIVE STORAGE

storage. A computer-controlled crane is particularly useful where random-lot storage or order picking is required. In such a system the computer memory stores each item's location and the device is programmed to pick merchandise with minimum travel.

Live Storage—Gravity Flow Rack

Gravity-flow racks, unlike conventional static shelving, slope from the back (feed-in side) toward the front (picking side). The flow rack will vary in depth (front-to-back) from as little as four or five feet to as much as 10, 12 or even 20 feet.[2]

Gravity-flow racks are made in several different styles for ease-of-flow from back-to-front. The original flow racks were made from fairly conventional shelving with sufficient tilt so that the products fed into the back would slide through to the front. Guide rods were sometimes used.

Subsequently, gravity-flow racks were made with small nylon wheels, then with wheel tracks. The use of wheels enhances the "flow" characteristics of the rack and thus reduces the required angle of declination. Reduced decline is desirable because it allows the elevation of the feed-in side of the rack to be lower.

You can achieve higher order-filling rates using gravity-flow racks because of two factors: (1) If the back-up or reserve cases can be located behind the pick location instead of to the side or on top of it, then the actual pick surface can be greatly reduced; (2) flow rack allows you to consolidate a greater number of pick facings within a limited number of running feet of aisle. Thus, an order filler standing in front of the rack can select a greater number of items without walking further down the aisle, so travel between picks is minimized.

Here is an example of how gravity-flow racks work. Assume a particular product has a case dimension of 12 inches wide by 24 inches deep by 12 inches high. The broken-case (less than full-case) demand for this product ranges from three-and-a-half to four cases. If four

---

[2] From Vol. 17, No. 8 of Warehousing and Physical Distribution Productivity Report, by W.B. "Bud" Semco, president of Semco, Sweet & Mayers, Los Angeles, CA. ©Marketing Publications, Inc., Silver Spring, MD.

cases are put on conventional shelving (no more than 24 inches deep) two lineal feet of shelving and at least 27 inches of shelf height (allowing for working clearance) are needed to accommodate the four cases.

With gravity-flow racks, however, a single facing, 12"×15" high, provides access to the first case. The additional three cases are lined up behind the front case, thus saving about 75 percent of the rack's facing area. If the "lay-back" type of gravity-flow rack is used, reaching into the open case is simplified and less overhead clearance is required.

When to Use Flow Racks

Gravity-flow racks are best for fast-moving products, and when the order-filling requirements allow the order filler to pull from a number of pick positions with little walking. Properly designed, with intelligent item placement and careful item selection, flow racks should increase order-picking speed considerably.

Since order-picking speed depends not only on the number of lines ("picks" or "hits"), but also the space between lines, the speed of order filling is improved by reducing the facing of each line. Thus, all facings should be made with the short side of the carton forward rather than the wide side. It is important to use the left-to-right span of each rack shelf to the utmost, and this requires careful initial arrangement of cartons.

Vertical space between shelves also is important. To waste as little space as possible, cartons of similar height should occupy the same shelf. For ease of picking, it's best to put tall cartons (packages) on the lower two shelves, and shorter packages on the upper shelves.

Order size also is a primary factor. Small orders that require a long walk between lines can reduce the effectiveness of a gravity-flow rack.

What Flow Racks Will Not Do for You

Flow racks are not the answer to all problems. Sometimes flow-rack installations do nothing to increase order-filling speed or save labor cost. On the contrary, they cost the warehouse operator more money and take up more space than conventional static shelving.

What's the difference between an effective flow-rack installation and one that wastes money and space? Typically, flow racks will provide little—if any—benefit unless the orders to be picked have enough lines in them to make the flow-rack technique work. For example, you frequently find ineffective flow racks in security areas where items in the small, confined area simply aren't fast-moving enough to warrant flow-racks. Generally speaking, items picked less than 20 times per month should not be in flow racks.

Two or three sections of flow-rack with one hit every four or five feet—three or four hits for the whole flow-rack system—will be picked at about the same speed as if the merchandise were placed in static shelving. While flow-rack picking *can* enhance order-picking speed, the flow-rack benefits depend on minimizing the space between hits, and having enough lines so the order filler can establish a fast pick-rate.

You must also consider an additional, though small, cost—the difference in shelf-loading time (and labor). Flow-racks take more time to load than shelves.

Any operator who installs a system to hold 100 percent of the items in the warehouse will have a cosmetically beautiful system with marginal utility. The best order-picking system is a hybrid, using flow racks for those items having high activity, and shelving or other systems for those with less activity or uncommon sizes.

Using the Carousel

The carousel saves money in the warehouse as it eliminates human travel, and it eliminates aisles, since the merchandise is brought directly to a picking station.[3] Typically, the carousel is in the shape of an elongated O, and several can be placed side by side with a minimal amount of space between each of them.

---

3   From T.A. Ewers, Kalmar A.C., Columbus, Ohio, first published in Vol. 19, No. 11 *Warehousing and Physical Distribution Productivity Report* ©Marketing Publications, Inc., Silver Spring, MD.

The best use of carousel equipment is when the boxes or products to be picked are quite small, and there are many SKUs (stockkeeping units). Items such as auto parts, plumbing supplies, paper products, nuts and bolts and pharmaceuticals are well suited for this kind of equipment. The carousel is particularly effective for picking of less-than-case quantities.

The carousel is excellent for items which are too small for gravity flow racks, items with a width of less than three inches; or items of unusual length or height.

Stock Layout in a Carousel

Once the decision is made to use carousels, how should merchandise be arranged in them? The most common stock arrangement uses a pre-determined pick list sequence, which thus minimizes the amount of rotation of the carousel itself. However, one of the virtues of such equipment is a control which automatically moves to the next required bin by the shortest distance. Thus, the machine will move either clockwise or counterclockwise, whichever distance is shorter to retrieve the desired item.

For maximum speed, product should be arranged in order of picking volume, with the very fastest movers in the same section of the carousel. Another option is to put the fastest movers in the middle area bins, with slow movers in the highest and lowest bins which require more stooping and reaching from order pickers.

Many carousel systems are linked to a computer, and the computer tracks information to control product arrangement. Most installations involve multiple carousels, and computer controls are particularly valuable for grouping in this situation. When a multiple carousel is used, frequently one operator can work with two carousel systems, since picking from one can be accomplished while the second unit is rotating to position the item for the next pick. One manufacturer offers a minicomputer control system capable of controlling and monitoring an inventory of up to 100,000 stockkeeping units.

Justifying the Carousel

Investment in carousel systems is paid back primarily by reducing travel time for picking. A second but important cost reduction is in storage space. In many cases, picking labor may be reduced by 30 to 40 percent. In one installation, storage space was reduced from 12,000 square feet to 4,000 square feet. The picking effort which had required 24 people on two shifts was accomplished with only nine people on one shift.

The carousel is sometimes selected when the user wants to integrate the picking system with a computer. Computer control is easier with carousels than with flow racks, and expanding the system is, too, because of its changeable shelves and the ability to increase the number of bins.

On the other hand, cost per-SKU in a carousel is higher than in a flow rack, and for fast-moving merchandise the carousel picking method will frequently be slower than a flow rack. If you think you may have to move the equipment to another building in the near future, consider the costs of disassembly and reinstallation, which will be higher with carousel than with flow rack.

While manufacturers and dealers may occasionally create the impression that the two systems are competitive, the experienced user will recognize that they really are not. For certain installations, the carousel is clearly superior, and the same is true of flow racks. As a user, you should be aware of the advantages of both systems, recognizing that there are relatively few times when one system is not clearly better than the other for a given application.

Conveyor Systems

Conveyor systems are employed in manufacturing and are frequently used to handle the interface between a production plant and a plant warehouse. There are seven factors that should be taken into account when selecting a conveyor:

1. Product or material to be handled.
2. Its outside measurements and physical characteristics.
3. How many and what flow rates are involved?

PRACTICAL HANDBOOK OF WAREHOUSING

4. What is the conveyor to accomplish specifically?
5. How large an expenditure is justified?
6. Is the product fragile, or is noise a problem?
7. Are there any other restrictive factors involved, such as space available or atmospheric conditions?

Figure 42-4 shows 18 different kinds of conveyors and ways in which they can be used.

Types of Conveyors

The hand-pushed monorail consists of a single overhead rail that may be a standard I-beam or a rail incorporating a contoured, har-

Conveyor Application Guide Chart

	Transportation	Assembly	Process	Accumulation	Sorting	Warehousing	Shipping Terminal	In Process Storage	Dispatching	Order Picking	Storage Handling	Live Storage	Horizontal Flexibility	Vertical Flexibility
Hand Pushed Monorail		O	O	O			O			O			●	
Powered Travel Monorail	◐		O				O	O					●	
Overhead Conveyor	●	●	●		O		O	◐				◐	●	●
Power And Free Conveyor	◐	●	●	●	◐		●	●	O		●	●	●	●
Overhead Tow Conveyor	●			O		●	●	◐	●		O	●	O	
Tow Conveyor	●	O	O	◐		●	●	◐	●		O	●	●	O
Roller Conveyor	O	O	O			◐	O	O		O	◐	O		
Live Roller Conveyor	◐			O	◐		◐	O	●	◐	◐	O	◐	
Belt Conveyor	●	◐	O		O	◐	◐		O	O	O		◐	
Single Strand Chain On Edge*	●	●	●										◐	
Single Strand Chain On Flat*	●	●	●											O
Two Strand Chain Conveyor	◐	◐	◐										◐	
Fixture Conveyor		●	O											
Slat Conveyor	●	●	O											O
Roller Flight Conveyor	O			●							◐			
Sorter Conveyor				◐	●	◐	●	◐	●				O	O
Storage And Retrieval Machine						◐	O	●	◐	●	●	●	●	●
Driverless Vehicles	●					●	O	O	●		◐		●	

* Handling four wheel dollies

O Fair   ◐ Good   ● Best   Compiled by Jervis B. Webb Co.

SOURCE: Distribution/Warehouse Cost Digest, Vol. 9, No. 19, © Marketing Publications, Inc., Silver Spring, MD.

Figure 42-4

EQUIPMENT: STORAGE AND LIVE STORAGE

dened lower flange on which a trolley runs. The load is carried on one or more trolleys. A similar unit with power can be adapted with dispatching controls, block systems and other automatic features.

The power-and-free conveyor consists of a trolley conveyor from which is hung a second set of rails to carry the free trolleys and the loads. A chain engages and pushes the free trolleys below. Several different types of mechanisms are available that permit the power chain and free trolleys to be coupled together. These allow stop-start control of one load independently of all other trolleys and loads on the conveyor. This important feature makes it possible to change horizontal and vertical direction easily so the free trolleys may be switched into alternate paths, and also allows the use of drop or lift sections. Power-and-free conveyors can be used to sort and provide live and in-process storage.

The dragline conveyor, or tow conveyor, is a power-and-free system turned upside down. The power chain under the floor is engaged by dropping pins mounted on the front of four-wheeled carts through the chain slot. Track switches allow alternate routes or destination spurs.

Skate-wheel conveyors are gravity conveyors of two or more rails with cross-shafts on which skate wheels are mounted. Skate-wheel conveyors can be set up on portable stands with power boosters (belt conveyors) at intervals.

Gravity roller conveyors have tube-type rollers mounted on the fixed-axle shafts instead of skate wheels. Though the roller conveyor is more expensive, it also is more flexible for handling many types of packages. Some are powered and are then called live rollers.

The difference between the belt conveyor and the belt-driven roller conveyor is that the belt runs on top of the rollers. Belt conveyors are widely used in order-assembly operations. Because they provide a continuous moving, flat surface, they are more suitable for order assembly when small or odd-shaped packages are to be handled.

Chain-on-edge conveyors are used for handling four-wheel dollies. This type of conveyor is usually found transporting dollies and loads through manufacturing operations.

The two-strand chain conveyor is vertically flexible and often is used for dipping loads, going up into ovens, or other process operations.

The slat conveyor has steel or wooden slats fixed between the strands generally at every pitch. These are used mainly to handle heavier loads, or in cases where the unit load could cause damage to a belt due to projections on the bottom of the load.

By putting tilting slats on a single or double-strand chain conveyor, it is possible to sort at fast speeds. The tilting slats tilt on command, dumping the load into specific slides, chutes, run-out conveyors, bins or wherever it is required.

The AGVS Alternative

As the price and availability of automatic guided vehicle systems improves, such systems will replace many existing conveyors. AGVS, described in Chapter 43, combine the automation of a conveyor with the flexibility of a mobile vehicle. While the route the vehicle follows is a fixed path, it is a great deal easier to change that path with an AGVS than with a conveyor.

Summary

The equipment just described is capable of both improving cube utilization for storage and in some systems moving that stored material to a more convenient spot. Much of it is justified through savings in both space and labor.

43

FORKLIFTS AND OTHER MOBILE EQUIPMENT[1]

Widespread use of the forklift truck had revolutionized warehousing practices before the middle of the 20th century, enabling warehouse operators to justify erecting sprawling one-floor buildings to replace the more compact multi-story structures used in the past. Many kinds of mobile equipment are used in warehouses today, but most are variants of the common forklift truck.

Choosing a Lift Truck

An almost infinite variety of lift trucks is available, nearly all designed to take full advantage of the cube in a modern warehouse and to handle products in a variety of configurations and packaging. Both electric and internal-combustion engine models of all types are available, with the latter using either propane, diesel fuel, or gasoline. For indoor warehousing uses, propane is considered to be the superior internal combustion fuel because of its relatively clean burning and odor-free qualities.

Another design variable is in the lift. Low-lift trucks are designed to transport palletized loads a few inches off the ground. Other

---

[1] The author is indebted to William C. Portman of Portman Equipment Company for advice in preparation of this chapter. Substantial help was received in the latest revision from Bill Thomas of Advanced Handling Systems and Rick Jahnigen of OKI Systems, Inc. Lee P. Thomas, management consultant, also provided assistance.

trucks have a mast capable of very high lifting, but the high mast may not collapse enough to permit the truck to move through low doorways or enter vans or boxcars. Some masts have a combination of low and high lift characteristics, sometimes referred to as triple- or quadruple-stage masts that extend three or four times, somewhat like a telescope.

SOURCE: Clark Equipment Co.

FORKLIFTS AND OTHER MOBILE EQUIPMENT

In specifying lift trucks, the first step is to examine the buildings in which the equipment will be used. The condition of the floors and the load limits of each floor are of critical importance. Equally important is the overhead clearance minimum and maximum, considering the lowest doorway through which the truck might pass as well as the highest bay in which merchandise may be stacked. Width as well as height of doorways must be measured. If storage rack or other storage equipment exists in these buildings, be sure that the storage equipment is compatible with the mobile equipment.

Certainly the safe speed of a forklift is more important than maximum speed. The safest way to move a loaded lift truck is in reverse because it affords the driver good visibility. Therefore, reverse as well as forward speed capability should be measured.

Before going into the many variations of lift trucks, consider the basic functions which the truck will perform. Figure 43-1 shows the lift truck operational elements which are essential to warehouse operations. Establish your own warehouse priorities.

Figure 43-2 is a checklist showing the features and services to consider before final decision on the specifications of a lift truck.

Fork Lift Truck Elements

- Travel Forward
- Travel Reverse
- Accelerate
- Stop
- Run-In Land Load
- Run-Out Empty
- Run-In Pick-Up Load
- Run-Out Loaded
- Turn Right Forward Stop
- Turn Right Reverse Stop
- Turn Left Forward Stop
- Turn Left Reverse Stop
- Hoist Up Empty
- Hoist Down Loaded
- Hoist Up Loaded
- Hoist Down Empty
- Tilt

Figure 43-1

Lift Truck Attachments

The first powered warehouse trucks were limited to the use of forks to lift loads on pallets. A variety of attachments was then developed for handling special commodities. Many of these are designed to handle cargo without pallets, sometimes by pushing and pulling loads stacked on fiberboard slip sheets. Carton clamps lift a unitized load by applying side pressure from vertical clamp blades. Other attachments are specifically designed for handling major appliances, drums, bales, or rolls of paper or floor covering. Attachments for unitized loads will be further described in Chapter 44.

Order Picking Trucks

Some lift trucks are designed to elevate the picker and permit

---

Fork-lift Checklist

___ Serviceability and maintenance experience
___ Spare parts availability
___ Availability and cost of maintenance contract
___ Quality of dealer service
___ Load rating, empty
___ Load rating, with attachments
___ Travel speed, both forward and reverse
___ Lifting speed, both empty and loaded
___ Mast height, both raised and lowered
___ Overall width of truck
___ Turning radius for right angle stack
___ Overall length with load
___ Maximum height to which load may be elevated
___ Weight of vehicle, both empty and loaded
___ Type of power used
___ Visibility, maneuverability
___ Safety features
___ Stacking stability
___ Availability of desired attachments
___ Operator safety and comfort
___ Operator training support

Figure 43-2

him to pick the order and control the truck from a platform. These are used most safely in a narrow aisle with a guidance mechanism so that the driver need not steer the truck. However, the driver does control forward and backward movement, as well as lifting and lowering the platform to position for comfortable picking heights at each level of the storage racks.

Tow tractors are useful when distances in the pick line are great, or when substantial capacity is needed. A walkie tractor is used when distances are short; this is usually a low-lift pallet jack designed to carry one or two loaded pallets. This machine can be modified to allow the operator to ride on it if distances are greater. Still greater distances may be covered with a sit-down towing tractor or an automatic guided vehicle.

Justifying Narrow-Aisle Equipment

As operators have sought ways to save space in warehouses, lift truck manufacturers have produced variations on the truck which allow it to operate in narrower aisles.[2] The oldest and still most commonly used mobile lift truck found in warehouses is the "sit-down" counterbalanced fork lift truck, referred to here as the common lift truck. Two variations on it are stand-up trucks which gain stability by using two outriggers extending in front of the driver to provide balance in handling a load. Three other variations have a mechanism which allows the load to be turned without turning the entire vehicle.

Each of the variations costs more than the common lift truck. How do you justify paying a premium for equipment? Here are the six types of mobile lifts to consider:

(1) Common lift truck
(2) Single-reach truck
(3) Double-reach truck
(4) Turret truck
(5) The Waldon truck
(6) The Drexel truck.

---

2   From Vol. IV Number 1 of Warehousing Forum ©1988, The Ackerman Co.

(1.) A 3,000-pound capacity *common lift truck* is the most popular type found in warehouses. This truck, which costs about $20,000 with battery and charger, requires a 12-foot aisle to allow a right-angle turn to stack merchandise in rows facing the aisle. It is very versatile. In addition to stacking goods in storage, it is used for unloading from vehicles, hauling loads from staging areas to storage bays, picking orders and loading vehicles. Many operators use this truck for virtually all the work done in the warehouse. It is relatively easy to learn to operate, and because it is so commonly used, it can be readily bought or rented from a wide variety of lift-truck dealers.

(2.) A first step in saving space is the *single-reach truck,* which can be operated in an 8-foot aisle, or 4 feet less than the aisle required for the counterbalanced type. The single-reach truck is priced in the neighborhood of $25,000.00, or about $5,000.00 more than the common lift truck. The truck uses a scissors-reach mechanism which moves the fork carriage forward into the storage pile, and somewhat greater skill is needed for the operator. Because the single-reach truck is commonly used in many grocery distribution warehouses, it is usually easy to find replacements in dealers' rental fleets. However, the truck is less versatile than a common lift truck, because its smaller outrigger front wheels are not designed for crossing dock plates and entering truck bodies. Unlike the common lift truck, the single-reach truck is available only with electric power. The user must therefore have other equipment for loading and unloading trucks and rail cars.

(3.) The *double-reach truck* has a scissors-reach mechanism which extends twice as far as the single-reach truck. This allows a warehouse operator to store two pallets-deep. The two-deep rack configuration allows twice as much product to be retrieved from the same aisle, but it is effective only where nearly every item in storage has a volume of at least two pallets. If this is not so, then the operator must either waste space or block one pallet with another pallet containing a different item. This wastes time, as the rear pallet must be unburied for shipping. The double-reach truck requires an aisle of 8.5 feet or about 6 inches more than the single-reach truck. It costs about

$5,000.00 more than the single reach or about one and one-half times the price of the common lift truck.

(4.) The *turret truck* has a fork carriage which rotates and eliminates the need to turn the entire truck. Turret trucks are available either with a conventional cab or with a "man-aboard" cab which allows the operator to ride up with the load. Turret trucks usually require a guidance system for safe operation in narrow aisles, and with the guidance system, the equipment cost is $65,000-$80,000 per unit, depending on lift height. The design of the turret prevents the use of a tilt mechanism for the mast. Therefore in high stacking (over 22 feet) the truck cannot be effectively operated unless the warehouse has a "super-flat" floor with variances of not more than $1/16$ inch. It is difficult, expensive and sometimes impossible to change conventional floors to "super-flat" floors. The turret operates in an aisle of about 5.5 feet, or 6.5 feet less than the common lift truck.

Two competitive alternatives to the turret truck are produced by Drexel and Waldon. (5.) The *Waldon* truck, like the turret, is equipped with a fork carriage which swivels. (6.) The *Drexel* swivels the entire mast rather than the fork carriage, but it swivels only to the right. Therefore the operator must turn the truck around to select from the opposite side of the aisle. Unlike the turret truck, both the Drexel and Waldon have a tilt mechanism to compensate for imperfections in the floor. Either of these trucks can be acquired for about $60,000. Either can be used without a guidance system, though they are more productive with the guidance system because aisle widths can be minimized and the driver can operate without the need to steer.

With a price range of $20,000 to over $80,000, a buyer must justify paying a premium for narrow-aisle trucks. If you operate a 200,000-square-foot warehouse with 40,000 square feet devoted to work aisles, the use of the very narrow aisle equipment might halve the aisle space needed, or save an additional 20,000 feet. Assuming that you need at least two narrow aisle trucks, this means that you would be spending up to $60,000 extra per-truck (the cost of the guided turret truck versus the common lift truck) or an extra $120,000 to save 20,000 square feet of space. This equals $6.00 per square foot, which is certainly less than the construction cost of additional warehouse space today. On the other hand, consider the fact that the turret truck is usually not used for loading and unloading trucks, which means that you may need extra equipment for this task. Also, the turret may not be as fast to operate as the common lift truck. If the turret is slower, the space saved in the warehouse must be balanced against the extra time used to operate the truck.

An order picking machine runs between angle irons in a narrow aisle.

FORKLIFTS AND OTHER MOBILE EQUIPMENT

Also consider the possibility that a super-flat floor may not be available. If stack heights are over 22 feet, conventional floors may prevent the use of the turret truck, regardless of cost. The Drexel or Waldon, because they operate in a similar aisle width without the need for a super-flat floor, appear to be most easily justified. Unlike the turret, these trucks can be used for the full variety of warehouse tasks.

Strategic Considerations

As you consider purchase of narrow-aisle equipment, you must place a value on alternative costs of space. If the choice is between new construction or increasing the capacity of an existing warehouse, the value of narrow-aisle equipment is greatly increased.

If new construction is required, you will face additional operating costs such as real estate taxes, new financing and increased utility bills.

The faster the product turns, the more important it is to have equipment which will operate quickly and easily with a minimum of product damage.

These variables should be considered:

- Cost of trucks.
- Availability of service for the trucks.
- Aisle width required.
- Wire guidance costs.
- Operating efficiency, speed and maneuverability of the trucks.
- Back-up equipment in case of breakdown or need for augmentation.
- Value of space saved by using narrower aisles.
- Operator training costs.

Automatic Guided Vehicle Systems

Automatic guided vehicle systems (AGVS) have been used in manufacturing plants for decades, but their use in warehouses is much more recent. AGVS have moved into the warehouse thanks to their improved technology and relatively lower cost. Furthermore, when the cost of new real estate is far higher than the cost of some "used" real estate, AGVS are justified by their capability of making an obsolescent building practical.³

There are two kinds of automatic guidance systems: a vertical system which automatically places stock selectors at the right spot to move into high-rack positions; and the more common AGVS, which is a horizontal system with path selection following a wire buried in the warehouse floor. It is the horizontal system that will be described here.

A signal transmitted by the wire buried in the floor activates the electronics in the AGV. Centralized computer systems can program

---

3   Significant contributions to the AGVS material came from Tom Ewers of Kalmar AC Handling Systems, Bob MacEwan of Clark Lift of Columbus, and Russ Gilmore of Autocon in Dayton, Ohio. Portions appeared in Vol. 20, No. 6, Warehousing and Distribution Productivity Report, Marketing Publications, Inc., Silver Spring, MD.

Guided vehicles improve utility of an older building.

the AGV to follow a specific path by deactivating certain wires and activating others. Sometimes the same vehicle can be steered automatically—following the wire path—or manually—off the path. One guided vehicle may carry one unit load, or a whole train of trailers may be attached to an automatically-guided tractor. By using the tractor-trailer system, up to 50,000 pounds of cargo can be horizontally transported behind the guided vehicle.

There are four main types of AGVs:

1. A towing vehicle designed to haul a string of pallet trucks or trailers.

2. A unit-load transporter designed to carry one individual load of up to 12,000 pounds. These vehicles are designed to accept loads delivered from a guided fork truck or from a powered or non-powered conveyor or load stand.

3. An automatic guided pallet truck, similar in appearance to the conventional jack used for order selection in grocery chains. These pallet trucks can handle up to four pallet loads or a total of 6,000 pounds.

4. A light-load transporter, a smaller vehicle designed to handle lighter loads such as parts or mail.

The heart of the AGVS is an onboard microcomputer which controls and monitors vehicle functions, giving each vehicle the ability to travel independently and automatically to a programmed destination.

Typically, the guide path is a wire loop which is embedded in a groove cut into the concrete floor. When the system is installed, the groove, $1/8''$ wide $\times$ $3/8''$ deep, is cut with a diamond concrete saw; the wire is placed in this small trench, the groove is filled with epoxy resin which dries to the same color and hardness as the original concrete. In one facility, a 1,500-2,000-foot path was installed in less than a week. Installation can be done either by the vendor or the user, and does not need to affect the day-to-day operations of the warehouse.

If layout changes occur in the warehouse, the paths can be altered. When a path is changed, the wire from the old path does not have to be removed.

In operating the system, a warehouse worker takes a load from the home area, places it on the guided vehicle, moves the vehicle onto the guide path, keys in the destination address, and pushes a start button. Once this is done, the computer takes over. The vehicle moves to the designated drop location, deposits the load at a specific address, and then returns to the home area for a new assignment.

The operator's control panel typically contains five function keys. Two are the start and stop keys, the third is a keyboard used to enter the description address, a fourth is a switch to move to manual operation instead of automatic, and the last is one which cancels an erroneous entry.

In newer, more sophisticated systems, the AGVs are controlled through a central queuing station in which they are automatically programmed to make certain stops along their paths. At the beginning of the day, the AGV enters the queuing station and is programmed by the central computer for a whole day's jobs. When the jobs are finished, the AGV is programmed to return to the queuing station for its next duties. At any point during the day a warehouse employee can step into the central control station and reprogram an AGV for another task.

The chief advantage of an AGVS to a warehouse is that the guidance system eliminates travel time by warehouse workers. A value can be put on this travel time through the use of generally accepted warehouse standards. In a warehouse standard developed originally by the U.S. Department of Agriculture, a time of 7.7 minutes is needed for each 1,000 feet of travel. If one assumes that the fully burdened cost of a warehouse worker is $25.00 per hour or $.42 per minute, this means that each 1,000-foot trip costs $3.23. If the $35,000 microprocessor is to be paid back in three years, it will have to eliminate 684 miles of travel per year in order to be justified.

Larger warehouse operations with long distances to travel will find it possible to eliminate at least this amount of travel. For example, if there is a 300-foot travel from storage rack to loading dock, the elimination of 27 round trips per day will cut somewhat more than 684 miles of travel per year.

FORKLIFTS AND OTHER MOBILE EQUIPMENT

Another advantage of AGVs is their flexible use in different warehouse layouts. There are buildings available today at bargain prices because their layout makes them of marginal value for conventional warehousing. In some cases, these are older buildings with docks that are not convenient to storage areas. Others may be multi-story buildings that cannot be operated economically with conventional equipment. Through the use of AGVs, possibly in combination with vertical conveyors, inbound loads can be automatically dispatched to a storage address without intervention by people. In such cases, the use of AGVs will allow the economical use of a building which would be quite uneconomical with conventional handling equipment.

In addition, a benefit that is frequently overlooked is the ability of AGVs to save space. Because tracking of the vehicle is so precise, lanes between loads can have as little as four-inch clearance between passing vehicles.

System Integrators

While most forklift trucks are still purchased from franchised dealers, a new kind of vendor has appeared on the scene called a systems integrator. This is a supplier who helps the user bring together the electronic and mechanical technologies needed to reach specified warehousing goals. The systems integrator is goal- and product-oriented, frequently using a mix of equipment to achieve those goals. The integrator can be involved in selling storage equipment, mobile equipment, software, and computer hardware.

What to Look For in Mobile Equipment

The critical element in making a purchase decision for mobile equipment is economics, but a cost study is not limited to the purchase price.

One forklift truck advertisement shows how the cost of fuel, maintenance and operator's salary in six years (10,000 hours) will exceed the purchase price of the truck by more than 1,000 percent. The real sales message is that it is false economy to purchase materials handling equipment on the basis of lowest initial price. Reliability must be measured both in terms of maintenance cost and risk of downtime. As handling systems grow in sophistication, the risk of downtime becomes ever greater. In measuring reliability, warehouse management should use some of the contingency plans described in Chapter 17. The conventional lift truck's greatest "reliability" is that it can be easily replaced by another lift truck. Some conveyors and cranes cannot. What will you do when the equipment breaks down? The time to consider this question is before the equipment is ordered—not after the operation is running at capacity.

In choosing mobile equipment, you must also anticipate future requirements and whether the equipment can be adapted or profitably resold when market conditions change your product or the way in which you handle it.

44

APPROACHING AUTOMATED HANDLING

Planning for warehouse automation is no different from planning for any other kind of capital improvement in the warehouse or in the production plant. Basically, the decision involves a weighing of costs and benefits, as well as considering the risk of failure *versus* the gain if the program is a success.

The first step in planning is to get the best possible estimate of the total cost of an automated installation, and then to compare that cost with the benefits and to consider the risk factors which could threaten the benefits. In measuring costs, consider both before-tax and after-tax cost consequences.

Three common risks should be considered:

- What changes in products handled could make this system obsolete? How many anticipated product changes can be handled by adapting the equipment?
- What are the contingency options in the event of major breakdowns, power failure, work stoppage, fire, earthquake or other disaster?
- What are the potentials for removal and resale of the equipment when it is no longer needed?

Any measurement of benefits should consider both economic and non-economic. In an automated handling system, economic benefits are usually measured in improved handling speed per-case. These are frequently measured in a savings in seconds per-case, and these savings are extended to a total dollars per-year. By comparing the dollars per-year saving with the cost of the automation investment, a pay-back in years for the automation investment is calculated. Remember to consider both the before-tax and after-tax effect.

In some installations, space saved is a primary economic benefit, and the value of space saved is sometimes measured by the avoidance of new construction. For example, partial warehouse mechanization costs $120,000.00 and saves enough space to allow storage which would otherwise require 20,000 square feet of additional space. This means that the cost equals $6.00 per square foot, certainly less than the construction or purchase of warehouse space today.

Some automated systems are justified by non-economic benefits—no need for labor in a severe labor shortage, an improvement in shipping accuracy and avoidance of errors, or a reduction in warehouse damage.

When considering the cost/benefit relationships, it is well to look at decisions reached by others, particularly in the early days of developing automated warehouse systems. One of the first such systems in the U.S. was installed at the finished goods warehouse of a manufacturer of frozen bakery items. The ultimate justification for the equipment was not based on cost savings, but rather on human factors. Because of the hostile environment of the freezer, the manufacturer had difficulty in recruiting workers. The cost of mechanizing the function was justified because it allowed the manufacturer to get a job done that workers did not want to perform. Other early applications of automation in warehousing have taken place in situations where expansion of a conventional warehouse was impossible. Faced with the alternative of enormous costs for a conventional structure, some warehouse operators may choose a high-rise automated system because the inflated cost of the alternative makes the high-rise attractive. In other cases, tax policy has been the benefit prompting the investment to be made. The most common example is the rack-supported building. Depreciation regulations for taxation in the U.S. sometimes allow the useful life of most of the building to be calculated as the relatively short life of the storage rack.

Without question, some early decisions for automation were

motivated primarily by desire for prestige, somewhat like the early computer installations. Many automated installations contain observation balconies for plant visitors, a clear sign that management considers the installation to be a showcase as well as a warehousing tool.

Strategic Planning

A major investment in automated warehousing should be coordinated with the long-range strategic plan for your company. Here are some of the strategy questions to be considered:

- How will the inauguration of automated warehousing affect marketing within the company?
- Will the installation provide the company with a significant competitive advantage?
- Will the system be compatible with other developments in the company over a five- to ten-year planning period (number of years depends upon equipment payback)?
- What are the potentials for expanding such systems in other places in the company?

The strategic plan must consider the shelf life of the products to be warehoused in the automated system. If corporate planning involves significant change in the size and shape of products to be moved through the system, then you must ensure that the new system has the ability to adapt to this change.

The most important strategic consideration is to ensure that you are automating the right things in the right way. In the past, some materials handling automation has been both costly and unnecessary. In some cases, the energy should have been spent in *eliminating* the need for materials handling, not *automating* it. Therefore, as you consider the strategic questions, look at whether everything is being done in the simplest way. Simplifying before automating can save significant investment dollars.

Ergonomics

One approach to materials handling automation is simply to arrange work so that it requires less effort from people. Ergonomics in warehousing calls for arranging work stations so people can do their jobs comfortably. Sometimes this takes only minor capital investments. For example, it is easier to handle cases when they come chest-high than to reach down to the floor or overhead to grasp them. By using a forklift adjusted the right height, it is possible to keep a pallet at a height which allows it to be loaded more comfortably. As a result, the loading is done significantly faster and with less effort and fatigue. Opportunities to use ergonomics exist all over the warehouse, and this factor should be carefully considered as automation is planned.

Robotics

Robot handling equipment is now used in some warehouses to stack cases on pallets. While palletizing machines have long been available, many of them are relatively inflexible, requiring extensive setup time if the size and shape of the package to be palletized changes. The robot palletizer is slower but more flexible than a palletizing machine. It can handle odd-shaped packages and can adapt

Ergonomics makes warehouse labor easier.

to different sizes of packages. For an installation with a reasonable variety of cases to be handled, the robot palletizer is a good alternate to a palletizing machine. Where palletizing must be done in a hostile environment, such as a freezer, the machine may be a good alternate to hand labor.

Conventional Warehouse System Vs. High Rise

An ASRS or automated storage and retrieval system handles either pallet loads, or other loads in pans or trays, on a unit-in, unit-out basis.[1] The ASRS is frequently installed in a high-rise rack-supported building where the storage racks provide the structure for the building walls and roof, and a skin of metal is attached to the rack framework. The ASRS cranes may also be installed in a conventional building. The rack-supported structure may also be coupled with either conventional forklift handling or narrow-aisle high-reach lift trucks.

The trade-off factors to be considered are the following:

1. Building cost.

—a. Land acquisition and land preparation

—b. Floor

—c. Structure

—d. Insulation, heat/cooling

—e. Fire protection

2. Storage equipment

3. Materials handling equipment

4. Back-up systems for materials handling equipment and maintenance of that equipment.

5. Flexibility to take current package and pallet-size variety and for future changes

6. Security, damage and lot number control

7. Computer hardware/software requirements.

---

[1] This and the next five subsections are by Lee P. Thomas. They appeared in Vol. 2, No. 10 of Warehousing Forum, ©The Ackerman Co.

The Building

Land costs favor the ASRS because its cube content is achieved through additional height on a smaller footprint. A smaller parcel of land should be required and fewer yards of earth will need to be moved to prepare the site. These factors may be especially critical when it is necessary to construct the maximum storage capacity on an existing site. A land locked existing site or excessively priced adjacent land will favor the high rise building. The ASRS system can be designed to over ninety feet height. If maximum storage on minimum land is the highest priority it favors the rack supported building and ASRS, as does lower land preparation cost.

Floor cost for a new structure would be a stand off as far as cost per square foot of floor is concerned unless the conventional building being considered is under twenty-six feet. The higher working level for storage in either an ASRS or a high rise conventional building will dictate a high class super level floor.

Narrow aisle conventional lift trucks working above the twenty-six foot level require an increasingly level floor and the storage rack they serve also need to be leveled. An ASRS crane system also needs a leveled floor and its racks must be level. Shimming the crane rail to a level orientation and also the racks is a possible alternative to a super level floor for an ASRS being installed in an existing building. Super floor leveling is not a concern in a conventional operation under 26 feet height.

Structure of the building is a unique factor when a conventional building is being compared to a rack supported building. The comparison must have careful thought.

- The cost of a rack supported structure will be significantly more than a conventional building.
- A rack supported building provides all the storage rack for the interior of an ASRS facility. In a conventional building the storage racks are an additional expense to be considered. However, rack layout in a conventional building can be revised and a rack supported building does not offer this advantage.
- Tax policy favors the racks in a rack supported building, since

rack is considered a part of a total machine with the useful life of mechanical equipment rather than building. Ten year write-off vs. thirty-five year write-off is a significant advantage for some firms when the support structure of the building can be written off fast. Professional tax advice should be sought to evaluate your own particular solution.
- Rack equipment from an existing facility is not usable in a rack supported building, though existing racks may be transferred to a conventional new building.

Insulation, heating and/or cooling favors the high rise structure because of the chimney effect of a tall building with a small footprint and a smaller roof surface. In heating or cooling, the roof area is a major source of heat loss or gain. Also, the smaller floor surface in an ASRS will be less costly to insulate for a frozen storage warehouse.

Fire protection installation cost may also favor the ASRS in that a fire wall is not a likely requirement for a small footprint high rise building. Sprinkler installation cost on a per cubic foot basis should be relatively equal. In a conventional facility, intermittent sprinklers built into storage racks will make rack relocation more expensive. In the high rise, rack cannot be moved in any case.

Storage Equipment

Pallet racks, drive-in racks, flow racks and pallets are examples of the storage equipment usually required to make a conventional building-shell function as an efficient warehouse. The rack-supported building requires no pallet racks in addition to its structure. It requires some additional flow-racks for less-than-pallet quantity picking and this type of facility is very likely to require dedicated or "slave" pallets or containers to complete the ASRS system. To transfer pallets or racks from an existing facility to a new rack-supported ASRS facility is much less feasible than transferring them from one conventional operation to another.

Rack tear-down and set-up cost and transportation cost should be considered in total-cost calculations when existing equipment is to be used.

Materials Handling Equipment

An ASRS crane unit serving only one aisle typically costs the equivalent of four to six counterbalanced lift trucks or two turret lift trucks. If more than one aisle is to be served by one crane, a transfer car may be used to move from aisle to aisle. The transfer car will have an additional cost equal to two or more conventional forklifts. One crane unit typically may handle 25 units in or 25 out per-hour, or forty units per-hour if loaded both inbound and outbound, 20-in 20-out. Comparing the crane's handling capacity per-hour to the handling capacity of its cost equivalent in forklifts is one point in comparing systems. Conversely, it should also be noted that each forklift must have an operator while a crane may be computer-controlled; the labor consideration is clearly significant.

Back-up Systems and Maintenance

One crane serving one or more aisles has no back-up system.

Handling capacity per equivalent handling cost

One crane's cost equals four fork lifts' cost. One transfer car for a crane equals two fork lifts' cost.

Below is a graph of handling capacity per equivalent handling cost based on:

Crane unit = 40 units in & out per hour
Fork lift = 15 units in & out per hour

Units per hour (in & out loaded both ways)

Crane unit: 40 units per hour

Crane and transfer car: 40 units per hour

Four F/L = crane cost: 60 units per hour

Six F/L = crane & car: 90 units per hour

Figure 44-1

When the crane is down for mechanical, electronic or power failure, the ability to store and retrieve is suspended. However, with forklifts there are usually additional trucks in the building and an alternative is available in case of breakdown. Furthermore, short-term rentals are usually available on short notice.

An ASRS crane system is likely to be tied closely to a computer inventory-control system and therefore may lose working time if the computer goes down. Forklift handling systems are often less dependent on the computer.

Maintenance cost for a forklift will be $3.00 to $4.00 per running-hour, compared to crane maintenance cost of under $1.00 per operating-hour. Cranes typically are less prone to breakdown than lift trucks. Forklift maintenance costs and downtime in zero-degree temperatures increase and battery cycles shorten, which is not the case with an ASRS.

Also to be considered, at least one lift truck must be used with a crane unit to bring unit loads from truck or rail car to the crane and/or to load outbound freight. A crane alone is not a complete handling system; it requires a lift truck with it.

Security

The ASRS has great advantages in warehouses where human failure can cause a problem. An unmanned ASRS will not cause damage, will not pilfer inventory and will seldom pull the wrong stock number. The ASRS warehouse storing valuable merchandise or drugs may need only minimum perimeter security. A burglar could neither find nor retrieve a drug or high-value product. During operating hours, no one can safely walk in the crane operating aisles so internal pilferage exposure is reduced in an ASRS facility.

Computer Requirements

While most ASRS crane units are controlled by computer, this is not an absolute necessity.

Some Japanese warehouse crane units are run without computerization. These units are low-cost because they are slower, with reduced horsepower. They are cost justified as a replacement for a forklift and are applied in systems as small as 400 total pallets. There appears to be a market niche for these lower-cost machines.

An ASRS crane unit is an expensive but an efficient handler of full units. Picking in less-than-full pallet quantity however, is not possible with the unmanned ASRS. For this activity in any volume, a separate flow rack or pick line is required within the facility.

An unmanned ASRS will cost an amount equal to four to six standard forklifts, but each of those forklifts would require an operator with an annual cost around $25,000 in wages and benefits.

Though a high-rise rack-supported building will cost significantly more than a conventional warehouse structure, it will fit on a smaller piece of land and requires less land preparation. Heating, refrigerating and fire-protection costs for the high-rise building will usually be less. Finally, the rack-supported building includes the cost of the storage racks that are an additional equipment cost for conventional construction. Because the rack is part of a machine, a tax advantage of some significance may be available to you.

Of course each firm's situation is different. However, with the foregoing information in mind, you may proceed to gather your own costs to reach an orderly choice between conventional warehouse or ASRS.

Coping with Change

Perhaps the toughest question to be answered in evaluating mechanized systems is whether a particular system can cope with anticipated change. Shocking examples of failure to do so are found in many warehouses. One highly-mechanized system for picking and shipping drug and pharmaceutical products was installed in an older building originally designed for tire storage. Dock doors in the building were rebuilt to fit the loading of low-bed retail delivery trucks. Conveying systems and aisle layouts were especially designed for the drug operation. After a relatively short period of use, the wholesale drug company had labor and financial problems that caused it to go out of business. At this point, the automated equipment was worth more than the value of the building. Furthermore, the high cost of removing this equipment made relocation an uneconomical alternative. Neither the dock doors nor the storage layout were adaptable to general warehousing purposes. As a result, a high percentage of this investment in automation was lost.

What might have been done differently? In this instance, specialized equipment was used when more standardized systems were available. Systems with wide application in grocery distribution were rejected in favor of equipment especially designed for drug products. No consideration for future relocation of the equipment was made, making it impossible to move the system economically.

The Future of Automated Systems

Investment in an automated system is a decision that balances the return on investment against the risks involved. And frequently, the savings from an automated system will not arise from materials handling operations within the warehouse, but rather from improved control over materials handling operations, which brings a faster reaction to increased service levels and sales. Automated systems probably will not be adopted by any company that does not forecast long-term market growth. One of the major risks—technical unreliability—has been considerably reduced as automated systems have matured.

The circumstances in which the risks are minimal and the benefits at their greatest can be summarized as follows:

1. An expanding market.

2. An expanding market serviced from specific sites as a result of rationalization.

3. The need to improve and maintain competitive service, to retain or increase market share.

4. High retained profits.

5. When successful experience can be transferred, with considerably reduced risks, to other company plants—suggesting a large multiple-plant organization as a prerequisite.

6. Where logistics costs are identifiably high—again favoring

multi-location component manufacturing and/or high-throughput assembly from a large range of component parts.

7. A top management and financial management structure that understands and accepts the risks of technical innovation and has the ability to motivate its management and staff to welcome innovation. This is difficult to achieve unless it is against a background of expanding sales to guarantee job security.

8. An understanding that adopting advanced technology is in many cases a function of the market share achieved. That market share is a function of price and service, which is a function of the technology of the production and distribution system. Less departmentalized decision-making and more integration are required if the true potential of technology is to be achieved.[2]

Conclusions

The enormous risk and desire for prestige that marked early applications of warehouse automation have been tempered by improving reliability, lower costs and greater acceptance. Yet, automated systems, particularly those involving high-rise installations, are most popular in parts of the world where the price of land is extremely inflated or where conditions limit the ability to expand the warehouse laterally. A decision to invest in automation involves strategic planning considerations, and automation should not be implemented without extensive planning. There also are steps toward partial automation or mechanization that should be considered before a plan to automate is completed. The payback can be calculated on the basis of individual items. A likely result of such analysis is to automate part of the warehouse, while leaving the rest of it on a manual system. Such separation should be done on the basis of speed of movement, whether or not items require repack, and segregation of items that do not fit in the system. Automated systems are most likely to be adopted by relatively large organizations that have an expanding market.

---

2 From "Automated Storage and Retrieval Systems," an article in "Logistics Today" by John Williams, director of the National Materials Handling Centre (National Materials Handling Centre, Cranfield, England, September 1982).

45

UNIT LOADS

There are three basic materials handling principles that underlie the concept of unitization. They are:

1. For maximum economy, handle units as large as can be practically lifted and moved.

2. Materials should be handled as few times as possible.

3. Mechanical equipment should replace manual labor wherever possible.

Unitization is governed by these four variables affecting the movement of goods within a warehouse:

- Movement by a fixed path (conveyors, towlines, wire guided paths),
- Movement by a variable path (forklift truck).
- Distance moved.
- Frequency of movement.

A materials handling operation that involves movement of small quantities of light-weight goods over short distances, or the movement of packages of limited size from fixed origin to fixed destination, may well be conducted more cheaply by conveyor rather than by being unitized. The value of unitization is enhanced in direct proportion to the increases in distances, in quantity of goods, in frequency of movement and in variability of paths.[1]

---

[1] Most of this chapter was written by Bob Promisel, Management Consultant. It first appeared as Vol. 18, No. 1, Warehousing and Physical Distribution Productivity Report, 1983 Marketing Publications, Inc., Silver Spring, MD.

Pallets

As the forklift truck became commonly used during World War II, there was a need to unitize freight for efficient handling with the forklift. Stacking cases on wooden pallets was the first unitizing method developed.

Wood was chosen as the commodity with which to build the movable platform called a pallet primarily because wood is one of the lowest-cost commodities available in most industrialized countries. In the early years of lift trucks and pallets there was little attempt at standardization. One of the first countries to develop a standard pallet was Australia.

In the U.S., efforts to standardize were made under the auspices of an organization known as the American National Standards Institute (ANSI). The first national standard for pallets was published in 1959. It later was revised and then subdivided into the following three standards:

1. MH1.1.2-1978 Pallet Definitions and Terminology.
2. MH1.4-1977 Procedures for Testing Pallets.

Imperial (inches)	Hard Metric (mm)	Hard Metric (inches)	Soft Metric (mm)
24 × 32	600 × 800	23.64 × 31.52	609 × 812
32 × 40	800 × 1000	31.52 × 39.40	812 × 1016
32 × 48	800 × 1200	31.52 × 47.28	812 × 1219
36 × 42	900 × 1060	35.46 × 41.75	914 × 1066
36 × 48	1060 × 1200	35.46 × 47.28	1066 × 1219
40 × 48	1000 × 1200	39.40 × 47.28	1016 × 1219
42 × 54	1060 × 1370	41.75 × 53.96	1066 × 1371
48 × 60	1200 × 1500	47.28 × 59.10	1219 × 1523
48 × 72	1200 × 1800	47.28 × 70.90	1219 × 1828
36 × 36	900 × 900	35.46 × 35.46	914 × 914
42 × 42	1060 × 1060	41.75 × 41.75	1066 × 1066
48 × 48	1200 × 1200	47.28 × 47.28	1219 × 1219

Figure 45-1

3. MH1.2.2-1975 Pallet Sizes.

The size definitions shown in the standard are listed in Figure 45-1.

Parts

The following parts are commonly used to make a pallet:

Deck boards—Boards forming the top and bottom decks of the pallet.

Stringers or Runners—The wooden runners to which deck boards are fastened. These serve as a spacer between the top and bottom decks to allow the forklift forks to enter.

Blocks—Square or rectangular wooden parts used to replace stringers in some four-way pallets.

When defining pallet sizes, dimensions are usually stated in inches, with the length designated before the width. The width is the dimension parallel to the deck boards.

Types of Pallets

As classified by use, there are three general groups of pallets: expendable, general purpose and special purpose. The expendable pallet is designed to be used just once, though some of them can be used for more than one trip if handled with great care. The prime requisite for expendable pallets is low cost, with the ideal expendable pallet being the cheapest device available that will allow a forklift truck to pick up and deposit a unit load.

A general-purpose pallet is one designed for repeated use with a variety of cargoes. The most popular general-purpose pallet used in the U.S. today is the 48" x 40" size (1.2 meters x 1 meter) originally specified by the Grocery Manufacturers of America (GMA).

Special-purpose pallets are designed for a specific product, with their design based on an agreement between pallet manufacturer and user.

The two most common designs for pallets in use today are two-way entry and four-way entry. The two-way pallets allow forks to enter only from two opposite sides. The four-way pallet allows entry on all four sides, either through two notches cut in each of the stringers (or runners) or by using blocks as a base for the pallet deck.

"Single-face" and "double-face" describe the pallet's deck. The single-face pallet, the most common, has only a top deck, the top surface. The double-face pallet has two decks and can be reversed. A full deck on both top and bottom of the pallet provides greater load stability for high stacking.

The most commonly-used construction is the flush-stringer pallet, in which the outside stringers or blocks are flush with the ends of the deck boards.

Wing pallets divide into single-wing and double-wing types. On a single-wing pallet, the top deck overhangs the stringers, while the bottom deck boards are flush with the stringers. In a double-wing pallet, both the top and bottom decks overhang the outside stringers. Wing pallets are usually designed for use with narrow-aisle lift trucks or equipment.

For the convenience of pallet users and manufacturers, six categories of pallets have been designated:

—Type 1 - Single-face, non-reversible pallet.

—Type 2 - Double-face, flush stringer or block, non-reversible pallet.

—Type 3 - Double-face, flush-stringer or block, reversible pallet.

—Type 4 - Double-face, single-wing, non-reversible pallet.

—Type 5 - Single-face, single-wing, non-reversible pallet.

—Type 6 - Double-face, double-wing, reversible pallet.

Types of Wood Employed

The best pallets are constructed of hardwood, but softer woods can be used. These include aspen, basswood, buckeye, cottonwood and willow. There are nine heavy hardwoods that are occasionally used: beech, birch, hackberry, hard maple, hickory, oak, pecan, rock elm and white ash. Pallets made with this type of wood have the greatest strength, but they are difficult to drive nails into and difficult to dry.

UNIT LOADS

A variety of fasteners is used to hold the pallet together. The most common type of fastener is the threaded nail, but auto nails, staples and nuts and bolts are sometimes used.

Performance Specifications

When pallets are designed for use with hand-pallet jacks, the space between the boards on the bottom deck must be specified by the user so it will match the specifications of the pallet jack.

Testing of a pallet usually involves its stiffness or deflection of the deck boards or other parts of the pallet. The American Society for Testing and Materials has published its ASTM Standard D1085-73 for testing pallet strength.

Box Pallets or Pallet Containers

Some pallets are more than a platform; they are in fact a complete container. In agriculture, a pallet container is typically used to transport fruits and vegetables from the field to the cannery or processing station. Sometimes the box pallet is used to move washed fruit directly to the supermarket. Pallet containers may also be used to transport small parts or units from storage directly to an assembly line.

There are three commonly-used types of pallet containers: the pallet bin, with wooden sides but no top; the pallet box, with a top, and the pallet crate, with open or slotted sides and ends, and no top. Pallet containers may be either expendable or non-expendable, collapsible or non-collapsible. Some are quite specialized, such as the "Gaylord" box used for bulk storage and handling of raw plastics.

Some wood pallets are specifically designed for use in automatic storage and retrieval systems. These are called "slave" pallets and they must be manufactured to very close tolerances in order to function in an automated system—at sometimes double the cost of an ordinary pallet.

Slip Sheets

A predecessor to the present slip sheet was a fiberboard pallet developed by the Mead Corp. Originally intended for heavy bags of low-cost materials, such as titanium dioxide, the early Mead pallets were designed to be lifted either with a special slip-sheet handling device or with sharp chisel forks or tines. The development of slip-sheet handling equipment came with the slip sheets themselves, with Clark Equipment pioneering this with the "Pul-Pac."

Slip sheets are usually made of solid fiberboard, corrugated board or plastic in a variety of thicknesses ranging from .035 inch to .15 inch.

Performance Specifications

Slip-sheets must have adequate tensile strength, tear resistance, bursting or puncture resistance, stiffness and moisture resistance. They also are graded according to moisture content and the quality of the scoring of the one or more tabs used by the gripper bar of the push/pull attachment. If a fiberboard slip sheet lacks sufficient moisture, for example, it will tear or have premature tab failure. The scoring of the tab is an important part of the product, since the objective of a quality slip sheet is to permit a 90-degree fold with a minimum disturbance to the fibers holding the sheet together.

Stiffness of the sheet is of critical importance for the storage of bagged or odd-shaped materials. Some slip sheets are used in wet conditions, as with produce or frozen foods, so the ability of the sheet to resist moisture is important.

The following factors should be considered in determining slip-sheet specifications:

—1. The number of times the unitized load is to be handled by push-pull equipment, i.e., five pulls, etc.

—2. Unit-load weight, composition and stability.

—3. Moisture exposure.

—4. Experience of lift truck operators.

—5. Requirements of receivers.

—6. Case size, product characteristics and packaging.

—7. Material handling equipment used.

UNIT LOADS

Building and Handling Unit Loads

An interlocking pattern is typically used when cases are unitized on a pallet or slip sheet. The pattern should provide the greatest amount of interlock for load stability, and the edges of cases should always be even with the edges of the pallet or slip sheet, leaving any necessary voids in the center of the unit load. For slip sheets, edges of cases should be ¼" to ½" away from the score line.

Many warehouse users keep a log of unit-load patterns, sometimes referred to as "tie and high" instructions, to be certain that all receivers unitize loads in the same way.

Adhesives and other stabilizers are sometimes used to hold the unit load together. The best pallet adhesive is one with high shear strength and low tensile strength. This allows cases to be removed by lifting them straight up, but does not allow the cases to slide sideways. Another stabilizing method is to use string or tape around the top layer of cases.

Plastic wrap in the form of either stretch-wrap or shrink-wrap is increasingly popular as a means of stabilizing loads. The shrink-wrap is applied as a loose bag around the load, which is then sent through a heat tunnel, shrinking the plastic around the load.

Stretch-wrap is usually applied either by rotating the load on a carousel or putting the wrapper on a roller and walking it around the unit load. Stretch-wrap has grown in popularity because it does not require the heat tunnel needed for shrink-wrap applications.

Loading Unit Loads on Trailer or Rail Cars

Avoid decking unit loads inside a trailer or rail car. Where possible, deck the loads on the dock and then move them into the rail car or truck—this method is both safer and faster.

Because vertical alignment inside trailers is often difficult, it is desirable to use a side-shifting mechanism on a lift truck. The nose of some trailers is rounded, so that alignment of the loads in the front of the trailer is most difficult.

Some loads consist of a mixture of unit loads and loose cases.

When this happens, the unit loads should be in the center of the rail car or trailer, with side voids filled in by loose cases.

Because substantial dollar investments are at stake, it's important that you analyze carefully the relative merits of pallets and slip sheets. Increasing numbers of warehouse users are adopting both techniques.

Equipment to Handle Unit Loads

The wood pallet can be handled with an ordinary forklift truck or pallet jack. More complex attachments are sometimes needed for the other unit-handling methods. The original Pul-Pac slip-sheet handling device has been further refined to handle greater volumes. This includes the use of a load-transfer device to provide low-cost transfer of merchandise from a slip sheet to a pallet, or the reverse.

The carton-clamp truck is used as a means of lifting a unit load without either a pallet or a slip sheet. In the early years, when this technique was first introduced, extensive damage was done by users who did not understand the proper pressure settings for carton clamps. More recent experience has been more successful.

A Case Example of Slip-sheet Handling

Norman E. Kuhl, Packaging Division Vice President of the Adolph Coors Company has described the conditions which caused his company to adopt a slip-sheet handling system in 1970.[2]

The extensive pallet repair required was one major concern. Another was the fact that the wooden pallet didn't sit consistently level in shipping. Coors was just beginning to convert to the aluminum can, and there were considerable problems with nails forcing their way up through the cans. When the pallet wasn't perfectly level, rocking occurred during transit and this caused the cans on the bottom layers to

---

2   This section from Norman E. Kuhl, Packaging Division Vice President, Adolph Coors Company. It first appeared as Vol. 19, No. 11, Warehousing and Physical Distribution Productivity Report, 1984 Marketing Publications Incorporated, Silver Spring, MD.

UNIT LOADS

be crushed. Sometimes splintered deck boards on the pallets caused dents and damage to cans.

It was difficult to predict the pallet-pool size on a given day. It seemed that either there were 50 thousand pallets stored in the warehouse, or supply was down to the last few. The entire operation hinged around being able to take beer from a line, put it on a pallet and load that pallet of beer. Running out of pallets was a production tragedy.

Coors decided to test the slip-sheet system with an actual shipment of its canned beer.

Since the fork trucks and the slip-sheet attachments required were rather special, Coors rented some fork trucks and some slip-sheet attachments. Two units were put in the field at two different distributor locations and one unit was kept at the brewery. In the course of the next three months 100 car loads of beer were shipped to distributors on a variety of slip sheets.

Coors had a problem. Shipping cold beer in insulated boxcars caused humidity. The beer cans would sweat and the paper slip-sheet would become moisture-laden. In every case the paper slip sheets worked alright at the brewery, but upon arrival at the distributorship they were no longer strong enough—or in many cases not even in one piece—and the distributors were back to unloading beer by hand.

A polypropylene plastic slip sheet was developed, patented and substituted for paper slip sheets. The plastic slip sheet weighs about five pounds and costs substantially more than the paper slip-sheet did. The high purchase-cost made reuse seem attractive, so the least-damaged sheets were washed and repressed; the remainder were shredded for remanufacture.

Re-evaluating the total program indicated self-manufacture of the slip-sheet was the most economic course. Self-manufacture was calculated to be one-fifth the cost of purchasing from vendors. It also allowed Coors to produce a one-trip sheet, which eliminated the sorting, washing and pressing costs associated with multi-trip sheets.

An effort was made to try to justify the program economically. The pallets cost between eight and ten dollars; it cost about $2.50 to repair each pallet that was returned (about one third of them). A slip sheet, however, cost (at the most) 60 cents a sheet. With some continued development, Coors was subsequently able to get two or three trips out of that sheet. Considerable savings could be made just on slip sheet versus pallet costs alone.

An even bigger savings was available in the freight costs to retail distributors. At that time, 80 percent of the beer was being shipped by rail. The maximum freight rate was established at 90,000 pounds. Regardless of how much more was put in the car, the cost per-case was still the same. If pallets were eliminated in the boxcars, an additional number of cases of beer could fit in the car.

Maximum freight rates have been established at the 130,000 pound- and 150,000 pound-level. This brings considerable savings to the railroad because they can supply fewer boxcars per million barrels of beer shipped out of the brewery, and Coors' distributors pay less money per-case in receiving their beer by rail.

The same savings, which approached a penny a case, held true for distributors receiving beer by truck. The pallets weighed about 75 pounds and with 20 pallets per-trailer, there is a saving of 1,500 pounds in pallets. Using slipsheets allowed another 1,500 pounds of beer per trailer.

The Potential Is Great

A standard materials handling text lists 27 advantages of unit loads, as opposed to only 11 disadvantages.[3] Indeed, the potential for cost savings through a unit-load program is great. The opportunities for reduced costs and improved control are such that every warehouse manager should reexamine the unit load subject carefully.

---

3   Apple, James M., *Materials Handling Systems Design*. (New York:John Wiley & Sons.)

46

DEALING WITH DAMAGE

Losses from accidental damage, casualty or pilferage are not only costly to industry, but can easily wipe out the productivity gains made elsewhere in the company. Loss and damage in the distribution system is a major source of waste in American industry today. And there is no indication that this waste is declining. The following national statistics should stimulate your interest in loss control:

1. "... As much as 3 percent to 5 percent of all products handled are damaged";

2. "... The costs of property damage range from 5 to 50 times the cost of personal injuries";

3. "... Product theft and pilferage costs have been estimated to exceed $160 billion per year";

4. "... Vandalism and arson lead to losses in excess of $600 million per year";

5. "... More than 40 percent of businesses having fires do not resume operation and of those that do, 60 percent go out of business within one year of reopening."[1]

If loss control programs are to be effective, formal audits must be performed regularly to see if current procedures and operations are effective in limiting losses. Banks perform such audits with a similar objective. An audit team should be appointed and given the responsibility of developing a hardnosed audit program.

---

1   J.A. Tomkins and Prof. John A. White, *Facilities Planning,* John Wiley & Sons, Inc., 1980.

Causes of Damage in Physical Distribution

Product damage in distribution usually results from three sources: mishandling in transportation, failure of packaging, or damage done in the warehouse. Even in the best of warehouses there will be some damage.

The packaging engineer usually develops a compromise between the lowest-cost carton and one which is so strong as to be completely damage-proof. *(See Chapter 7)*. If the compromise moves too far toward low price, the resulting package will not protect against the normal amount of wear and tear found in all physical distribution systems. The warehouse operator has a unique opportunity to report on the results of the packaging engineer's work. When damage is discovered, the warehouseman should aid in discovering the cause of the loss.

Causes of External Damage

The warehouse operator ultimately has two concerns: controlling the damage caused within the facility, and identifying and controlling that which occurs outside. Warehouse workers must understand that external damage not identified at the receiving dock will subsequently be charged to the operators of the warehouse. In some warehouse operations, all damage found on receipt is rejected to the delivering carrier, but in most the warehouse operator takes damaged goods in for repacking or repair.

Since most external damage occurs in transportation, we list six typical causes of transportation damage:

1. Faulty loading practices.
2. Abrasion from shifting within the vehicle.
3. Shock damage from bouncing or sudden starts and stops in transit.
4. Leaks in the cargo van or boxcar.
5. Transfer handling damage.
6. Improper blocking and bracing in a rail car.

Disposal of Carrier Damage

When goods are damaged by common carrier, the warehouseman's damage-control procedure might include these seven steps.

1. Segregate all products in which damage is evident or suspected, including both damaged merchandise and damaged packaging. On those inbound loads where a carrier representative is not present (rail cars or unattended trucks), take photographs of the damage while the freight is still in place on the carrier vehicle.

2. The carrier's bill of lading should be noted with the quantity and nature of suspected damage. If a carrier representative is available, he initials this notation as his acknowledgement of it.

3. Warehouse receiving tallies should contain similar notations.

4. In a public warehouse, the owner of the goods should be notified.

5. After tagging for identification, each piece of damaged merchandise should be moved to a segregated area.

6. After inspection, merchandise that has suffered damage to packaging only should be recartoned and returned to stock. In some cases minor repairs to merchandise can be made at the warehouse; more severely damaged merchandise should be shipped to a repair center or scrapped.

7. Finally, the total cost of dealing with the damage should be calculated and reported. This cost includes handling, storing and processing damaged merchandise. The report used is an OS&D (Over, Short and Damage) report.

Concealed Damage

Certainly the most vexing problem in controlling damage in physical distribution is concealed damage. No one learns that the product is damaged until it gets to destination and the packaging is removed. Frequently, this is after the product has moved through all the distribution channels and is in the hands of the ultimate consumer.

The damage is concealed because the outside of the container

ADDRESS ____________________

OVER, SHORT AND DAMAGE REPORT

ACCOUNT: Nonesuch Company	CARRIER: Manyties RR
	CAR #: MT 2468
ORIGIN: Blissfield, Minn.	DATE UNLOADED: 5/20/84
SEALS: K-13467 & 8	B/L #: Z-15379
ROUTE: MTRR	F/B # OR W/B #:

NO. UNITS BILLED	NO. UNITS RECEIVED	DESCRIPTION	NO. OF UNITS: OVER	SHORT	DAMAGED	RESULTS OF RECONDITIONING: GOOD STOCK	REJECTED	LOST
2000	1995	Wigit, NOIBN		5				
					14		14	

CAUSE OF DAMAGE:

IMPORTANT - NOTICE TO CARRIER

THIS REPORT IS NOTICE THAT A FORMAL CLAIM WILL BE FILED. ALL SALVAGE ITEMS AND/OR REJECTED GOODS ARE BEING HELD FOR YOUR ACCOUNT SUBJECT TO A CHARGE FOR STORAGE IF NOT PICKED UP WITHIN 48 HOURS. WE ASSUME NO RESPONSIBILITY FOR LOSS OF WEIGHT, SHRINKAGE, LEAKAGE, OR CONDITION OF MERCHANDISE.
PLEASE ARRANGE FOR PROMPT DISPOSITION.

NOTICE TO CUSTOMER

FOLLOWING PAPERS ARE ATTACHED:

- [X] INBOUND FREIGHT BILL
- [ ] ORIGINAL BILL OF LADING
- [ ] INVOICE # ______
- [ ] OTHERS

CARRIER CERTIFICATION OF INSPECTION

I HEREBY CERTIFY THAT I HAVE EXAMINED THE ABOVE DAMAGED SHIPMENT AND FIND AS REPORTED HEREON, AND THAT I HAVE BEEN FURNISHED A COPY OF THIS REPORT.

INSPECTOR'S SIGNATURE ______

NAME OF CARRIER MTRR

DATE INSPECTED 5/30/84

BY: J.M.

1-003-1069

CARRIER RECEIPT

RECEIVED 14 UNITS AS ITEMIZED IN REJECTED COLUMN ABOVE

NAME OF CARRIER MTRR

SIGNED ______

DATE 5/30/84

ORIGINAL

OS&D reports document the exceptions as to count or conditions of the goods received on a single shipment. This form serves to notify the owner of the goods and the carrier of the exceptions.

offered no hint that its contents had been damaged. When this happens, the party discovering the concealed damage returns it for credit, and a claim to recover this loss goes back through the system. The delivering carrier, after determining that damage did not take place in the consignee's own handling operation, may attempt to involve prior carriers or prior warehouse operators in a sharing of the claim. Sometimes, a compromise is adopted to spread the claim cost among the several parties who might have caused the damage.

Three kinds of evidence are helpful in dealing with a concealed damage claim: the delivering carrier's bill of lading, photographic evidence of the damage and an invoice indicating the value of the item or its cost of repair.

Warehouse Damage

Warehouse damage is inevitable. Any warehouseman who claims to be running a "damage-free" facility is either suffering from delusion or is untruthful. The forklift driver who never damages anything is probably working below optimum speed. Normal handling and storage functions will cause a certain amount of damage in any operation.

U.S. Department of Agriculture officials studied warehouse damage and found 66 causes. Of these, six accounted for 50 percent of the damage. These six are listed below, with the damage rate per-100,000-cases-shipped:

Dropping cases in aisles	16.1 cases
Protruding nails in pallets	15.8 cases
Damage by forks of lift trucks	14.4 cases
Damaged during storage	14.4 cases
Damaged in filling racks	12.8 cases
Damaged in removing from 2nd level rack slot (order picking)	8.6 cases
Total	82.1 cases per 100,000 shipped[2]

2 Distribution Warehouse Cost Digest, Vol. 15, No. 1. Marketing Publications, Inc., Silver Spring, MD.

Warehouse damage is most prevalent at corners of stacks, in narrow aisles, or at the face of a load (through a collision in stacking). Faulty palletizing or dropped packages may also cause warehouse damage. If merchandise s piled above its packaging limitations, the stack will collapse. Shifting may cause the corners of boxes in a unit load to be misaligned; this in turn reduces the stack strength and may cause the pile to collapse. High humidity also is a frequent cause of package failure, since most corrugated materials will be weakened by moisture. Roof or sprinkler leakage may soak packages and cause them to fail. Occasionally, cross contamination of stored goods can be caused either by odor transfer or by chemical reaction.

By definition, warehouse damage is the result of human failure. Such damage is minimized through training, proper maintenance of handling equipment, constant communication of correct storage and handling practices, and knowledge of the sources of chemical reaction or cross contamination.

Reducing Warehouse Damage

Following these seven steps will reduce damage in many warehouse operations:

- Use storage racks for bagged materials to eliminate damage caused by leaning stacks.
- Prevent nail damage by covering pallet faces with light material such as a sheet of plywood.
- Buy pallets with high nail-retention.
- Don't use forks that are too long and protrude on other side of a pallet.
- Maintain warehouse training programs.
- Remove damage as soon as it is discovered.
- Maintain proper aisle widths.[3]

A great deal of the damage found in most warehouses is ultimately caused by carelessness. Greater concern can be developed

---

3   Ibid.

through keeping score on the results of damage control and communicating those results to the warehouse employees concerned. Highlighting the damage ratios described in this chapter is a vital part of developing this concern. These ratios should be publicized at employee meetings, or used as the basis for contests, with prizes awarded for the best record in controlling damage.

Recooperage and Repair

When cases of packaged goods are damaged, it is frequently possible to salvage much of the damaged shipment by removing the inner cartons that are broken and replacing them. This process is frequently referred to as recooperage and must be done meticulously. Damaged food products are likely to attract insects or rodents. Products that are stained, but otherwise undamaged, are still considered unsalable. Fumes or odors from damaged products could cause contamination to good merchandise stored near them. Liquids can spill and stain other containers. For this reason, any repackaging operation must be designed to handle the product and cleanly separate good stock from bad.

Some product damage in a warehouse can be repaired so perfectly that it is undetectable. Major appliances, such as refrigerators or washing machines, sometimes suffer minor dents or scratches that can be repaired by using processes similar to automotive body work. Metal office furniture can be repaired with equal ease. Again, quality control in such repair is absolutely essential, since the consumer who receives an apparently new product that is actually damaged will be an extremely unhappy customer.

At times it is more economical to sell damaged goods to a salvage firm than to engage in either recooperage or damage repair. However, damage disposal must conform to your marketing policy. If slightly damaged product is allowed to be sold at a discount to salvage firms, common carriers, or warehouses handling the product, the ability to market that same product at list price might be impaired. For this reason, many manufacturers strictly control the salvage and disposal of all damage, even products judged to be a total loss.

Protecting Damaged Merchandise

One of the toughest problems in any warehouse storing damage-prone products[4] is to be sure that powders or liquids do not leak from damaged containers and stain or spoil other merchandise in the warehouse. With many products, the consequences of leakage can cause a great deal of additional damage.

Some warehouse operators have been very innovative in developing systems to control secondary damage.

When damage of sensitive materials is discovered, it is important that the damaged cases immediately be removed and segregated from good stock. At times, this removal is not easy, because it could involve depalletizing for which there is no labor available. One answer is the development of a drip pan, a piece of sheet metal a little larger than a pallet with corners turned up and sealed to hold powder or fluid. In this way, leakage from the damaged pallet is held in the drip pan and not allowed to migrate through the warehouse.

Another answer is the use of fiber or metal drums as a temporary repository for damaged cases or bags. Like the pan, the drum keeps the spillage from causing further damage.

One showcase warehouse handles nothing but carbon black, one of the most sensitive products from a standpoint of secondary damage. Carbon black is the same material used as a toner in photocopy machines, and bags of this material, when broken, can cause substantial discoloration of goods stored nearby. In this warehouse, every lift truck is equipped with a portable vacuum as well as broom and dust pan. When a leaker is discovered, cleaning and controlling the leakage has higher priority than completing the shipment. By placing the emphasis on damage control, the carbon black warehouse maintains superior housekeeping with a dirty product.

Storage of Damaged Products

Damaged products stored in a warehouse are an embarrassment, so most warehouse operators keep their damage storage where it's not

---

4   From Volume I, No. 1 of Warehousing Forum, ©The Ackerman Company.

DEALING WITH DAMAGE

seen by visitors. Unfortunately, however, management doesn't see it either, and the accumulation of damaged goods is overlooked.

The best way to control the cost of storing damage is to keep down the quantity in the warehouse. And the best way to control the quantity is to store it in a conspicuous spot.

Because damage is unpleasant, management as well as warehouse workers may try to sweep it under the rug. Yet, dealing with damage is part of good warehouse discipline—and remember that the warehouse operator now has a golden opportunity to submit his diagnosis of its cause. Because damage always represents waste, improved control represents a primary way to improve overall warehouse productivity.

Loss and Damage Guidelines

The primary method of measuring loss-control performance is by conducting a warehouse operations audit. But before making an audit, be sure that the following conditions exist:

1. Aisles are clear, well-marked and sufficiently wide to allow equipment to maneuver.

2. Special guards are used to protect racks and columns, and guided aisles have guide rail entries.

3. Block stacking is not leading to load crushing or leaning stacks, and loads do not overhang the pallet.

4. Floor-loading, rack-loading and structure load limits have not been exceeded by recent changes in material or equipment.

5. Equipment is not being over-loaded or operated at excessive speeds.

6. Front-to-back members are installed in selective pallet racks to prevent loads falling through openings.

7. Masts on industrial trucks fit easily through doorways and passageways.

8. Safety screens are located beneath overhead conveyors.

9. Hazardous and flammable materials are well-marked and are stored, handled, and transported according to regulations.

10. No sagging load beams and no bent upright trusses are present in the storage racks.

11. Materials handlers have undergone formal training in the use of handling equipment.

12. Preventive maintenance programs are used for all handling equipment.

Damage Control Ratios

Using damage-control ratios is a useful damage consciousness-raising exercise. Here are some examples:

$$\text{Damaged loads ratio} = \frac{\text{Number of damaged loads}}{\text{Number of loads}}$$

$$\text{Inventory shrinkage ratio} = \frac{\text{Inventory investment verified}}{\text{Inventory investment expected}}$$

Other statistics that can be used are "Percentage of Accident-Free Operators," and "Number of Accidents per Industrial Truck-Operating Hours."

Damaged Loads Ratio

The damaged loads ratio provides a measure of the loss due to poor handling. The ratio should be computed at each stage of flow throughout the system, beginning with receiving and ending with shipping. A load is defined as damaged if there are *any* damages to it.

It is not likely that individual equipment operators will maintain accurate damage reports, even if they are not at fault. The time required to prepare the reports and a lack of interest in the task combine to create an atmosphere that is not conducive to reliable reporting of damaged loads.

Random samples should be made periodically to ascertain the damage percentage on loads being staged in receiving, in-bound inspection, packing and shipping. Additionally, when perpetual inventory audits are performed, a damage report should be prepared.

For example, in a receiving department pallet loads are staged awaiting movement to storage. An audit of the loads is performed with the following results:

Number of loads 64
Number of damaged loads 4
Number of cases 670
Number of damaged cases 12

$$\text{Damaged loads ratio} = \frac{4 \text{ Damaged loads}}{64 \text{ Total loads}} = 6.24\%$$

$$\text{Damaged cases ratio} = \frac{12 \text{ Damaged cases}}{670 \text{ Total cases}} = 1.79\%$$

Inventory Shrinkage Ratio

The inventory shrinkage ratio provides a measure of the loss of material due to pilferage, errors in record keeping, errors in material shipments to customers, and misplacement of inventory in storage. The ratio compares what was found in a physical inventory of the material in storage with what was believed to be there, based on inventory records.

Unfortunately, one does not always know why differences exist, if there are differences. In fact, even if there are not differences in the physical inventory and the inventory records, there is no guarantee that losses have not occurred. Furthermore, offsetting errors can cause the verified value of overall inventory investment to equal the expected value. A sample computation of an inventory shrinkage ratio follows:

A physical audit yielded an inventory value of $2,500,000. Inventory records indicate the investment in inventory was $2,550,000.

$$\textit{Ratio of actual recorded inventory value} = \frac{\$2,500,000}{\$2,550,000} = 98\%$$

Therefore, there was a two percent "shrinkage" due to losses, errors, or both.[5]

A Management Checklist for Damage Control

Use the questions listed below as a checklist to make sure no areas of damage control are being overlooked.

- Is there a daily report of repackaging showing units processed, quantity rejected and quantity returned to stock?
- Are the boundaries of the recooperage areas delineated by a fence, a painted stripe or by stacks of stored merchandise?
- Is there an inventory of packaging material available, stored for easy retrieval?
- Is carrier damage separated from warehouse damage, reported by carrier identification number (car number) and date received?
- Are spills from all damaged goods cleaned up, and is remaining damage protected from potential infestation or cross-contamination?
- Are there ample trash receptacles for disposal of spilled products in the repackaging area?
- Are corner guards (made of sheet metal or plywood) used to protect stacks of stored merchandise at aisle and dock intersections?
- Is there a pallet repair program, and are pallet repairs up-to-date?
- Are cracks, wear spots or holes in the warehouse floor repaired so that the floor remains smooth?

---

5 Prof. John A. White, "Yale Management Guide to Productivity," Industrial Truck Division of Eaton, Yale & Towne.

Part IX

HANDLING OF INFORMATION

47

CLERICAL PROCEDURES

The clerical section of a distribution center is frequently the most undervalued part of the system. It often receives an inordinately small amount of attention from management. While entire magazines, books and trade shows inform management about stacker cranes, storage racks and other hardware associated with the storage and handling functions, relatively little attention is given to the most critical function of all—the clerical operation.

Most warehouse operations that are in trouble can trace the difficulty to malfunctions in the clerical operation. This is especially important because of the "middle-man" role of warehousing in the typical distribution system. Prompt and accurate communication for both shippers and consignees using the warehouse is critical to the success of any warehouse operation.

While major improvement in warehouse materials handling methods occurred during and after World War II, the technological change in office procedures has been even more pronounced. The prime cause was the introduction of electronic data processing equipment, followed by its miniaturization and substantial cost reduction.

The emphasis on computer-oriented information systems has placed greater responsibilities on the clerical personnel at the warehouse. The interface between the warehouse and the user's computer has become increasingly important. Since clerks now routinely feed information directly into warehouse users' computers, the potential for disruption through clerical error is greatly increased.

Any warehouse clerical system should be designed around a profile of the user's needs. The operator must know what kinds of information are needed, and when.

Customer Service Standards

Order-response time is the most important customer service requirement that affects clerical operations. This is the time lapse between the moment when the warehouse is notified to ship and the time when the goods must be physically out the door, or the time when they must be in the hands of the consignee. Order-response time is the most critical service factor at most distribution centers. Yet, the allowable time may vary substantially in different industries, or even between different customers in the same industry. Customer service standards in a distribution center for pharmaceuticals, for example, will be different from those of a furniture warehouse.

In public warehousing, it is common to establish a standard order-response time. A typical standard in this industry is the following:

- All orders received by noon are shipped on the same day.
- All orders received after noon are shipped on the following day.

Location of Clerical Center

The clerical function need not be handled at the warehouse, since today's data communications equipment provides almost unlimited location options. Clerical work can be performed at a remote location, with bills of lading sent to the shipping dock via teletype or facsimile transmission. Receiving information can be reported via the same means of transmission and perpetual inventories can be maintained at a location outside the warehouse.

With today's capabilities in processing and transmission of information, a warehouse can be controlled from a clerical center which is miles away from the storage building. Yet, most managers want the clerical function under the same roof, primarily since clerical personnel frequently participate in inventory checks and other activities that involve the cooperation of both the office and materials-handling employees.

Office Environment

The atmosphere and appearance of an office play a vital role in its effectiveness. When the office is isolated from the warehouse, an insular attitude may develop among the clerical personnel. Since effective customer service requires complete cooperation between the clerical and materials handling personnel, poor office location can deter teamwork.

Security for the office area is frequently overlooked. Today's office contains electronic equipment small enough to be easily carried out. Few warehouse offices handle cash or negotiable securities, but they do handle documents and records that are difficult or impossible to replace. Back-up systems stored separately should be available to replace records that may be lost or destroyed.

Since the warehouse office is usually the first place customers are received, it should be a showcase for the entire distribution operation. A pleasant work environment also will pay off in improving office morale and efficiency. Ample lighting, effective use of colors and control of temperature and noise levels must all be considered in developing an effective work environment.

Computer printers create noise problems unknown in offices a few years ago. These machines should be isolated or surrounded with sound-dampening enclosures. A carefully-planned office layout is the best means of controlling noise levels. As more computers use the quiet laser printers, noise promises to become less of a problem in the future.

Organizing the Clerical Function

There are two methods of organizing work in a warehouse office—either by clerical function or by customer account.

In the clerically-oriented system certain individuals perform specific tasks for every user of the warehouse. For example, one individual might be in charge of preparing receiving reports. Another prepares all the bills of lading, and still another has duties limited to inventory record-keeping. The clerically-oriented system permits a person with limited skills to be employed in repetitive work. In effect, it is a variation of a manufacturing assembly line. Just as the assembly line worker will do nothing but install wheels, the receiving clerk will do nothing but prepare receiving reports. As in an assembly line, such repetition can be tedious, but the boredom can be relieved by rotating responsibilities and cross-training personnel on different jobs.

In public warehousing individual users or customers are sometimes referred to as "accounts." Most public warehouses will assign certain clerks to handle one or a group of accounts. That clerk will perform all of the clerical functions in connection with specific accounts. These may include receiving reports, bills of lading and supervision of inventory records. When the customer calls, there is one person in the office who maintains a hands-on familiarity with all the clerical functions of that account. The same practice can be used in a private warehouse, though in this case the "account" is the outbound shipment consignee. The account-oriented system requires greater versatility and skill on the part of each clerk in the system. On the other hand, it affords a more personalized approach to a customer who wants to talk with someone who has total familiarity with that customer's transactions.

Interaction with Marketing

One activity of the clerical department in most warehouses is interacting with the users' sales staff. Most warehouse users have recognized the desirability of separating warehouse workers from sales personnel in order to maintain maximum efficiency of both. Yet, the demands of sales personnel must be balanced with the shipping capacity of the warehouse in order to develop a customer service system that is both predictable and realistic.

For example, if warehouse shipping capacity is 20 truckloads per day and shipment requests for 25 truckloads are received for a given day, some compromises must be made. So the clerical coordinator's job is to merge the demands of salesmen with the capabilities of the warehouse. When releases exceed capacity, the coordinator will develop a schedule approved by the user to designate which orders will be shipped and which will be deferred. This avoids friction and the time lost if sales personnel communicate directly with the warehouse personnel.

Manual and Electronic Data Processing

In the pre-computer era warehouse inventory control systems consisted of hand-posted ledgers and manually-prepared shipping and receiving documents. Visible record card files were developed to make it fast and easy to use inventory records. These well-designed systems allowed the user to see low stock or low activity at a glance. Although the cost of computing equipment has been reduced drastically in recent years, there are still times when a visible-record manual system is the most cost-effective way to control inventory.

With today's array of computers, a wide degree of automation is available for inventory control, word processing, or communication. One important advantage of computers is their ability to speed order-response time. Nonetheless, there have been faulty computer installations in which order-response time has been slowed rather than expedited.

Batch-processing is a means of increasing data processing efficiency by first accumulating a group of orders and then processing all of them at one time. Batch processing uses the computer more efficiently, but it will do so at a sacrifice in order-response time unless the batches are processed frequently.

JIT (Just in Time systems—see Chapter 35) and faster response times have substantially reduced the popularity of batch processing.

The need to increase clerical capacity is a frequent motive for changing to an automated system. Some automated systems can handle far more line items and orders per-day than manual systems.

A fringe benefit of an automated system is the requirement for rigid standardization of clerical procedures. Enforcing the discipline necessary to operate computers has side benefits in most warehousing operations, if an adequate level of standardization can be maintained.

Control of Automation

Automated systems are never as automated as they first appear.

Correcting the system when it develops trouble is one drawback. When incorrect information is produced, someone must be able to quickly diagnose and correct the problem. This is not easy to do in a complex system, and failure to do so can have disastrous results in a warehouse operation. Keeping back-up data is a necessity for reconstructing the pieces when the system goes awry.

Loss of inventory control is a disaster in any warehouse clerical system. This could occur if inventory records are destroyed and no back-up records are available. Under such circumstances, the only recourse would be a physical count of the inventory. Such failures might be caused by deliberate tampering with data processing systems. Good computer security is therefore very important.

Measuring Clerical Costs

The best standard for comparison between various clerical systems is a cost per-unit. This may be expressed as a cost per-bill of lading, a cost per-billing line, or a clerical cost per-unit shipped. By developing a clerical unit-cost for both existing and proposed systems, it is easy to determine whether a new system is justified.

Though more sophisticated systems may be installed for non-cost reasons, such as improved response time or accuracy, a true cost of these should still be determined. In this way management can establish trade-offs between a proposed new system and an existing one.

Management Information

A bonus benefit of computerized clerical operations in the warehouse is the development of management information that can aid in controlling warehouse operations. The operations manager can gain information on productivity and labor expended for each warehouse account. Payroll reports help to control overtime and measure materials handling productivity. In a public warehouse, cash flow is controlled through reporting on aging of accounts receivable.

Clerical systems have changed as much as, or more than, any other aspect of warehousing. Regardless of the proliferation of equipment, ultimately the system depends on well-motivated personnel.

48

COMPUTER HARDWARE

With the advent of the industrial forklift truck, the question became not whether to palletize but how pallets should be configured and what applications were best suited to them. In today's electronic age, the issues are parallel, not whether to use computers but rather which ones to use, how they should be configured, and where they can be most productive.[1]

A capsule review of computer hardware development looks like this:

- 1950: no broad business applications.
- 1960: used by large companies primarily in accounting.
- 1970: present in medium-sized companies in operational applications (order entry, inventory control, bookkeeping).
- 1980: increasingly visible in small businesses and much broader application (dispatch, location systems, batch recall, labor standards, word processing, office automation).
- 1990: more common than the typewriter in most businesses and present on the shop floor.

Rapid advances in computer hardware are moving computers from the narrow applications of a generation ago to broad and diverse applications today. We do not speak of the company's computer today but rather the computerization of the company. This marks the shift in computer architecture from a single central-processing center to in-

---

[1] Most of this chapter is from an unpublished article by John T. Menzies, Terminal Corporation, Baltimore, MD.

dividual "desk top" computers bringing intelligence and independence close to the job. Connectivity between computers is all-important today. A few year ago computers were used almost exclusively for data processing. Office automation and personal productivity are now of equal importance and must be considered in hardware selection and configuration.

The nature of warehousing makes it a natural fit for computer applications in these areas:

- Repetitive work such as receiving reports and bills of lading.

- The necessary retention and organization of vast amounts of information including quantities, weights, descriptions, locations, storage characteristics, etc.

- The recording and retention of receiving information by bar code scanning or optical character recognition (OCR).

- The rapid receipt and transmission of information by using electronic data interchange (EDI).

- The timely processing of information to allow for inventory status reporting, work planning, productivity reporting, etc.

While the business of warehousing represents a natural application for computers, it also presents a number of challenges. The warehouse is a classic "real time" situation. Decisions must be made quickly and they must be right, particularly in the era of minimal inventories and JIT (Just In Time) service. The warehouse is often where a business comes in contact with very many other businesses. This requires great diversity in terms of information and the form it takes. The closer the computer comes to the work area the more likely it will be to encounter a hostile environment in terms of temperature, dirt and abuse. Computers, while offering great power, have not always been flexible in their application. At the warehouse floor level, they must be able to deal with new items and changes in packing without advanced notification. Localized computer processing power allows jobs to be done on the warehouse floor which a few years ago would have seemed beyond the level of staff skills and training.

COMPUTER HARDWARE

Micro, Mini, Mainframe

The difference between microcomputers, minicomputers, and mainframe computers is hard to discern, save in price. A Fortune magazine article differentiated the classes by indicating that microcomputers cost less than $12,000, minicomputers cost between $12,000 and $700,000, and mainframes cost more than $700,000. Some microcomputers can emulate mainframes with software developments. A microcomputer can serve as a work station (terminal) linked to a central processing computer which may either be another microcomputer, a minicomputer, or a mainframe. Or a microcomputer may be one of a few or many hundred microcomputers which are linked together by a network allowing information and programs to be shared. Today hardware is customized to satisfy the application needs of the user. This may be done by a system which stands alone, is closely linked to other computers, or periodically connects with other computers. Distinctions between systems speak more to size than function and the overlap is so great as to make them virtually irrelevant, so far as most applications are concerned.

While there are different classes of computers, their operating characteristics to the user are similar. All must have a method of input. This input may be by keyboard, bar-code scanner, optical character recognition (OCR), mouse, or some other device such as a portable terminal. All process the information they receive in the same way. This may involve manipulation or calculations. The product of the process is output of some type which may be stored, displayed on a computer monitor, printed, or electronically transferred. Some form of storage is the final common characteristic of computer systems. The storage may take the form of a disk drive (fixed or removable), tape, or electronic memory in the processor. All computer hardware functions revolve around input, processing, output, and storage.

Computer Jargon

Data processing, like warehousing, has developed a language specific to the profession. The language may seem intimidating, but most of the concepts are straight-forward. The concepts that they represent may become more meaningful if we roughly compare them to things we are used to:

- A bit and byte are units of information and might be compared to a case and a pallet?
- A "K" is a measure of information capacity or storage. Perhaps 10 K is like 10,000 square feet?
- Megahertz refers to a computer's speed - is 20 megahertz so much different in concept than 80 miles per hour?
- ASC II is a standard computer code - in many ways like the GMA pallet standards.

Figure 48-1 lists some of the commonly used terms and what they describe.[2]

Portable Data Collection Terminals

Recently technology has provided a number of devices that can assist in the timely and accurate collection of data. One of these devices is the Portable Data Terminal (PDT). The PDT is a device that an operator can carry and collect data at its source e.g. the receiving dock, storage location, shipping dock. These devices usually have a key pad (alpha and numeric) and a screen that will display one or more lines of data. Many of the PDTs can be equipped with bar-code scanners. The operator keys in the appropriate data such as bill of lading number, product code, serial number, location, and quantity. If a bar-code scanner is attached, some of the data may be scanned directly from a bar-coded label.[3]

Since the data in the PDT is in a computer-compatible form, it never has to be key-entered into another system. The data may be passed directly to a computer from the PDT, eliminating the possibility of more errors being generated by re-keying data.

Usually one of two methods is employed to transfer data from

---

2 Excerpt from Glossary of Black Box Catalog, a Micom Company, Pittsburgh, PA.
3 This material was originally written by Jim Allen, President of BarCode Resources, for a publication of Warehousing Education and Research Council.

the PDT to the host computer. The first method is a direct batch-transfer which requires the operator periodically to connect the PDT to the host computer by cable and transfer the batch of data.

The second method is for the PDT to transfer data to and from the host computer by radio signal. Both the PDT and the host computer have radio transmitters and receivers attached. As data is entered into the PDT, it is immediately transmitted to the host. Upon receipt, the host will verify and store the data on the computer. If errors are detected, the host will send the appropriate response back to the PDT operator. This type of system operates in a real-time mode.

The most prevalent applications of the PDT today are in recording l) physical inventories; 2) material receipts; 3) warehouse storage locations; 4) material shipments and 5) order picking. In all of these activities, significant labor saving can be realized. In addition, the result of more accurate data collection can affect many intangibles such as customer service and scheduling. In most applications, a PDT data collection system can provide a payback of less than one year.

Purchasing Hardware

The miniaturization of computers and their use outside the once-traditional data processing functions have led to changes in the way hardware is purchased. Computer hardware was traditionally purchased from a few manufacturers. Today the sources include the same manufacturers, value-added resellers (VARS, companies which sell software applications as well as the hardware to run them), computer stores, and mail-telephone order companies. Because hardware in any one warehouse may come from different sources, there is an additional burden of integrating it with other systems and insuring reliability.

Hardware Development

There are significant trends developing within the industry. These trends provide some guidance as to what will be appropriate in hardware and what to be aware of.

Communication and or connectivity is an important trend. This may be between computers of the same make and class (micro, mini, mainframe) or computers of different makes and different classes. A centerpiece of this trend is IBM's SAA architecture which advances this firm's plans to allow all its equipment to interconnect. IBM projects that in 1991, 90-95 percent of its machines will be in networks. The move to connectivity is evident among hardware manufacturers. Electronic data interchange (EDI) is an approach to communication by agreeing to standard message codes. Links may be established between businesses through EDI or networks. In-house connections take place through local area networks (LAN) which connect similar computers or computers of different types, or by a multi-user system which represents the traditional approach to distributing data processing. Architecture can be mixed and the same computer can be connected to a LAN and to a multi-user system.

Another trend is towards "intelligent" terminals. Terminals used to rely on minicomputers or mainframe computers for their data-processing power. More and more of these terminals now have local processing power (they are in fact microcomputers) but are also gateways to larger and more powerful systems if needed. This trend seems likely to continue due to the amount of local processing power needed to support popular applications such as spread-sheets and graphics-oriented programs.

Alternative forms of input will be important. Connectivity between computers is an important way to avoid redundant keystroking of data which is already in a system. Bar-code scanning, and optical character recognition offer the potential of eliminating manual input. Nearly all consumer items are now marked with bar codes and there are growing industrial and warehouse applications of bar code technology. It can be expected that most warehouse operations will have to be able to integrate this technology into their operations, including scanned input to inventory systems and bar-code label output to meet down-stream requirements. Voice data-entry is a fledgling technology which we may expect to expand.

Graphics will become increasingly important as a way of communicating with computers and also as output. The human brain is a very good processor of images. One of the primary applications for super computers is the creation of graphic output rather than numeric output because scientists better recognize trends and anomalies when data are presented in picture form rather than as raw numbers. More down-to-earth graphic applications will be seen in our industry. Image processing is now being used to supply proof of delivery in the trucking industry. Under this approach, a graphic representation of the signed proof of delivery is retrieved from a computer storage medium rather than searching a bank of file cabinets for the hard-copy proof.

Departmental computing will increase as applications which previously had to depend upon a centralized EDP staff and mainframe support can now be processed on smaller systems which stand alone and or tie-in with other systems.

The blurring which has already begun between classes of systems will continue. Technological advances and ready communications between systems will likely make the differentiation between micro, mini, and mainframe computers increasingly ambiguous.

Price competition for lower-end systems will continue to cause an increase in data-processing power for fewer dollars. Competition will also mean there is danger of systems becoming obsolete in terms of new product development and service should some companies fail to make the grade.

The first response to changes taking place in the industry must be to stay aware of, and attuned to, the changes.[4]

Applications in Public Warehousing

Inventory control is the most widespread application of computers in public warehousing. It is, however, only one of many possible applications.[5]

The computer allows a public warehouse manager not only to

---

4 The last two subchapters are by Menzies, op. cit.
5 From "Micro Computers — Don't Trust a Computer You Can't Lift!" by James L. Allen, Milwaukee Commercial Storage, Inc., from Warehousing Review, 1982, American Warehousemen's Association, Chicago, IL.

analyze costs but to isolate them by individual customer, or by particular operation in a sequence of operations. Such analysis can be used to measure profitability and performance of current rates. A second category of uses involves strategic planning functions within the warehouse. This includes forecasting changes in volume and business levels, as well as the easy ability to ask "what if" questions. The computer can be used to simulate what happens if warehouse volume increases by 10 percent, for example.

A third category of use is in control. This involves the accounting aspect, including all the basic accounting functions.

Yet another area in which computers can be used is determining the status of individual orders. Because of the search speed of the computer, individual orders can be checked more rapidly than by hand. The computer can be used to keep records of each item in stock, where it is located in the warehouse, how it was shipped, where it was shipped, etc. Again, the advantage of using computer equipment for this task is the speed and ease of information retrieval.

Other auxiliary applications that have been developed for the warehousing industry include those that assist in warehouse layout. Often, shifts in product, customer, volume and product mix require that the layout of certain areas within the warehouse be changed to make the operation most cost-effective. Using the microcomputer to analyze this layout change would make it easier and also help determine the most efficient layout based on particular product characteristics.

For equipment maintenance scheduling, the microcomputer could keep records of all equipment used and note the maintenance schedule, as well as keep track of when maintenance was done. It could then generate reports showing the costs of ownership and maintenance of various equipment. Similar records and information for building maintenance can be of great help.

COMPUTER HARDWARE

Figure 48-1
A Glossary of Computer Terms

ANSI (American National Standards Institute)—The principal standards development body in the U.S.A. ANSI is a nonprofit, nongovernmental body supported by over 1000 trade organizations, professional societies and companies. U.S.A.'s member body to ISO (International Standards Organization).

ASCII (American Standard Code for Information Interchange)—Pronounced asky. A seven-bit-plus-parity code established by ANSI to achieve compatibility between data services.

asynchronous transmission—Transmission in which time intervals between transmitted characters may be of unequal length. Transmission is controlled by start and stop bits at the beginning and end of each character.

baud—Unit of signaling speed. The speed in baud is the number of discrete conditions or events per second. If each event represents only one bit condition, baud rate equals bps. When each event represents more than one bit (e.g. dibit), baud rate does not equal bps.

bisynchronous transmission (BSC)—A byte-or character-oriented IBM communications protocol that has become the industry standard. It uses a defined set of control characters for synchronized transmission of binary coded data between stations in a data communications system.

bit (binary digit)—Contraction of "binary digit," the smallest unit of information in a binary system; a one or zero condition.

BPS (bits per second)—Unit of data transmission rate.

buffer—A temporary storage device used to compensate for a difference in data rate and data flow between two devices (typically a computer and a printer); also called a spooler.

bus—A data path shared by many devices (e.g. multipoint line) with one or more conductors for transmitting signals, data or power. In LAN technology, a bus is a type of linear network topology.

byte—A binary element string functioning as a unit, usually shorter than a computer "word." Eight bit bytes are most common. Also called a "character."

character—Letter, numeral, punctuation, control figure or any other symbol contained in a message.

PRACTICAL HANDBOOK OF WAREHOUSING

clock—Shorthand term for the source(s) of timing signals used in synchronous transmission. More generally: the source(s) of timing signals sequencing electronic events.

console—The device used by an operator, system manager or maintenance engineer to monitor or control computer, system or network performance.

CPU (central processing unit)—Portion of a computer that directs the sequence of operations and initiates the proper commands to the computer for execution.

dip switches—Switches for opening and closing leads between two devices. When a lead is opened, a jumper wire can be used to cross or tie it to another lead.

dumb terminal—Both hard copy and VDT type ASCII asynchronous terminals that do not use a data transmission protocol and usually send data one character at a time.

emulation—The imitation of a computer system, performed by a combination of hardware and software, that allows programs to run between incompatible systems.

EPROM—Read-only, non-volatile, semi-conductor memory that is erasable via ultra-violet light and reprogrammable.

firmware—A computer program or software stored permanently in PROM or ROM or semi-permanently in EPROM.

full duplex (FDX)—Simultaneous, two-way independent transmission in both directions (4 wire).

half duplex (HDX)—Transmission in either direction, but not simultaneous (2 wire).

handshaking—Exchange of predetermined signals between two devices establishing a connection. Usually part of a communications protocol.

header—The control information added to the beginning of a message—either a transmission block or a packet.

hertz (Hz)—A measure of frequency or bandwidth; the same as cycles per second.

hexadecimal number system—The number system with the base of sixteen. In hexadecimal, the first ten digits are 0-9 and the last six digits are represented by the letters A-F.

COMPUTER HARDWARE

interface—A shared boundary defined by common physical interconnection characteristics, signal characteristics and meanings of interchanged signals.

interleave—To send blocks of data alternately to two or more stations on a multipoint system, or to put bits or characters alternately into the time slots in a TDM (time division multiplexor).

LAN (local area network)—A data communications system confined to a limited geographic area (up to six miles or about ten kilometers) with moderate to high data rates (100 Kbps to 50 Mbps). The area served may consist of a single building, a cluster of buildings or a campus type arrangement. The network uses some type of switching technology, and does not use common carrier circuits—although it may have gateways or bridges to other public or private networks.

LED (light emitting diode)—A semiconductor light source that emits visible light or invisible infrared radiation.

mainframe—A large scale computer system that can house comprehensive software and several peripherals.

mean-time-between-failures (MTBF)—The average length of time that a system or component works without failure.

menu—The list of available software functions for selection by the operator, displayed on the computer screen once a software program has been entered.

mnemonic code—Instructions for the computer written in a form that is easy for the programmer to remember. A program written in mnemonics must be converted to machine code prior to execution.

modem (modulator-demodulator)—A device used to convert serial digital data from a transmitting terminal to a signal suitable for transmission over a telephone channel, or to reconvert the transmitted signal to serial digital data for acceptance by a receiving terminal.

multiplexor—A device used for division of a transmission facility into two or more subchannels, either by splitting the frequency band into narrower bands (frequency division) or by allotting a common channel to several different transmitting devices one at a time (time division).

packet—A group of bits (including data and call control signals) transmitted as a whole on a packet-switched network. Usually smaller than a transmission block.

parity check—The addition of noninformation bits that make up a transmission block to ensure that the total number of 1s is always either even (even parity) or odd (odd parity); used to detect transmission errors.

port—A computer interface capable of attaching to a modem for communicating with a remote terminal.

PROM (programmable read only memory)—Nonvolatile memory chip that allows a program to reside permanently in a piece of hardware.

protocol—A formal set of conventions governing the formatting and relative timing of message exchange between two communicating systems.

RAM (random access memory)—Semiconductor read-write volatile memory. Data stored is lost if power is turned off.

redundancy check—A technique of error detection involving the transmission of additional data related to the basic data in such a way that the receiving terminal, by comparing the two sets of data, can determine the probability that an error has occurred in transmission.

response time—The elapsed time between the generation of the last character of a message at a terminal and the receipt of the first character of the reply. It includes terminal delay and network delay.

ROM (read-only memory)—Nonvolatile semiconductor memory manufactured with predefined data content, permanently stored.

RS-232—Interface between data terminal equipment and data communication equipment employing serial binary data interchange.

serial transmission—Data transmission in one direction only.

software—A computer program or set of programs held in some kind of storage medium and loaded into read/write memory (RAM) for execution. (Compare with firmware and hardware.)

synchronous transmission—Transmission in which data bits ar sent at a fixed rate, with the transmitter and receiver synchronized. Synchronized transmission eliminates the need for start and stop bits.

TWX (teletypewriter exchange service)—A teletypewriter dial network owned by Western Union. ASCII coded machines are used.

X-on/X-off (transmitter on/transmitter off)—Control characters used for flow control, instructing a terminal to start transmission (X-on) and end transmission (X-off).

Excerpt from Glossary of Black Box Catalog, a Micom Company, Pittsburgh, PA.

49

SOFTWARE FOR WAREHOUSING OPERATIONS

Advances in computers and computer software have changed the warehousing industry as they have so many other service industries.[1] Today, one need know practically nothing about computer programming to make very effective use of a computer. Though the capacity of hardware in relation to its size has increased exponentially, this larger capacity allows use of software that is increasingly simple or "user friendly."

Warehouse software must address five fundamental types of operations: warehouse floor operations, customer service, management information, transportation, and electronic data interchange (EDI).

Warehouse Floor - the system software must assist the user in defining storage locations; select appropriate, efficient locations to store and pick product; track movement of product from one location to another; direct replenishment of pick lines; allocate product according to FIFO, LIFO or lot-specific rules; maximize opportunities for bulk picking, pick-line utilization and other productivity-enhancing methods; track equipment-time and labor for all activities; and support cycle counting and physical inventory efforts.

Customer Service - the system software must process shipping orders and generate purchase orders; maintain historical information

---

[1] Most of this chapter is from an article by W.G. Sheehan, V.P., Distribution Centers, Inc.

for status inquiry and product recall purposes; print bills of lading, receipts and purchase orders; and provide immediate access to information to respond to customer, vendor and carrier inquiries.

Transportation - the system software must identify freight consolidation opportunities, specify route sequence loading, process claims, and track the status of shipments.

Electronic Data Interchange (EDI) - the system software must conduct real time or batch communication of shipping orders, purchase orders and status reports between the warehouse computer and those of customers, vendors and carriers.

Warehouse Floor

As an information field is built for each storage location, you can detail the area of the building, row, section and storage rack slot number. If your warehouse has various types of storage, your records can indicate whether the item is stored on the floor, in a bin, on a pallet, in a drawer, shelf or flow rack. Where both bulk and pick-line locations exist, the difference will be indicated.

When new merchandise arrives, the system searches for an appropriate storage location. If all existing locations for the same item are full, the system will select a new one from the inventory of empty slots. Thus, the manual search for new locations, which is time-consuming and error-prone, is eliminated in most cases.

Storage locations must be well-defined in a manner that corresponds with definition of product-storage requirements. For each location, the system needs to "know" the *category, size, type, status, and environmental characteristics. Category* will prohibit mixing of incompatible products (i.e., soaps and food products and assist in logically assigning put-away locations; *size* should typically be indicated by stock keeping unit or pallet count, corresponding to units used in shipping orders; *type* could be rack storage or floor storage, in a pick line or in bulk storage; *status* would indicate whether or not the location is full; and if your warehouse has cooler or freezer space, *the environmental characteristics* would indicate whether or not the location is in such an area.

SOFTWARE

Such a system can greatly improve productivity, prevent honeycombing, and assure proper FIFO (First In, First Out) or LIFO (Last In, First Out) rotation by allocating product to orders properly. Productivity is aided by identifying bulk-picking opportunities and using pick lines to minimize travel on the warehouse floor. Honeycombing can be prevented by cleaning out locations with small quantities of products, making those locations available as soon as possible for consolidation of bulk storage or putting away new receipts. Pickers can be directed to the oldest or newest product by lot number or by warehouse location. Pick lines can be used to fill the greatest number of orders possible before requiring replenishment.

The most up-to-date information possible must be provided to the warehouse floor and collected from the warehouse floor. Having pick-lists, receiving tallys, put-away sheets, and other documents with handwritten information returned to the office for data entry is the traditional approach. Some systems then queue that information for processing in a "batch." Somewhat better are systems that update the inventory as transactions are key-entered from the warehouse documents. In today's just-in-time (JIT) systems and in other systems with high-velocity inventories, neither approach may provide information that sufficiently is up-to-date.

Information can be collected from the warehouse floor using radio frequency (RF) portable data terminals. As products are received, put away, picked, shipped and rewarehoused, the warehouseman keys or scans information into the portable terminal instead of writing it down. The inventory is updated immediately. With bar code-labeled product and scanning equipment attached to the portable terminals, even greater efficiencies and fewer information errors can be realized.

Physical inventories and cycle counts are expensive to conduct. The system must support efficient approaches to both processes. Systems with continuous cycle-counting procedures may make periodic, total physical inventories unnecessary. The system should provide preprinted inventory tickets for counting each location and an efficient method of entering counts. Bar-code labels on count tickets can speed entry and practically eliminate data entry errors.

Finally, the system should track labor hours and equipment utilization time as transactions are processed. With manual input, additional information must be handwritten on warehouse documents and key-entered. With bar code readers, practically no additional keystrokes are required; much of the information can be generated automatically by the system as transactions occur.

Customer Service

Customer service managers must have immediate access to historical information about receipts and shipments to be able to respond to client inquiries and provide support for product recalls. Up-to-date inventory status and location information must be on-hand to process shipping orders and provide timely, accurate instructions to the warehouse. Receipts, picking documents, bills of lading and purchase orders must be produced on demand.

In a just-in-time operation, the type of information collected in the warehouse must be transmitted and processed immediately in order to support very precise delivery requirements. JIT requires both greater speed and greater accuracy in recording inventory and movement.

In the clerical services department of the warehouse, perhaps the most important hardware feature is the use of scanners to read bar coded labels on merchandise as it moves over the receiving and shipping docks, or packages being counted for an inventory. Computer generation of such labels eliminates many opportunities for error. Furthermore, similar errors in reading labels are greatly reduced. *(See Chapter 50.)*

One by-product of reading labels is the ability to create transaction histories for various aspects of the warehouse operation.

Documents needed for receiving or shipping (warehouse receipts or bills of lading) should be available as a by-product of the warehouse control system.

SOFTWARE

Management Information

In an ideal system, measurements of labor and equipment time should be a by-product of the inventory system. If information is gathered from documents returned from the warehouse floor, it must include start and stop times, employee identification numbers, and equipment identification numbers. That information is tied-in with shipment, receipt and product information.

If radio frequency terminals are used, each activity is reported to the computer *as it happens,* and the system can determine and record the elapsed time for each activity.

Space utilization information can be developed automatically from the status and size of each storage location. Product activity reports can be printed from historical receipt and shipment information. Information of that nature might only be required periodically (i.e., weekly), while an analysis of labor and equipment productivity might be made daily.

Although most management information regarding space, labor and equipment utilization should be available in reports from the main inventory system, there is a variety of special operating and analytical programs tailored to rate development, manpower requirements forecasting, activity-standards development, and space analysis in the warehouse.

Transportation

Based on destination zip codes and the characteristics of each product available from the inventory data base, the system should be capable of analyzing shipments scheduled for one or several customers, on one or more scheduled ship dates, in order to identify freight consolidation opportunities.

Route-sequence-loading may be required by a carrier picking up several orders. Either the loader can sort the bills of lading into route-sequence order, then attempt to locate the correct products for each shipment; or, the system can print pick-sheets and staging instructions to establish the best route-sequence order for consolidated shipments freight charges. This practice will help minimize time required to load a shipment.

The system also should support freight claims processing by providing documentation about each shipment.

The automatic detection of weight overages and shortages, which is a feature of many warehousing software systems, provides benefits in both receiving and in warehouse security. Such automatic detection provides a check on the accuracy of packing clerks, since packing errors can be detected by weight variances before the boxes leave the shipping dock. The predicted box weight for every repacked order is automatically calculated before the packing is completed. Once the box is weighed and its label scanned, the computer compares the actual with the projected weight. When used for full truckloads, a weight check also can prevent accidental overloading of truck axles.

Electronic Data Interchange (EDI)

A warehouse software system should support electronic data interchange of shipping orders, purchase orders, receipt and shipment confirmations, shipment status reports, and inventory status information. Freight invoicing, payments and other transactions can also be exchanged electronically. Information should be transmitted and received in a standardized format.

When a warehouse operation is fully integrated with manufacturing or sales activities of a customer, "batches" of this type of information may have to be transmitted several times a day or every few minutes to be timely enough. Realtime EDI can be made when necessary, but only at great expense. Batch transmissions are more practical, less costly, and can be scheduled to occur automatically without requiring a computer operator's attention.

EDI capability is fast becoming a standard requirement for distribution services. As partnerships develop among warehouses, customers and vendors, all benefit by the freshness of information possible by using EDI.

SOFTWARE

Cost Analysis Programs

Until the advent of the microcomputer, warehouse cost calculation programs to support the management decision process were too technical and expensive for the typical public warehouse operation.

The American Warehousemen's Association has sponsored the development of operating and analytical programs for the specific needs of the public warehouse industry. AWA has also developed a series of microcomputer programs that enable a user to make accurate calculations for warehouse cost control and productivity. Using account information, these programs calculate and analyze storage and handling costs to the limit of accuracy of the input information.

The storage-rate analysis program produces an overview of space utilization—either for a prospective account or for a current user—and illustrates the effects of any operational changes on the overall cost picture. The output correlates basic cost information, summarizes the demands for warehouse space use, and evaluates the percentage of warehouse utilization. A manager can easily analyze the current situation and project the impact of operating changes. The handling-rate analysis program evaluates the activity required by an account and helps the operator price that activity.

Another program called *Standard Time Values for Public Warehouse Handling* has been developed by AWA to systematize the setting of performance standards for warehouse labor. The *AWA Program* uses a system of engineered values corresponding to the elements of work involved in any warehouse function. This program combines the selected job elements to compute very accurate standard time values. By comparing this standard level of productivity with the actual level of productivity attained, the manager can focus on the areas needing improvement.

Since this calculation is quickly done on the microcomputer, any warehouseman can use this accurate engineering tool to evaluate handling efficiency. Management control devices such as time and production reporting, methods improvement, work assignment and manpower control procedures are available through this software.[2]

Program Costs

Consider not only the initial licensing or purchase costs, but also the cost of modifying and enhancing the system. The most important part of the process should occur before a single software system is considered—and that is defining your requirements.

The Requirements Definition

Carefully analyze your existing system, whether it's completely manual, partially-automated, or fully-automated. Examine each procedure to determine (and formally document) the objective of the procedure, how often the procedure is required, what information is required and how much information is normally input for that procedure, how that information is used, what printed and displayed output formats are produced and "why it's done this way." The individuals or departments who provide input to and receive output from each procedure should also be documented.

Listing advantages and disadvantages of current procedures can be helpful in setting objectives for the new system software. New warehouse systems are normally implemented to save money or to improve control. You should set more detailed objectives for the new system.

When all the documentation is complete, the individuals who provide inputs to and receive outputs from each procedure should "sign off," agreeing the documentation is accurate and complete and that all their requirements have been defined. This will help assure a lasting commitment to the new system.

The Role of the User

Of critical importance when evaluating software is the input

---

[2] This section was written by John Doggett, American Warehousemen's Association, Chicago, IL.

from the ultimate users—those who will operate the hardware and make use of the output of the system. These users must participate in not only the requirements definition but also the evaluation. The users of the *present* system are the real experts in what might be needed in the *new* system. Technical language should be avoided. By providing this participation and doing so in terms understandable to the user, there is much greater likelihood that the sense of user ownership will develop, which ensures a lasting commitment to the system. A common cause of user resistance to automated systems is a sense of loss of control. This resistance can be overcome by participating in the planning of the system.

The user is frequently in the best position to identify previous problems as well as to suggest special features and report formats that might make the system more useful.

Finding Software Sources

The next step is to identify sources of warehouse operations software. It is best to consider, as a minimum, four to five sources.

The systems definition should be reduced to a checklist which is the outline of your requirements when asking software houses for bids on the job. Sometimes flow charting the system will help. *(See Fig. 49-1.)* By being able to standardize the competitive bids, a more logical and rational comparison can be made among the various vendors. The request to the software supplier should include some description of the using company and its operations. The systems definition should be put into questionnaire-type format. Then develop a checklist to evaluate the alternatives. The request for bids should provide a description of the hardware that will be used with the software. The systems requirement for warehousing should include all the warehouse functions. *(See Figure 49-2.)* The requirement needs must be in sufficient detail for the software vendors to comprehend what is required and yet be simple to understand, because ultimately, it is simplicity that you want in a system. An example of how this might be worded is shown in Figure 49-3.

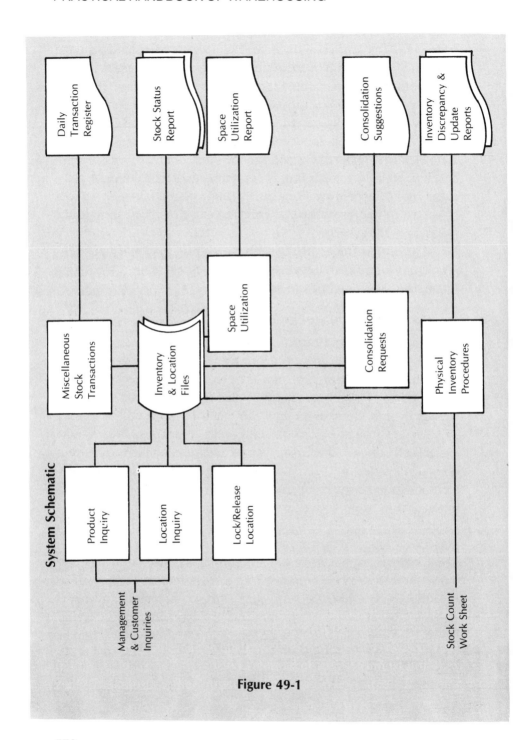

Figure 49-1

SOFTWARE

When you outline software requirements for management controls, consider the following areas:

Labor scheduling
Productivity measurement
Space utilization
Operations analysis
Budgeting
Financial controls
Inventory management—stock status, locations, reorder
Physical inventories

Security or access control is another key point within the request proposal. You should describe the documentation that should accompany the software. Also specify the communications network, including any on-line requirements or planned use of batch processing.

As part of the software evaluation, you should define your criteria. Those most commonly considered are:

Warehouse Functions

Receiving

Order Picking
  Zone
  Bulk
  UPS

Shipping
  Backorders
  Advanced orders

Reports, Documents
  Bills of lading
  Receiving reports
  etc.

Management Controls
  Labor scheduling
  Productivity measurements
  Space utilization
  Operations analysis
  Budgeting
  Financial controls
  Inventory management
  Stock status
  Locations
  Reorder
  Physical inventories

Figure 49-2

- Software responsiveness
- Hardware reliability
- Simplicity of operation
- Economics
- Vendor stability and support

The selection criteria might be ranked as essential, very desirable, desirable, etc. Sometimes, software may be available "off the shelf" to meet the user's needs. However, it is not at all unusual to modify existing software to "customize" the system. Part of the selection process if considering off-the-shelf software should be to specify those changes needed and to determine the amount of time and cost it would take to make them.

Evaluation

Four or five software suppliers should be selected and they should be asked to document how their systems will satisfy your requirements. Require demonstrations of systems that meet most of the requirements. Your choices should be narrowed to two or three systems after the demonstrations.

During the demonstrations, consider the quality of documentation supplied with the systems. Quality and utility of documentation is not only a necessity, it demonstrates the maturity of the supplier's product—don't consider a system for which documentation is not usable, "isn't necessary," or "will be available soon." Cost is always a major consideration in the selection process. Beware of putting too much emphasis on cost, however; what appears to be the more expensive system may be less expensive to use over a period of three to five years. Consider not only the initial licensing or purchase cost but also the cost of modifying and enhancing the system, monthly support costs, training costs, the costs of implementation.

The system should be implemented using a standard programming language, and the source code should either be provided with the system or in an escrow arrangement which assures its availability should the vendor no longer support the system.

SOFTWARE

The vendor should have a proven record of successful implementation and support for the system, and should provide several references. The references will probably be the vendor's most satisfied clients, but their opinions of the vendor's support, expertise in warehousing operations, and responsiveness are important. Prepare a standard list of questions and compare the answers.

Training should be provided with the system. It should be geared to everyday users, not just computer experts in your organization. Insist on an initial measurement of trainees' ability to operate the system successfully, with additional training obligated should the results be less than satisfactory.

Demonstration

An important part of software selection is receiving a demonstration of the program. It is worth the trouble to outline exactly what you expect in a demonstration. You may want to prepare "dummy" data. If you do not prepare the demonstration outline, you can be sure the vendor will take charge. You will see what he wants you to see, not necessarily what you may need to see.

Certain things should be looked for when evaluating software in a demonstration. These include unusual time delays between data input and feedback received through the VDU or printer, responsiveness (particularly using your data base), compatibility with other software systems you may already have, the ease of use and the documentation. Each function of the software package should be readily understandable and, of course, have adequate documentation. To correct problems, instructions should be immediately displayed on the screen. This is the type of feature that is important to the user, as opposed to documentation that is highly technical and can only be found in a thick and difficult-to-understand volume.

Other indicators of good software will be the documents or information produced. This should be available as hard copy reports or display on the VDU. In addition, the data input and the format required for input are important, as is the layout of the master file, what is included, how readily it can be updated, etc.

Warehouse Information System
A Sample Definition

I. Overall Scope
The system will have the ability to process receipts, generate order picking documents, and shipping documents that will contribute to increased efficiency and operational control. The system also will provide reports facilitating management control in the area of workload planning, measuring labor and space utilization, facilitating stock rotation, and providing reports for productivity and operational analysis.

II. Item Control numbers
The system will handle batch numbers, lot numbers, serial numbers, and code dates. This will be an option on each item; some items will require control, hence record keeping of one or more of the above items.

VII. Type of Storage
The types of storage handled by the system will be:
A. Standard Rack
B. Drive-Thru Rack
C. Drive-In Rack
D. Gravity Rack
E. Floor
F. Mixed
G. Cooler
H. Red Label

IX. Receipts Processed
A. Receipts Pending
If information about a receipt is available prior to the actual time of receipt, that information may be entered into the system. It may also be modified prior to the actual receipt. The information kept on an expected receipt would include the following:
1. Account or purchase order number
2. Receipt type (rail or truck)
3. Document number
4. Number of units expected

Figure 49-3

5. Weight expected
6. Expected delivery date
7. Carrier name
8. Car/trailer number
9. Receiving Warehouse
10. All line items expected in receipt with item code and quantity for each line item.

B. Receipts Without Notice

The information on the packing list will be entered into the system when the truck or rail car arrives. This information will include that listed above plus date and time of the receipt, the sequential receipt number (receiving tally number) any line item control numbers relating to the receipt such as lot numbers, product date codes, dates, serial numbers, number of pallets, and/or type received, etc.

C. Storage Locations

After entering receipt data, the system will print a receiving report including the locations where the product is to be stored. The locations assigned to the items will be determined by an algorithm based on item movement. It will include the type of storage required that will be part of the item master information file and the type of storage available within the locator system developed for the warehouse. It also will attempt to store an item adjacent to the most recently received of the same item. The system will be capable of assigning the storage locations for some receipts and not assigning them for others, depending on timing, number of line items, the receipt, and other factors. If the system assigns a storage location, the space assigned is reserved at that time and will not be assigned to other incoming product.

D. Confirmation

After the determination of storage location, the system will then proceed to process receipt confirmation. This will be a confirmation that the product shown in the receipt data has been put away and is available for use. At this time the actual locations in which the items were placed will be entered into the system data file, if they are different than those assigned by the system or if the locations were manually assigned. If a product is placed in a location other than the one assigned, the one assigned location space is "unreserved" and available for use.

SOURCE: W.G. Sheehan

Flexibility and Control

Documentation must include adequate detail for the user to understand what is happening (functional documentation). System documentation is in fact essential for "in-house" programming and explaining the internal workings of the hardware.

Software being evaluated should contain some system of security control—what data can be input and changed, who can get data from the system, what data are available to what people, etc.

The Vendor

Other considerations when evaluating software alternatives are the vendor's financial strength, vendor support in your area, and whether you will be provided updates as modifications are made to the hardware. Also of importance will be the matter of software ownership. Will you own the software outright, or will there be limitations placed on your use?

As regards vendor support, look at what kind of maintenance agreement is provided with the software. Will you have to pay if problems arise? How easy is it for the vendor to troubleshoot when problems arise and how timely will the corrections be?

Training

Purchase of the software should also entitle the user to a certain amount of training from the vendor. This training should be oriented toward the day-to-day user and not just the technical personnel within the organization. In addition to this, it is also important that representatives of the vendor be available at your site during the installation to provide an immediate response to problems. It might also be necessary to augment the vendor's user documentation with procedures or policies of your own. If this is the case, that documentation should be prepared before the training (feedback on the accuracy of documentation or procedures should be reviewed and incorporated in it). To lose the user's commitment to the system because of sketchy or inadequate training is a self-defeating exercise.

Finally, ask yourself the following questions -

SOFTWARE

1. What changes will have to occur in current procedures and policies to use the system? What will those changes cost? Will those changes improve customer service or productivity?

2. What changes (if any) will have to be made to the system before it can be implemented? What will those changes cost?

3. Where is the nearest source of service and support? Is the vendor the only source of that service and support?

4. Will the vendor guarantee the system to meet documented requirements and fix inadequacies free?

5. What computer system hardware and system software will be required to use this system?

6. Is the system compatible with other, related application systems already implemented?

7. Does initial training seem adequate and is additional training available at a reasonable cost?

8. Will the vendor provide on-site assistance during implementation of the system or install it as a complete, turnkey package?

9. Are all terms of the licensing or purchasing agreement proposed by the vendor acceptable?

10. How will problems be reported to the vendor, and fixes received from the vendor, and what turnaround time is guaranteed for fixes?

50

ELECTRONIC IDENTIFICATION

In the whole field of warehousing, perhaps no technical development has shown greater potential than electronic identification. There is more than one way to "machine-read" information from a package, but today the most common method is bar coding. The concept of bar coding has been around for decades, but it is just in recent years that its full potential in warehousing has been realized.

Bar coding improves warehouse operations in several ways. First, electronically reading and copying a symbol from bar coding is both faster and more accurate than any manual processing. A government study showed that scanning bar-code systems was 75 percent faster than entering the information through typing, and 33 percent faster than entering the information on a ten-key data board. Keypunchers had an accuracy of 98.2 percent, but in a study of over 1.25 million lines of bar code symbol data an accuracy level of 99.9997 percent was achieved.

Bar Codes—What They Are and How They Work

Bar codes are a grouping of lines, bars and spaces in a special pattern. This pattern can be read by a machine, which communicates with people or other machines. The bar code itself can be applied when the carton or package is manufactured and printed. The cost of applying or printing the code at that stage is virtually nil, since package lettering has to be printed anyway. The code can also be applied on-site, perhaps at the receiving dock of the warehouse, by a wide variety of small printing machines.

Once applied, the code can later be read or "captured" by a bar-code reader or scanner. Several kinds of scanners of various sizes are available. Used in the warehouse, this method of electronic identification offers many advantages, and there seems little doubt that new applications will be developed in the years to come. At the receiving dock, a code can be read and the identification used accurately to update the inventory. *(See photo.)* Since many inventory errors occur because of mistakes in receiving, this application has great potential for improving accuracy. As the product is identified, a determination can be made as to its best storage location, and immediate instructions relayed back to the receiver to indicate the best storage address for the in-bound item. When a physical inventory or a location check is made, electronic identification will ensure that items are identified correctly as they are counted or located. When it is time to ship, automatic identification drastically reduces the possibility of shipping error through misidentification of goods. Furthermore, if every item

ELECTRONIC IDENTIFICATION

is scanned as it moves past a loading conveyor, count can be verified. In those operations where broken case order picking is involved, bar code identification can be particularly valuable in verifying that the right number and assortment of inner packages are included in each master carton. Finally, electronic identification can be used as input for automated sorting or picking equipment. As the identification of each package is determined, instructions can be given to the sorting equipment.

Currently, the different bar codes in use are: Code 3 of 9, Interleaved 2 of 5, Code 128, Code 93, Codabar, UPC (The Universal Product Code), and the Plessy Code. *(See Figure 50-1 for an analysis of Code 3 of 9.)*

Scanners

Machines that read or "capture" the data contained in a bar code are called bar code readers or scanners. They are classified as follows:

- Portable scanners
- Laser scanning guns
- Fixed beam stationary scanners
- Moving beam stationary scanners.

Scanners are mobile and can be carried to a receiving area, a racking operation or a bulk storage area. They can be used on a lift truck anywhere in the facility. A clerk can communicate with someone who is out of the office through a visual display terminal (VDT).

The bar codes are read by shining a beam of light on the code. A hand-held terminal can have a pen-type wand for contact reading (to read, the wand must touch the bar code), or can be equipped with a laser gun for distance reading. Scanners use either a moving or fixed beam. Portable scanners are lightweight (two to three pounds), are programmable and have visual display capability. Their principal use is for capturing data on packages not moving on a conveyor. Portable scanners can be used for order selection, inventory counting, cycle counting, location product verification, and stock replenishment.

Stationary scanners can be positioned on the side of conveyors to scan bar codes of passing cartons and transmit the information to a computer. Stationary scanners can also be positioned above conveyors. Whatever their fixed position, these scanners use grid patterns of light to pick up bar code signals and transmit the data to any point for decoding.

Stationary scanners can accurately read bar codes on objects traveling at high speeds. For variable carton heights, the overhead scanner is equipped with a flip-lens photo eye that adjusts to the various carton heights.

One advantage of stationary scanners is that they can read bar codes without someone having to feed the material through a given point. This is also done through the flip-lens photo eye, which consists of two overhead scanners. One is a single-line laser; the other sends a multi-directional pattern of laser light in the form of "X's" that scan the carton. No specific placement of the carton is required, because

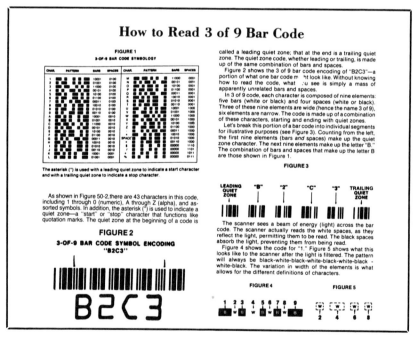

Figure 50-1

ELECTRONIC IDENTIFICATION

Figure 50-2

with the cross-weave pattern the bar code symbol can be read in any position

Contact and Non-Contact Readers

Contact readers can be portable or stationary. Both types come with a variety of options, including keypads and displays. The aperture size (opening) of a wand determines how much reflected light the reader will "see." Take care to match the wand resolution to the code density. A high-resolution wand being used to read a low-density symbol might "see" an ink spot as a bar or a void as a space. Alternatively, a low-resolution wand may not be able to recognize a narrow bar in a high-density code.

The angle at which the wand is positioned to the surface of the bar code symbol (45-90 degrees) and the speed at which the wand is moved across the symbol (three to sixty inches per second) are also key parameters affecting the first-pass read rate.

The major benefits of contact readers are reductions in the number of clerical errors in reading data, reductions in the labor and paperwork required to process data, faster and more accurate inventory taking, and enhanced efficiency of forms and document-tracking control.

A drawback of contact readers is that the surface which the pen or wand is reading must be very clean. A dust particle can miscue the reading being transmitted to the computer. Labeled products positioned in warehouse racks or shelves tend to accumulate dust.

Non-contact readers include fixed- and moving-beam scanners as well as portable laser guns. Fixed-beam scanners are usually stationary and use a fixed beam of light to read bar-code symbols. Moving-beam scanners are also usually stationary but use a moving beam of light to scan and decode the bar-code symbols.

Moving-beam scanners can read bar codes without the intervention of human operators. The material—a carton, tote pan, pallet, container or drum—is transported past the scanner on a conveyor belt. Scanners mounted on conveyors can read passing bar code symbols at speeds in excess of 1,400 scans per second by using a light spot. A simple movement of the beam over the bar code constitutes a "scan."

An advantage of using moving beam scanners is that bar code placement is not critical as long as the location is within its field of view (scanning height).

Through the use of moving beam scanners, management can optimize storage utilization, deliver material when needed to the approximate shipping point, speed up order-processing, and provide accurate and prompt billing.

Printing Bar Codes

The key to an automatic identification system is the bar code printing. If the printing is erratic, or the lines are not within a tolerance level of the scanners, the entire identification system may be worthless.

When you see automatic identification systems demonstrated at trade shows, you see how nicely the scanners work—but with label images that are perfect for recognition. Take those same labels and test them in a dirty environment, and you'll get a different reading or no reading at all.

Using Bar Coding in a Warehouse

Bar-code scanning in a warehouse works best when it is tied to the use of radio frequency terminals which transmit the image received of the bar code. Scanning has moved the power of the computer out of the office and onto the warehouse floor. It provides a means of fast and accurate data entry, which allows warehouse employees to collect data from anywhere in the building. Radio frequency terminals transmit the scanned data directly to a host computer. The computer then sends a response which provides the worker with the instructions needed to complete the current task and start the next one. Here are six examples of the sort of information exchanged in these response messages:

Next item to pick
Location to pick
Quantity
Verification of item picked
Verification of item for shipping
Staging or storage location for receipted container.

Scanning can transform an inventory management system from a batch-oriented system which merely chronicles what *did* happen, to a real-time system which directs what *will* happen.

Benefits

It is helpful to divide the benefits into two groups. The first group includes those benefits which enhance the value of the product. The second has benefits which reduce warehouse operating costs.

Four benefits which enhance the time-and-place value of a product are:

- Improved delivery services
- Improved customer service
- Improved quality and dependability
- New or improved information services

The ability to provide these services may well result in increased market share or support an increase in price.

Three benefits that tend to reduce warehouse operating costs are:

- Increased productivity
- Improved accuracy
- Improved space utilization

With scanning, annual physical inventories can be taken in a few hours with consistent accuracy to within .01%. Scanning with real-time inventory control will result in productivity gains of at least twenty percent. Receipting is an excellent example. A team member scans the inbound container, communicating all receipt information such as part number, quantity, lot number and other data to the host computer. The host computer checks all orders for an outbound order matching that receipt. If one is found, the system fulfills the order requirement and sends a message to the team member indicating where outbound freight should be delivered. If no order requirement exists for that item, then the system assigns it to a storage location based on variables such as container type, part number, product grouping and demand.

Improvements in accuracy are also quite dramatic, as are the benefits flowing from these improvements. Safety stocks can be reduced. Stockouts and the corresponding lost sales or lost production can be significantly reduced, if not eliminated. Stock rotation can be improved, thereby reducing product loss due to obsolescence or spoilage. Returns due to shipping errors are practically eliminated. Better inventory control allows product "loss" to be identified and eliminated.

Improvements in space utilization are gained in two ways. First, storage requirements can be reduced by increasing throughput. Second, a real-time system will permit better use of remaining storage space.

Productivity measuring and reporting accuracy can be dramatically improved because the system is designed to capture information about the work being performed. After selecting any terminal function or scanning activity, the team member must scan a bar code on his identification card. As the work continues, the computer stores information about how many units were handled and how much time elapsed during the handling process.[1]

Using the System

Here is how the system will work in the warehouse cycle. At the receiving dock, the bar codes for inbound materials are read and entered into the computer. The computer returns instructions on the locations to be used in storing inbound goods. Then the "put-away" function is completed, with the warehouse operator entering both stock location and Stock Keeping Unit (SKU) identification as each item is put into storage. These are verified to ensure that the specified put-away instructions are carried out accurately. The employee also reads his personal bar code identifier, so that a record is kept as to *who* put the merchandise away and how long the operation took. This allows identification of errors as well as tracking of individual productivity.

It is the order-picking function which gains the greatest advantages from the use of automatic identification. In most warehouses, order picking is labor-intensive and error-prone. Through an automatic identification system, the computer guides picking by selecting the shortest feasible path to pick the items on the list. The picker scans both the SKU identification and the warehouse location identification. The computer verifies that both are as specified, which provides a double verification that the right item was selected.

Once goods are picked, they are typically repacked in larger shipping containers. Packing has always been a manual operation, but computer analysis will provide information to speed the process. By considering the cubic volume of each product, the computer will indicate the number and size of repackaging boxes which will be needed in the operation. Automatic identification will be used to prepare the final packing list as items are placed in the reshipment box. Through

---

[1] This section is by Russell A. Gilmore III, President of Autocon, Inc. It first appeared in Volume 4, No. 2 of Warehousing Forum, ©The Ackerman Company.

the use of bar codes, the packing operation will be done more quickly and accurately than before.

In shipping, automatic identification, scales, and automated routing are all used to improve accuracy and save time.[2]

The Future of Electronic Identification

Perhaps the most interesting aspect of bar coding as an electronic identification system is that the cost of equipment has steadily come down, primarily because usage has gone up and thus created economies of scale. Today, bar coding is the commonly-used method of electronic identification. However, there are alternatives which could become popular at a later date. One of these is optical character recognition systems which are able to "read" ordinary printing rather than codes.[3]

As costs continue to fall, we are likely to see the day when electronic identification at a warehouse receiving dock is as common as lift trucks and dock plates. The dividend to every user of warehouses will be faster handling, fewer errors, and faster and more accurate transmission of inventory information.

---

[2] This section is by Morton T. Yeomans of Integrated Automation. It first appeared in Vol. 2, No. 11 of Warehousing Forum ©The Ackerman Company.

[3] This chapter is based on articles written by Bob Promisel, Management Consultant. Parts of this appeared in Vol. 20, Nos. 3 & 4 of *Warehousing and Physical Distribution Productivity Report,* ©Marketing Publications, Inc., Silver Spring, MD. R. Gilmore of Autocon provided photography and information about warehousing applications.

Appendix 1

UNIFORM COMMERCIAL CODE

ARTICLE 7 - Warehouse Receipts, Bills of Lading and Other Documents of Title

PART 1 - GENERAL

Sec. 7-101. *Short Title*

This Article shall be known and may be cited as Uniform Commercial Code—Documents of Title.

Sec. 7-102. *Definitions and Index of Definitions.*

(1) In this Article, unless the context otherwise requires:

(a) "Bailee" means the person who by a warehouse receipt, bill of lading, or other document of title acknowledges possession of goods and contracts to deliver them.

(b) "Consignee" means the person named in a bill to whom or to whose order the bill promises delivery.

(c) "Consignor" means the person named in a bill as the person from whom the goods have been received for shipment.

(d) "Delivery order" means a written order to deliver goods directed to a warehouseman, carrier or other person who in the ordinary course of business issues warehouse receipts or bills of lading.

(e) "Document" means document of titles as defined in the general definitions in Article 1 (Section 1-201).

(f) "Goods" means all things which are treated as movable for the purposes of a contract of storage or transportation.

(g) "Issuer" means a bailee who issues a document except that in relation to an unaccepted delivery order it means the person who orders the possessor of goods to deliver. Issuer includes any person for whom an agent or employee purports to act in issuing a document if the agent or employee has real or apparent authority to issue documents, notwithstanding that the issuer received no goods or that the goods were misdescribed or that in any other respect the agent or employee violated his instructions.

(h) "Warehouseman" is a person engaged in the business of storing goods for hire.

(2) Other definitions applying to this Article or to specified Parts thereof, and the sections in which they appear are:

"Duly negotiate." Section 7-501.

"Person entitled under the document." Section 7-403(4).

(3) Definitions in other Articles applying to this Article and the sections in which they appear are:

"Contract for sale." Section 2-106.

"Overseas." Section 2-323.

"Receipt of goods." Section 2-103.

(4) In addition Article 1 contains general definitions and principles of construction and interpretation applicable throughout this Article.

Sec. 7-103. *Relation of Article to Treaty, Statute, Tariff, Classification or Regulation.*

To the extent that any treaty or statute of the United States, regulatory statute of this State or tariff, classification or regulation filed or issued pursuant thereto is applicable, the provisions of this Article are subject thereto.

Sec. 7-104. *Negotiable and Non-Negotiable Warehouse Receipt, Bill of Lading or Other Document of Title.*

(1) A warehouse receipt, bill or lading or other document of title is negotiable

(a) if by its terms the goods are to be delivered to bearer or to the order of a named person; or

(b) where recognized in overseas trade, if it runs to a named person or assigns.

Any other document is non-negotiable. A bill of lading in which it is stated that the goods are consigned to a named person is not made negotiable by a provision that the goods are to be delivered only against a written order signed by the same or another person.

Sec. 7-105. *Construction Against Negative Implication.*

The omission from either Part 2 or Part 3 of this Article of a provision corresponding to a provision made in the other Part does not imply that a corresponding rule of law is not applicable.

PART 2—WAREHOUSE RECEIPTS: SPECIAL PROVISIONS

Sec. 7-201. *Who May Issue a Warehouse Receipt; Storage Under Government Bond.*

(1) A warehouse receipt may be issued by any warehouseman.

(2) Where goods including distilled spirits and agricultural commodities are stored under a statute requiring a bond against withdrawal or a license for the issuance of receipts in the nature of warehouse receipts, a receipt issued for

the goods has like effect as a warehouse receipt even though issued by a person who is the owner of the goods and is not a warehouseman.

Sec. 7-202. *Form of Warehouse Receipt; Essential Terms; Optional Terms.*

(1) A warehouse receipt need not be in any particular form.

(2) Unless a warehouse receipt embodies within its written or printed terms each of the following, the warehouseman is liable for damages caused by the omission to a person injured thereby:

(a) the location of the warehouse where the goods are stored;

(b) the date of issue of the receipt;

(c) the consecutive number of the receipt;

(d) a statement whether the goods received will be delivered to the bearer, to a specified person, or to a specified person or his order;

(e) the rate of storage and handling charges, except that where goods are stored under a field warehousing agreement a statement of that fact is sufficient on a non-negotiable receipt;

(f) a description of the goods or of the packages containing them;

(g) the signature of the warehouseman, which may be made by his authorized agent;

(h) if the receipt is issued for goods of which the warehouseman is owner, either solely or jointly or in common with others, the fact of such ownership; and

(i) a statement of the amount of advances made and of liabilities incurred for which the warehouseman claims a lien or security interest (Section 7-209). If the precise amount of such advances made or of such liabilities incurred is, at the time of the issue of the receipt, unknown to the warehouseman or to his agent who issues it, a statement of the fact that advances have been made or liabilities incurred and the purpose thereof is sufficient.

(3) A warehouseman may insert in his receipt any other terms which are not contrary to the provisions of this Act and do not impair his obligation of delivery (Section 7-403) or his duty of care (Section 7-204). Any contrary provisions shall be ineffective.

Sec. 7-203. *Liability for Non-Receipt or Misdescription.*

A party to or purchaser for value in good faith of a document of title other than a bill of lading relying in either case upon the description therein of the goods may recover from the issuer damages caused by the non-receipt or misdescription of the goods, except to the extent that the document conspicuously indicates that the issuer does not know whether any part or all of the goods in fact were received or conform to the description, as where the description is in terms of marks or labels or kind, quantity or condition, or the receipt or description is qualified by "contents, condition and quality unknown," "said to contain" or the like, if such indication be true, or the party or purchaser otherwise has notice.

Sec. 7-204. *Duty of Care; Contractual Limitations of Warehouseman's Liability.*

(1) A warehouseman is liable for damages for loss of or injury to the goods caused by his failure to exercise such care in regard to them as a reasonably careful man would exercise under like circumstances but unless otherwise agreed he is not liable for damages which could not have been avoided by the exercise of such care.

(2) Damages may be limited by a term in the warehouse receipt or storage agreement limiting the amount of liability in case of loss or damage, and setting forth a specific liability per article or item, or value per unit of weight, beyond which the warehouseman shall not be liable; provided, however, that such liability may on written request of the bailor at the time of signing such storage agreement or within a reasonable time after receipt of the warehouse receipt be increased on part or all of the goods thereunder, in which event increased rates may be charged based on such increased valuation, but that no such increase shall be permitted contrary to a lawful limitation of liability contained in the warehouseman's tariff, if any. No such limitation is effective with respect to the warehouseman's liability for conversion to his own use.

(3) Reasonable provisions as to the time and manner of presenting claims and instituting actions based on the bailment may be included in the warehouse receipt or tariff.

(4) This section does not impair or repeal . . .

NOTE: Insert in subsection (4) a reference to any statute which imposes a higher responsibility upon the warehouseman or invalidates contractual limitations which would be permissible under this Article.

Sec. 7-205. *Title Under Warehouse Receipt Defeated in Certain Cases.*

A buyer in the ordinary course of business of fungible goods sold and delivered by a warehouseman who is also in the business of buying and selling such goods takes free of any claim under a warehouse receipt even though it has been duly negotiated.

Sec. 7-206. *Termination of Storage at Warehouseman's Option.*

(1) A warehouseman may on notifying the person on whose account the goods are held and any other person known to claim an interest in the goods require payment of any charges and removal of the goods from the warehouse at the termination of the period of storage fixed by the document, or, if no period is fixed, within a stated period not less than thirty days after the notification. If the goods are not removed before the date specified in the notification, the ware-

houseman may sell them in accordance with the provisions of the section on enforcement of a warehouseman's lien (Section 7-210).

(2) If a warehouseman in good faith believes that the goods are about to deteriorate or decline in value to less than the amount of his lien within the time prescribed in subsection (1) for notification, advertisement and sale, the warehouseman may specify in the notification any reasonable shorter time for removal of the goods and in case the goods are not removed, may sell them at public sale held not less than one week after a single advertisement or posting.

(3) If as a result of a quality or condition of the goods of which the warehouseman had no notice at the time of deposit the goods are a hazard to other property or to the warehouse or to persons, the warehouseman may sell the goods at public or private sale without advertisement on reasonable notification to all persons known to claim an interest in the goods. If the warehouseman after a reasonable effort is unable to sell the goods he may dispose of them in any lawful manner and shall incur no liability by reason of such disposition.

(4) The warehouseman must deliver the goods to any person entitled to them under this Article upon due demand made at any time prior to sale or other disposition under this section.

(5) The warehouseman may satisfy his lien from the proceeds of any sale or disposition under this section but must hold the balance for delivery on the demand of any person to whom he would have been bound to deliver the goods.

Sec. 2-207. *Goods Must Be Kept Separate; Fungible Goods.*

(1) Unless the warehouse receipt otherwise provides, a warehouseman must keep separate the goods covered by each receipt so as to permit at all times identification and delivery of those goods except that different lots of fungible goods may be commingled.

(2) Fungible goods so commingled are owned in common by the persons entitled thereto and the warehouseman is severally liable to each owner for that owner's share. Where because of overissue a mass of fungible goods is insufficient to meet all the receipts which the warehouseman has issued against it, the persons entitled include all holders to whom overissued receipts have been duly negotiated.

Sec. 7-208. *Altered Warehouse Receipts.*

Where a blank in a negotiable warehouse receipt has been filled in without authority, a purchaser for value and without notice of the want of authority may treat the insertion as authorized. Any other unauthorized alteration leaves any receipt enforceable against the issuer according to its original tenor.

Sec. 7-209. *Lien of Warehouseman.*

(1) A warehouseman has a lien against the bailor on the goods covered by a warehouse receipt or on the proceeds thereof in his possession for charges for

storage or transportation (including demurrage and terminal charges), insurance, labor, or charges present or future in relation to the goods, and for expenses necessary for preservation of the goods or reasonably incurred in their sale pursuant to law. If the person on whose account the goods are held is liable for like charges or expenses in relation to other goods whenever deposited and it is stated in the receipt that a lien is claimed for charges and expenses in relation to other goods, the warehouseman also has a lien against him for such charges and expenses whether or not the other goods have been delivered by the warehouseman. But against a person to whom a negotiable warehouse receipt is duly negotiated a warehouseman's lien is limited to charges in an amount or at a rate specified on the receipt or if no charges are so specified then to a reasonable charge for storage of the goods covered by the receipt subsequent to the date of the receipt.

(2) The warehouseman may also reserve a security interest against the bailor for a maximum amount specified on the receipt for charges other than those specified in subsection (1), such as for money advanced and interest. Such a security interest is governed by the Article on Secured Transactions (Article 9).

(3) A warehouseman's lien for charges and expenses under subsection (1) or a security interest under subsection (2) is also effective against any person who so entrusted the bailor with possession of the goods that a pledge of them by him to a good faith purchaser for value would have been valid but is not effective against a person as to whom the document confers no right in the goods covered by it under Section 7-503.

(4) A warehouseman loses his lien on any goods which he voluntarily delivers or which he unjustifiably refuses to deliver.

Sec. 7-210. *Enforcement of Warehouseman's Lien.*

(1) Except as provided in subsection (2), a warehouseman's lien may be enforced by public or private sale of the goods in bloc or in parcels, at any time or place and on any terms which are commercially reasonable, after notifying all persons known to claim an interest in the goods. Such notification must include a statement of the amount due, the nature of the proposed sale and the time and place of any public sale. The fact that a better price could have been obtained by a sale at a different time or in a different method from that selected by the warehouseman is not of itself sufficient to establish that the sale was not made in a commercially reasonable manner. If the warehouseman either sells the goods in the usual manner in any recognized market therefor, or if he sells at the price current in such market at the time of his sale, or if he has otherwise sold in conformity with commercially reasonable practices among dealers in the type of goods sold, he has sold in a commercially reasonable manner. A sale of

more goods than apparently necessary to be offered to insure satisfaction of the obligation is not commercially reasonable except in cases covered by the preceding sentence.

(2) A warehouseman's lien on goods other than goods stored by a merchant in the course of his business may be enforced only as follows:

(a) All persons known to claim an interest in the goods must be notified.

(b) The notification must be delivered in person or sent by registered letter to the last known address of any person to be notified.

(c) The notification must include an itemized statement of the claim, a description of the goods subject to the lien, a demand for payment within a specified time not less than ten days after receipt of the notification, and a conspicuous statement that unless the claim is paid within that time the goods will be advertised for sale and sold by auction at a specified time and place.

(d) The sale must conform to the terms of the notification.

(e) The sale must be held at the nearest suitable place to that where the goods are held or stored.

(f) After the expiration of the time given in the notification, an advertisement of the sale must be published once a week for two weeks consecutively in a newspaper of general circulation where the sale is to be held. The advertisement must include a description of the goods, the name of the person on whose account they are being held, and time and place of the sale. The sale must take place at least fifteen days after the first publication. If there is no newspaper of general circulation where the sale is to be held, the advertisement must be posted at least ten days before the sale in not less than six conspicuous places in the neighborhood of the proposed sale.

(3) Before any sale pursuant to this section any person claiming a right in the goods may pay the amount necessary to satisfy the lien and the reasonable expenses incurred under this section. In that event the goods must not be sold, but must be retained by the warehouseman subject to the terms of the receipt of this Article.

(4) The warehouseman may buy at any public sale pursuant to this section.

(5) A purchaser in good faith of goods sold to enforce a warehouseman's lien takes the goods free of any rights of persons against whom the lien was valid, despite noncompliance by the warehouseman with the requirements of this section.

(6) The warehouseman may satisfy his lien from the proceeds of any sale pursuant to this section but must hold the balance, if any, for delivery on demand to any person to whom he would have been bound to deliver the goods.

(7) The rights provided by this section shall be in addition to all other rights allowed by law to a creditor against his debtor.

(8) Where a lien is on goods stored by a merchant in the course of his business the lien may be enforced in accordance with either subsection (1) or (2).

(9) The warehouseman is liable for damages caused by failure to comply with the requirements for sale under this section and in case of willful violation is liable for conversion.

PART 3—BILLS OF LADING: SPECIAL PROVISIONS

(Omitted)

PART 4—WAREHOUSE RECEIPTS AND BILLS OF LADING: GENERAL OBLIGATIONS

Sec. 7-401. *Irregularities in Issue of Receipt of Bill or Conduct of Issuer.*

The obligations imposed by this Article on an issuer apply to a document of title regardless of the fact that

(a) the document may not comply with the requirements of this Article or of any other law or regulation regarding its issue, form or content; or

(b) the issuer may have violated laws regulating the conduct of his business; or

(c) the goods covered by the document were owned by the bailee at the time the document was issued; or

(d) the person issuing the document does not come within the definition of warehouseman if it purports to be a warehouse receipt.

Sec. 7-402. *Duplicate Receipt or Bill; Overissue.*

Neither a duplicate nor any other document of title purporting to cover goods already represented by an outstanding document of the same issuer confers any right in the goods, except as provided in the case of bills in a set, overissue of documents for fungible goods and substituted for lost, stolen or destroyed documents. But the issuer is liable for damages caused by his overissue or failure to identify a duplicate document as such by conspicuous notation on its face.

Sec. 7-403. *Obligation of Warehouseman or Carrier to Deliver; Excuse.*

(1) The bailee must deliver the goods to a person entitled under the document who complies with subsections (2) and (3), unless and to the extent that the bailee establishes any of the following:

(a) delivery of the goods to a person whose receipt was rightful as against the claimant;

(b) damage to or delay, loss or destruction of the goods for which the bailee is not liable (but the burden of establishing negligence in such cases is on the person entitled under the document);

NOTE: The brackets in (1)(b) indicate that State enactments may differ on this point without serious damage to the principle of uniformity.

(c) previous sale or other disposition of the goods in lawful enforcement of a lien or on warehouseman's lawful termination of storage;

(d) the exercise by a seller of his right to stop delivery pursuant to the provisions of the Article on Sales (Section 2-705);

(e) a diversion, reconsignment or other disposition pursuant to the provisions of this Article (Section 7-303) or tariff regulating such right;

(f) release, satisfaction or any other fact affording a personal defense against the claimant;

(g) any other lawful excuse.

(2) A person claiming goods covered by a document of title must satisfy the bailee's lien where the bailee so requests or where the bailee is prohibited by law from delivering the goods until the charges are paid.

(3) Unless the person claiming is one against whom the document confers no right under Sec. 7-503(1), he must surrender for cancellation or notation of partial deliveries any outstanding negotiable document covering the goods, and the bailee must cancel the document or conspicuously note the partial delivery thereon or be liable to any person to whom the document is duly negotiated.

(4) "Person entitled under the document" means holder in the case of a negotiable document, or the person to whom delivery is to be made by the terms of or pursuant to written instructions under a non-negotiable document.

Sec. 7-404. *No Liability for Good Faith Delivery Pursuant to Receipt or Bill.*

A bailee who in good faith including observance of reasonable commercial standards has received goods and delivered or otherwise disposed of them according to the terms of the document of title or pursuant to this Article is not liable therefor. This rule applies even though the person from whom he received the goods had no authority to procure the document or to dispose of the goods and even though the person to whom he delivered the goods had no authority to receive them.

PART 5—WAREHOUSE RECEIPTS AND BILLS OF LADING: NEGOTIATION AND TRANSFER

Sec. 7-501. *Form of Negotiation and Requirements of "Due Negotiation."*

(1) A negotiable document of title running to the order of a named person

is negotiated by his endorsement and delivery. After his endorsement in blank or to bearer any person can negotiate it by delivery alone.

(2)(a) A negotiable document of title is also negotiated by delivery alone when by its original terms it runs to bearer.

(b) When a document running to the order of a named person is delivered to him the effect is the same as if the document had been negotiated.

(3) Negotiation of a negotiable document of title after it has been endorsed to a specified person requires endorsement by the special endorsee as well as delivery.

(4) A negotiable document of title is "duly negotiated" when it is negotiated in the manner stated in this section to a holder who purchases it in good faith without notice of any defense against or claim to it on the part of any person and for value, unless it is established that the negotiation is not in the regular course of business or financing or involves receiving the document in settlement or payment of a money obligation.

(5) Indorsement of a non-negotiable document neither makes it negotiable nor adds to the transferee's rights.

(6) The naming in a negotiable bill of a person to be notified of the arrival of the goods does not limit the negotiability of the bill nor constitute notice to a purchaser thereof of any interest of such person in the goods.

Sec. 7-502. *Rights Acquired by Due Negotiation.*

(1) Subject to the following section and to the provisions of Section 7-205 on fungible goods, a holder to whom a negotiable document of title has been duly negotiated acquires thereby:

(a) title to the document;

(b) title to the goods;

(c) all rights accruing under the law of agency or estoppel, including rights to goods delivered to the bailee after the document was issued; and

(d) the direct obligation of the issuer to hold or deliver the goods according to the terms of the document free of any defense or claim by him except those arising under the terms of the document or under this Article. In the case of a delivery order the bailee's obligation accrues only upon acceptance and the obligation acquired by the holder is that the issuer and any endorser will procure the acceptance of the bailee.

(2) Subject to the following section, title and rights so acquired are not defeated by any stoppage of the goods represented by the document or by surrender of such goods by the bailee, and are not impaired even though the negotiation or any prior negotiation constituted a breach of duty or even though any person has been deprived of possession of the document by misrepresentation, fraud, accident, mistake, duress, loss, theft or conversion, or even though a pre-

vious sale or other transfer of the goods or document has been made to a third person.

Sec. 7-503. *Document of Title to Goods Defeated in Certain Cases.*

(1) A document of title confers no right in goods against a person who before issuance of the document had a legal interest or a perfected security interest in them and who neither

(a) delivered or entrusted them or any document of title covering them to the bailor or his nominee with actual or apparent authority to ship, store or sell or with power to obtain delivery under this Article (Section 7-403) or with power of disposition under this Act (Sections 2-403 and 9-307) or other statute or rule of law; nor

(b) acquiesced in the procurement by the bailor or his nominee of any document of title.

(2) Title to goods based upon an unaccepted delivery order is subject to the rights of anyone to whom a negotiable warehouse receipt or bill of lading covering the goods has been duly negotiated. Such a title may be defeated under the next section to the same extent as the rights of the issuer or a transferee from the issuer.

(3) Title to goods based upon a bill of lading issued to a freight forwarder is subject to the rights of anyone to whom a bill issued by the freight forwarder is duly negotiated; but delivery by the carrier in accordnce with Part 4 of this Article pursuant to its own bill of lading discharges the carrier's obligation to deliver.

Sec. 7-504. *Rights Acquired in the Absence of Due Negotiation; Effect of Diversion; Seller's Stoppage of Delivery.*

(1) A transferee of a document, whether negotiable or non-negotiable, to whom the document has been delivered but not duly negotiated, acquires the title and rights which his transferor had or had actual authority to convey.

(2) In the case of a non-negotiable document, until but not after the bailee receives notification of the transfer, the rights of the transferee may be defeated.

(a) by those creditors of the tranferor who could treat the sale as void under Section 2-402; or

(b) by a buyer from the transferor in ordinary course of business if the bailee has delivered the goods to the buyer or received notification of his rights; or

(c) as against the bailee by good faith dealings of the bailee with the transferor.

(3) A diversion or other change of shipping instructions by the consignor in a non-negotiable bill of lading which causes the bailee not to deliver to the consignee defeats the consignee's title to the goods if they have been delivered

to a buyer in ordinary course of business and in any event defeats the consignee's right against the bailee.

(4) Delivery pursuant to a non-negotiable document may be stopped by a seller under Section 2-705, and subject to the requirement of due notification there provided. A bailee honoring the seller's instructions is entitled to be indemnified by the seller against any resulting loss or expense.

Sec. 7-505. *Indorser Not a Guarantor for Other Parties.*

The endorsement of a document of title issued by a bailee does not make the endorser liable for any default by the bailee or by previous endorsers.

Sec. 7-506. *Delivery Without Indorsement: Right to Compel Indorsement.*

The transferee of a negotiable document of title has a specially enforceable right to have his transferor supply any necessary indorsement but the transfer becomes a negotiation only as of the time the indorsement is supplied.

Sec. 7-507. *Warranties on Negotiation or Transfer of Receipt of Bill.*

Where a peron negotiates or transfers a document of title for value otherwise than as a mere intermediary under the next following section, then unless otherwise agreed he warrants to his immediate purchaser only in addition to any warranty made in selling the goods

(a) that the document is genuine; and

(b) that he had no knowledge of any fact which would impair its validity or worth; and

(c) that his negotiation or transfer is rightful and fully effective with respect to the title to the document and the goods it represents.

Sec. 7-508. *Warranties of Collecting Bank as to Documents.*

A collecting bank or other intermediary known to be entrusted with documents on behalf of another or with collection of a draft or other claim against delivery of documents warrants by such delivery of the documents only its own good faith and authority. This rule applies even though the intermediary has purchased or made advances against the claim or draft to be collected.

Sec. 7-509. *Receipt or Bill: When Adequate Compliance with Commercial Contract.*

The question whether a document is adequate to fulfill the obligations of a contract for sale or the conditions of a credit is governed by the Articles on Sales (Article 2) and on Letters of Credit (Article 5).

PART 6—WAREHOUSE RECEIPTS AND BILLS OF LADING: MISCELLANEOUS PROVISIONS

Sec. 7-601. *Lost and Missing Documents.*

(1) If a document has been lost, stolen or destroyed, a court may order

delivery of the goods or issuance of a substitute document and the bailee may without liability to any person comply with such order. If the document was negotiable the claimant must post security approved by the court to indemnify any person who may suffer loss as a result of non-surrender of the document. If the document was not negotiable, such security may be required at the discretion of the court. The court may also in its discretion order payment of the bailee's reasonable costs and counsel fees.

(2) A bailee who without court order delivers goods to a person claiming under a missing negotiable document is liable to any person injured thereby, and if the delivery is not in good faith becomes liable for conversion. Delivery in good faith is not conversion if made in accordance with a filed classification or tariff or, where no classification or tariff is filed, if the claimant posts security with the bailee in an amount at least double the value of the goods at the time of posting to indemnify any person injured by the delivery who files a notice of claim within one year after the delivery.

Sec. 7-602. *Attachment of Goods Covered by a Negotiable Document.*

Except where the document was originally issued upon delivery of the goods by a person who had no power to dispose of them, no lien attaches by virtue of any judicial process to goods in the possession of a bailee for which a negotiable document of title is outstanding unless the document be first surrendered to the bailee or its negotiation enjoined, and the bailee shall not be compelled to deliver the goods pursuant to process until the document is surrendered to him or impounded by the court. One who purchases the document for value without notice of the process of injunction takes free of the lien imposed by judicial process.

Sec. 7-603. *Conflicting Claims; Interpleader.*

If more than one person claims title or possession of the goods, the bailee is excused from delivery until he has had a reasonable time to ascertain the validity of the adverse claims or to bring an action to compel all claimants to interplead and may compel such interpleader, either in defending an action for non-delivery of the goods, or by original action, whichever is appropriate.

Appendix 2

AWA BOOKLET OF DEPOSITOR-WAREHOUSEMAN AGREEMENTS

STANDARD CONTRACT TERMS AND CONDITIONS FOR MERCHANDISE WAREHOUSEMEN

(Approved and Promulgated by the American Warehousemen's Association, October 1968)

Acceptance—Sec. 1

(a) This contract and rate quotation including accessorial charges endorsed on or attached hereto must be accepted within 30 days from the proposal date by signature of depositor on the reverse side of the contract. In the absence of written acceptance, the act of tendering goods described herein for storage or other services by warehousemen within 30 days from the proposal date shall constitute such acceptance by depositor.

(b) In the event that goods tendered for storage or other services do not conform to the description contained herein, or conforming goods are tendered after 30 days from the proposal date without prior written acceptance by depositor as provided in paragraph (a) of this section, warehouseman may refuse to accept such goods. If warehouseman accepts such goods, depositor agrees to rates and charges as may be assigned and invoiced by warehouseman and to all terms of this contract.

(c) This contract may be cancelled by either party upon 30 days' written notice and is cancelled if no storage or other services are performed under this contract for a period of 180 days.

Shipping—Sec. 2

Depositor agrees not to ship goods to warehouseman as the named consignee. If, in violation of this agreement, goods are shipped to warehouseman as named consignee, depositor agrees to notify carrier in writing prior to such shipment, with copy of such notice to the warehouseman, that warehouseman named as consignee is a warehouseman and has no beneficial title or interest in such property and depositor further agrees to indemnify and hold harmless warehouseman from any and all claims for unpaid transportation charges, including undercharges, demurrage, detention or charges of any nature, in connection with goods so shipped. Depositor further agrees that, if it fails to notify carrier as required by the next preceding sentence, warehouseman shall have the right to refuse such goods and shall not be liable or responsible for any loss, injury or damage of any nature to, or related to, such goods. Depositor agrees that all promises contained in this section will be binding on depositor's heirs, successors and assigns.

Tender for Storage—Sec. 3

All goods for storage shall be delivered at the warehouse properly marked and packaged for handling. The depositor shall furnish at or prior to such delivery, a manifest showing marks, brands, or sizes to be kept and accounted for separately, and the class of storage and other services desired.

Storage Period and Charges—Sec. 4

(a) All charges for storage are per package or other agreed unit per month.

(b) Storage charges become applicable upon the date that warehouseman accepts care, custody and control of the goods, regardless of unloading date or date of issue of warehouse receipt.

(c) Except as provided in paragraph (d) of this section, a full month's storage charge will apply on all goods received between the first and the 15th, inclusive, of a calendar month; one-half month's storage charge will apply on all goods received between the 16th and last day, inclusive, of a calendar month, and a full month's storage charge will apply to all goods in storage on the first day of the next and succeeding calendar months. All storage charges are due and payable on the first day of storage for the initial month and thereafter on the first day of the calendar month.

(d) When mutually agreed upon by the warehouseman and the depositor, a storage month shall extend from a date in one calendar month to, but not including the same date of the next and all succeeding months. All storage charges are due and payable on the first day of the storage month.

Transfer, Termination of Storage, Removal of Goods—Sec. 5

(a) Instructions to transfer goods on the books of the warehouseman are not effective until delivered to and accepted by warehouseman, and all charges up to the time transfer is made are chargeable to the depositor of record. If a transfer involves rehandling the goods, such will be subject to a charge. When goods in storage are transferred from one party to another through issuance of a new warehouse receipt, a new storage date is established on the date of transfer.

(b) The warehouseman reserves the right to move, at his expense, 14 days after notice is sent by certified or registered mail to the depositor of record or to the last known holder of the negotiable warehouse receipt, any goods in storage from the warehouse in which they may be stored to any other of his warehouses; but if such depositor or holder takes delivery of his goods in lieu of transfer, no storage charge shall be made for the current storage month. The warehouseman may, without notice, move goods within the warehouse in which they are stored.

(c) The warehouseman may, upon written notice to the depositor of record and any other person known by the warehouseman to claim an interest in the goods, require the removal of any goods by the end of the next succeeding storage month. Such notice shall be given to the last known place of business or abode of the person to be notified. If goods are not removed before the end of the next succeeding storage month, the warehouseman may sell them in accordance with applicable law.

(d) If warehouseman in good faith believes that the goods are about to deteriorate or decline in value to less than the amount of warehouseman's lien before the end of the next succeeding storage month, the warehouseman may specify in the notification any reasonable shorter time for removal of the goods and in case the goods are not removed, may sell them at public sale held one week after a single advertisement or posting as provided by law.

(e) If as a result of a quality or condition of the goods of which the warehouseman had no notice at the time of deposit the goods are a hazard to other property or to the warehouse or to persons, the warehouseman may sell the goods at public or private sale without advertisement on reasonable notification to all persons known to claim an interest in the goods. If the warehouseman after a reasonable effort is unable to sell the goods he may dispose of them in any lawful manner and shall incur no liability by reason of such disposition. Pending such disposition, sale or return of the goods, the warehouseman may remove the goods from the warehouse and shall incur no liability by reason of such removal.

Handling—Sec. 6

(a) The handling charge covers the ordinary labor involved in receiving goods at warehouse door, placing goods in storage, and returning goods to warehouse door. Handling charges are due and payable on receipt of goods.

(b) Unless otherwise agreed, labor for unloading and loading goods will be subject to a charge. Additional expenses incurred by the warehouseman in receiving and handling damaged goods, and additional expense in unloading from or loading into cars or other vehicles not at warehouse door will be charged to the depositor.

(c) Labor and materials used in loading rail cars or other vehicles are chargeable to the depositor.

(d) When goods are ordered out in quantities less than in which received, the warehouseman may make an additional charge for each order or each item of an order.

(e) The warehouseman shall not be liable for demurrage, delays in unloading inbound cars, or delays in obtaining and loading cars for outbound shipment unless warehouseman has failed to exercise reasonable care.

Delivery Requirements—Sec. 7

(a) No goods shall be delivered or transferred except upon receipt by the warehouseman of complete instructions properly signed by the depositor. However, when no negotiable receipt is outstanding, goods may be delivered upon instructions by telephone in accordance with a prior written authorization, but the warehouseman shall not be responsible for loss or error occasioned thereby.

(b) When a negotiable receipt has been issued no goods covered by that receipt shall be delivered, or transferred on the books of the warehouseman, unless the receipt, properly endorsed, is surrendered for cancellation, or for endorsement of partial delivery thereon. If a negotiable receipt is lost or destroyed, delivery of goods may be made only upon order of a court of competent jurisdiction and the posting of security approved by the court as provided by law.

(c) When goods are ordered out a reasonable time shall be given the warehouseman to carry out instructions, and if he is unable because of acts of God, war, public enemies, seizure under legal process, strikes, lockouts, riots and civil commotions, or any reason beyond the warehouseman's control, or because of loss or destruction of goods for which warehouseman is not liable, or because of any other excuse provided by law, the warehouseman shall not be liable for failure to carry out such instructions and goods remaining in storage will continue to be subject to regular storage charges.

Extra Services (Special Services)—Sec. 8

(a) Warehouse labor required for services other than ordinary handling and storage will be charged to the depositor.

(b) Special services requested by depositor including but not limited to compiling of special stock statements; reporting marked weights, serial numbers or other data from packages; physical check of goods; and handling transit billing will be subject to a charge.

(c) Dunnage, bracing, packing materials or other special supplies, may be provided for the depositor at a charge in addition to the warehouseman's cost.

(d) By prior arrangement, goods may be received or delivered during other than usual business hours, subject to a charge.

(e) Communication expense including postage, teletype, telegram, or telephone will be charged to the depositor if such concern more than normal inventory reporting or if, at the request of the depositor, communications are made by other than regular United States Mail.

Bonded Storage—Sec. 9

(a) A charge in addition to regular rates will be made for merchandise in bond.

(b) Where a warehouse receipt covers goods in U.S. Customs bond, such receipt shall be void upon the termination of the storage period fixed by law.

Minimum Charges—Sec. 10

(a) A minimum handling charge per lot and a minimum storage charge per lot per month will be made. When a warehouse receipt covers more than one lot or when a lot is in assortment, a minimum charge per mark, brand, or variety will be made.

(b) A minimum monthly charge to one account for storage and/or handling will be made. This charge will apply also to each account when one customer has several accounts, each requiring separate records and billing.

Liability and Limitation of Damages—Sec. 11

(a) The warehouseman shall not be liable for any loss or injury to goods stored however caused unless such loss or injury resulted from failure by the warehouseman to exercise such care in regard to them as a reasonably careful man would exercise under like circumstances and warehouseman is not liable for damages which would not have been avoided by the exercise of such care.

(b) Goods are not insured by warehouseman against loss or injury however caused.

(c) The depositor declares that damages are limited to _____, provided, however, that such liability may at the time of acceptance of this contract as provided in Section 1 be increased on part or all of the goods hereunder in which event a monthly charge of _____ will be made in addition to the regular monthly storage charge.

Notice of Claim and Filing of Suit - Sec. 12

(a) Claims by the depositor and all other persons must be presented in writing to the warehouseman within a reasonable time, and in no event longer than either 60 days after delivery of the goods by the warehouseman or 60 days after depositor of record or the last known holder of a negotiable warehouse receipt is notified by the warehouseman that loss or injury to part or all of the goods has occurred, whichever time is shorter.

(b) No action may be maintained by the depositor or others against the warehouseman for loss or injury to the goods stored unless timely written claim has been given as provided in paragraph (a) of this section and unless such action is commenced either within nine months after date of delivery by warehouseman or within nine months after depositor of record or the last known holder of a negotiable warehouse receipt is notified that loss or injury to part or all of the goods has occurred, whichever time is shorter.

(c) When goods have not been delivered, notice may be given of known loss or injury to the goods by mailing of a registered or certified letter to the depositor of record or to the last known holder of a negotiable warehouse receipt. Time limitations for presentation of claim in writing and maintaining of action after notice begin on the date of mailing of such notice by warehouseman.

Additional Terms and Conditions Applicable to This Contract and Rate Quotation.

Nothing entered hereon shall be construed to extend the warehouseman's liability beyond the standard of care specified in Section 11 above.

THE ABOVE QUOTATION IS HEREBY ACCEPTED AND TERMS HEREIN AGREED TO

_____
DEPOSITOR

By _____ Date _____ By _____

WAREHOUSE AGREEMENT

THIS AGREEMENT, Made and entered into this ____ day of ____, 19 __, by and between _____
(Company, City State)

hereinafter referred to as "DEPOSITOR," and _____
hereinafter referred to as "WAREHOUSEMAN."

WITNESSETH

WHEREAS, DEPOSITOR is desirous of obtaining and utilizing certain warehouse facilities and services in the _____ area; and

WHEREAS, WAREHOUSEMAN has certain warehousing facilities and services of the type and kind desired by DEPOSITOR located at _____ _____; and

WHEREAS, WAREHOUSEMAN desires to make said facilities and services commercially available to DEPOSITOR subject to the terms and conditions herein specified;

NOW, THEREFORE, for and in consideration of the mutual agreements, covenants and promises herein contained, it is hereby mutually agreed, covenanted and promised as follows:

Article I. Term of Agreement

The term of this Agreement shall commence on the date of its execution by the parties thereto and shall continue thereafter in full force and effect for a period of _____ and shall thereafter automatically renew on a month-to-month basis subject only to either party's right to terminate at any time by serving not less than thirty (30) days' prior written notice to that effect upon the other party, said notice to be effective upon receipt. This Agreement shall be deemed cancelled if DEPOSITOR does not store goods with WAREHOUSEMAN for any period exceeding one hundred and eighty (180) days.

Article II. Acceptance of Goods, Rates and Charges

During the term of this Agreement, and any extensions or renewals thereof, WAREHOUSEMAN agrees to provide for DEPOSITOR certain warehousing facilities and services described in this Agreement and attached Schedules "A" and "B" which are made a part hereof, and to accept and keep in a neat and orderly condition such goods described in Schedule "A" as from time to time may be tendered by DEPOSITOR. WAREHOUSEMAN further agrees to furnish sufficient personnel, equipment, and other accessories necessary to perform efficiently and with safety the services herein described. Rates and charges for public warehousing services are set forth in Schedule "A" and for extra services and conditions in Schedule "B." For any services not specified in Schedule "A" or Schedule "B," DEPOSITOR shall pay to WAREHOUSEMAN such consideration and compensation as may mutually be agreed upon. Consideration for WAREHOUSEMAN'S performance of this Agreement shall be paid to WAREHOUSEMAN by DEPOSITOR within _____ days after receipt by DEPOSITOR of WAREHOUSEMAN'S statement.

Article III. Shipping

DEPOSITOR agrees not to ship goods to WAREHOUSEMAN as the named consignee. If, in violation of this Agreement, goods are shipped to WAREHOUSEMAN as named consignee, DEPOSITOR agrees to notify carrier in writing prior to such shipment, with a copy of such notice to WAREHOUSEMAN, that WAREHOUSEMAN named as consignee is a warehouseman under law and has no beneficial title or interest in such property. DEPOSITOR further agrees to indemnify and hold harmless WAREHOUSEMAN from any and all claims for unpaid transportation charges, including undercharges, demurrage, detention, or charges of any nature in connection with goods so shipped.

Article IV. Tender for Storage

All goods tendered for storage shall be delivered at the warehouse in a segregated manner, properly marked and packaged for handling. DEPOSITOR shall furnish or cause to be furnished at or prior to such delivery, a manifest showing the goods to be kept and accounted for separately, and the class or type of storage and other services desired.

Article V. Storage Period and Charges

(A) Storage charges become applicable upon the date that WAREHOUSEMAN accepts care, custody and control of the goods, regardless of unloading date or date of issue of warehouse receipt.

(B) Unless otherwise stated in Schedule "A" hereof, storage rates and charges shall be computed and are due and payable as follows:

1. Goods received on or after the first (1st) day of the month up to and including the fifteenth (15th) day of the month shall be assessed the full monthly storage charge;

2. Goods received after the fifteenth (15th) day of the month up to and including the last day of the month shall be assessed one-half ($\frac{1}{2}$) of the full monthly storage charge; and

3. A full month's storage charge will apply to all goods in storage on the

first day of the next and succeeding calendar months. All storage charges are due and payable on the first day of storage for the initial month and thereafter on the first day of the calendar month.

Article VI. Termination of Storage, Removal of Goods

(A) Instructions to transfer goods on the books of WAREHOUSEMAN shall not be effective until said instructions are delivered to and accepted by WAREHOUSEMAN, and all charges up to the time transfer is made shall be chargeable to DEPOSITOR. If a transfer involves the rehandling of goods, it will be subject to rates and charges shown in the attached Schedule "A" or Schedule "B" or as otherwise mutually agreed upon. When goods in storage are transferred from one party to another through issuance of a new warehouse receipt, a new storage date is established on the date of such transfer.

(B) WAREHOUSEMAN may, without notice, move goods within the warehouse in which they are stored; but shall not, except as provided in VI(C), move goods to another location without prior consent of DEPOSITOR.

(C) If, as a result of a quality or condition of the goods of which WAREHOUSEMAN had no notice at the time of deposit, the goods are a hazard to other property or to the warehouse or to persons, WAREHOUSEMAN shall immediately notify DEPOSITOR and DEPOSITOR shall thereupon claim its interest in the said goods and remove them from the warehouse. Pending such disposition, the WAREHOUSEMAN may remove the goods from the warehouse and shall incur no liability by reason of such removal.

Article VII. Handling

(A) Handling rates and charges as shown in the attached Schedules shall, unless otherwise agreed, cover the ordinary labor involved in receiving goods at warehouse door or dock. Additional expenses incurred by the warehouseman in loading or unloading cars or vehicles shall be at rates shown in attached Schedules or as otherwise mutually agreed upon.

(B) WAREHOUSEMAN shall not be liable for demurrage, detention or delays in unloading inbound cars or vehicles, or detention or delays in obtaining and loading cars or vehicles for outbound shipment unless WAREHOUSEMAN has failed to exercise reasonable care and judgment as determined by industry practice.

(C) If detention occurs for which WAREHOUSEMAN is liable under Paragraph (B) of this Article, payment of such detention shall be made by DEPOSITOR to the carrier and WAREHOUSEMAN shall reimburse DEPOSITOR for such payment. WAREHOUSEMAN shall keep records concerning the detention of vehicles to assist DEPOSITOR in processing any objection to carrier's imposition of detention charges.

(D) DEPOSITOR shall be responsible for payment of all demurrage charges resulting from receipt by WAREHOUSEMAN of more than _____ carloads of DEPOSITOR'S goods received in any regular working day.

Article VIII. Delivery Requirements

(A) No goods shall be delivered or transferred except upon receipt by WAREHOUSEMAN of complete instructions properly signed by DEPOSITOR. However, when no negotiable receipt is outstanding, goods may be delivered or transferred upon instructions received by telephone, TWX, Dataphone or other agreed upon method of communication, but WAREHOUSEMAN shall not be responsible for loss or error occasioned thereby except as caused by WAREHOUSEMAN'S negligence.

(B) When goods are ordered out, a reasonable time shall be given WAREHOUSEMAN to carry out instructions, and if he is unable to because of acts of God, war, public enemies, seizure under legal process, strikes, lockout, riots and civil commotions, or any other reason beyond WAREHOUSEMAN'S reasonable control, or because of loss or destruction of goods for which WAREHOUSEMAN is not liable, or because of any other excuse provided by law, WAREHOUSEMAN shall not be liable for failure to carry out such instructions provided, however, that goods remaining in storage will continue to be subject to regular storage charges as provided for herein.

Article IX. Extra and Special Services

(A) Warehouse labor required for services other than ordinary handling and storage must be authorized by DEPOSITOR in advance. Rates and charges will be provided for herein or as mutually agreed by the parties hereto.

(B) Special services requested by DEPOSITOR will be subject to such charges as are provided herein or as mutually agreed upon in advance.

(C) Dunnage, bracing, packing materials or other special supplies used in cars or vehicles are chargeable to DEPOSITOR and may be provided at a mutually agreed upon charge in addition to WAREHOUSEMAN'S cost.

(D) By prior arrangement, goods may be received or delivered during other than usual business hours, subject to a reasonable charge.

(E) Communication expense including postage, teletype, telegram or telephone will be charged to DEPOSITOR if such expense is the result of more than normal inventory reporting or if, at the request of DEPOSITOR, communications are made by other than regular First Class United States Mail.

Article X. Bonded Storage

(A) A charge in addition to regular rates will be made for merchandise in bond.

(B) Where a warehouse receipt covers goods in U.S. Customs bond, such receipt shall be void upon the termination of the storage period fixed by law.

Article XI. Inbound Shipments

(A) WAREHOUSEMAN shall immediately notify DEPOSITOR of any known discrepancy on inbound shipments and shall protect DEPOSITOR'S interest by placing an appropriate notation on the delivering carrier's shipping document.

(B) WAREHOUSEMAN may refuse to accept any goods that, because of infestation, contamination or damage, might cause infestation, contamination or damage to WAREHOUSEMAN'S premises or to other goods in the custody of WAREHOUSEMAN and shall immediately notify DEPOSITOR of such refusal and shall have no liability for any demurrage, detention, transportation or other charges by virtue of such refusal.

(C) All notices required under paragraphs A or B of this Article shall be directed to the attention of _____.

Article XII. Liability and Limitation of Damages

(A) WAREHOUSEMAN shall be liable for loss of or injury to all goods while under his care, custody and control when caused by his failure to exercise such care in regard to them as a reasonably careful man would exercise under like circumstances. He shall not be liable for damages which could not have been avoided by the exercise of such care.

(B) In consideration of the rates herein, DEPOSITOR declares that said damages will be limited to _____.

Article XIII. Legal Liability Insurance

WAREHOUSEMAN shall maintain at its sole expense and at all times during the life of this Agreement a policy or policies of legal liability insurance covering any loss, destruction or damage for which WAREHOUSEMAN has assumed responsibility under the terms of Article XII. WAREHOUSEMAN further agrees to provide satisfactory evidence of such insurance upon request by DEPOSITOR.

Article XIV. Notice of Loss and Damage, Claim and Filing of Suit

(A) WAREHOUSEMAN agrees to notify DEPOSITOR promptly of any

loss or damage, howsoever caused, to goods stored or handled under the terms of this agreement. All such notices shall be directed to the attention of _____.

(B) Claims by DEPOSITOR must be presented in writing to WAREHOUSEMAN not longer than either sixty (60) days after delivery of the goods by WAREHOUSEMAN or sixty (60) days after DEPOSITOR is notified by WAREHOUSEMAN that loss or injury to part or all of the goods has occurred, whichever time is shorter.

(C) No action may be maintained by DEPOSITOR against WAREHOUSEMAN for loss or injury to the goods stored unless timely written claim has been given as provided in paragraph (B) of this Article and unless such action is commenced either within nine (9) months after the date of delivery by WAREHOUSEMAN or within nine (9) months after DEPOSITOR is notified that loss or injury to part or all of the goods has occurred, whichever time is shorter.

(D) When goods have not been delivered, notice may be given of known loss or injury to the goods by the mailing of a registered or certified letter to DEPOSITOR. All such notices shall be directed to the attention of _____.

Time limitations for presentation of claim in writing and maintaining of action after notice begin on the date of receipt of such notice by DEPOSITOR.

Article XV. Records

DEPOSITOR reserves the right upon reasonable request to enter WAREHOUSEMAN'S premises during normal working hours to examine and count all or any of the goods stored under the terms of this Agreement. WAREHOUSEMAN shall at all reasonable times permit DEPOSITOR to examine its books, records and accounts for the purpose of reconciling quantities and determining with WAREHOUSEMAN whether certain amounts are payable within the meaning of the Agreement.

Article XVI. Independent Contractor

It is hereby agreed and understood that WAREHOUSEMAN is entering into this Agreement as an independent contractor and that all of WAREHOUSEMAN'S personnel engaged in work to be done under the terms of this Agreement are to be considered employees of WAREHOUSEMAN and under no circumstances shall they be construed or considered to be employees of DEPOSITOR.

DEPOSITOR-WAREHOUSEMAN AGREEMENTS

Article XVII. Compliance with Laws, Ordinances, Rules and Regulations

(A) WAREHOUSEMAN shall comply with all laws, ordinances, rules and regulations of Federal, State, municipal and other governmental authorities and the like in connection with the safeguarding, receiving, storing and handling of goods.

(B) DEPOSITOR shall be responsible for advising WAREHOUSEMAN of all laws, ordinances, rules and regulations of Federal, State, municipal and other governmental authorities and the like relating specifically to the safeguarding, receiving, storing and handling of DEPOSITOR'S products.

Article XVIII. Notification of Product Characteristics

DEPOSITOR shall notify WAREHOUSEMAN of the characteristics of any of DEPOSITOR'S products that may in any way be likely to cause damage to WAREHOUSEMAN'S premises or to other products that may be stored by WAREHOUSEMAN.

Article XIX. Assignment

This Warehousing Agreement shall inure to the benefit of and be binding upon the successors and assigns of the parties hereto, provided neither party to this Agreement shall assign or sublet its interest or obligations herein, including but not limited to the assignment of any monies due and payable, without the prior written consent of the other party.

Article XX. Applicable Statutes

The parties understand and agree that the provisions of Article 7 of the Uniform Commercial Code as enacted by the State law governing this Warehousing Agreement shall apply to this Warehousing Agreement. (In the State of Louisiana the provisions of the Uniform Warehouse Receipts Act shall apply.)

Article XXI. Additional Terms and Conditions

(Nothing entered hereon shall be construed to extend WAREHOUSEMAN'S liability beyond the standard of care specified in Article XII(A) above.)

Article XXII.

This agreement constitutes the entire understanding between DEPOSITOR and WAREHOUSEMAN, and no working arrangements, instructions, or operating manuals intended to facilitate the effective carrying out of this Agreement shall in any way affect the liabilities of either party as set forth herein.

IN WITNESS WHEREOF, the parties hereto have duly executed this Agreement in duplicate the day and year first above written.

WAREHOUSEMAN
By: _____
DEPOSITOR
By: _____

Schedule A
Storage and Handling Rates and Charges

Attached to and made a part of Agreement dated the _____ day of __ _____, 19 ____, by and between _____ and _____

_____
DEPOSITOR
WAREHOUSEMAN

Schedule B
Accessorial Rates and Charges

Attached to and made a part of Agreement dated the _____ day of _____ _____, 19 ____, by and between _____ and _____

_____
DEPOSITOR
WAREHOUSEMAN

ADDITIONAL CONTRACT CONDITIONS AND CONSIDERATIONS

Averaging of Overages and Shortages

The concept of warehousemen and depositors averaging routine "overs" and "shorts" has for many years been considered valid on an almost universal basis, but the manner in which overages and shortages are averaged ranges widely. At one extreme we find simple letter agreements stating, without further explanation, that "overages and shortages will be averaged out," leaving to the parties the resolving of how this will be done, presumably upon the basis of individual facts in a situation and in a manner that is equitable to both. Depending upon the good faith of both parties, this appears to have worked fairly well. Somewhere in the middle, there are written attempts to set forth the understanding in more detail. At the other extreme, although admittedly not frequently, there are letter agreements in which the depositor states to the warehouseman simply, "You will pay us for all shortages and we will pay you for all overages."

For a long time, even though the obvious intent of both parties is that neither will receive an unearned "windfall" at the expense of the other, one area of the overages and shortages situation has generally been ignored by most warehousemen and depositors, and this has to do with the definition of the word "shortage." It appears that the customary intent of warehousemen and depositors is that averaging arrangements handle the normal relatively minor (in relation to total volume) overages and shortages that may be expected in the normal course of business due to clerical error, faulty inventory count, over or under shipments on the part of either depositor or warehouseman, and the like. Few, if any, understandings regarding the averaging of overages and shortages, however, place any quantity, percentage, or dollar limitation on the averaging arrangements.

It is not contended that a warehouseman has either right or title to overages. It would be, however, improper and unwise for a warehouseman to assume responsibility for a "shortage" with no limitation on the amount and without regard to the question of whether or not he has exercised the reasonable care required by law. In fact, there is the serious probability that the warehouseman's so doing might be considered as assuming a liability that would negate his warehouseman's legal liability insurance.

While it is recognized that, in many instances, accounting procedures of both depositor and warehouseman cannot allow adjustment from one inventory period to another "ad infinitum," equitableness demands that provision should be made for adjustment from one period to the next in the case of obvious miscount or proven clerical error. There have been numerous situations, at least one of which involved several hundred thousand dollars, in which a shortage was indicated at the end of an inventory period. Subsequent investigations disclosed that the depositor had never shipped the supposedly "short" product, that the supposed shortage never existed, and actually resulted from a clerical error or (as in the case of the large item referred to) computer malfunction. Similar situations have existed as to overages. Obviously, these possibilities should be considered in any effort to set forth an understanding with respect to this subject.

It would appear to be best to clearly define in detail the circumstances under which overages and shortages will be averaged, the method of settlement, and the method and basis for determination of values involved, and the following is intended to accomplish that definition:

1. Inventory Period. The period for settling stock differences shall be ——— months, each period ending on the ——of—————(month), or a date as near thereto as may be mutually agreed upon by Depositor and Warehouseman. Joint Depositor-Warehouseman physical inventories will be taken on those dates.

2. Additional physical inventories may be taken by Depositor's personnel at any time during normal working hours of Warehouseman, provided that such personnel must, for security and safety reasons, be accompanied by an employee of Warehouseman and, provided further, that Warehouseman shall be reimbursed on the hourly basis as set forth in Schedule "B" or as separately agreed, for time of Warehouseman's employee or employees so engaged.

3. If stock differences are found in any count, the Warehouseman will list gains as receipts, and losses as deductions, thus correcting the book record to agree with the actual stock on hand. These changes will be based upon counts agreed to and signed by Depositor's representative and Warehouseman's representative.

4. For purposes of determining the net balance on the Warehouseman's account, all debits and credits for the period shall be netted together without regard to commodity group.

5. If there be a net-money debit balance for the period covered, after netting across commodity lines, that amount shall be payable to Depositor by Warehouseman. Unless otherwise agreed, if there be a net-money credit on the account, the account shall be closed.

6. The dollar values used to establish the net-money balance shall be __ ____.

7. Carry-over of net debits or net credits from a settlement period to succeeding ones, or carry-back to preceding periods will not be made except under one or more of the following conditions:

   a. Proven miscount in physical inventory.

b. Proven clerical error by either Depositor or Warehouseman.

c. Located or recovered mis-shipment.

8. The maximum total payment by Warehouseman to Depositor for shortage in any one inventory period as defined in Paragraph 1, above, shall be $_____. Payment by Warehouseman for any greater net-money debit shall depend upon determination of whether, with respect to the shortage involved, Warehouseman has exercised the degree of care that a reasonably careful man would exercise under similar conditions (Article 7-204, Uniform Commercial Code).

Shrinkage Allowance

While not a universal practice, an increasing number of depositor-warehouseman understandings recognize that the nature, variety, and volume of products handled, and a reduction in the quality of packaging that causes carton failure in handling and in the stack may result in a small percentage of loss in both the depositor's own facilities as well as those of the warehouseman. These understandings, permitting a "shrinkage allowance," result from two considerations.

The first is recognition that the clerical and administrative time involved in making adjustments can cost more than the loss involved. The second is acceptance of the "reasonably careful man" concept of care set forth in the Uniform Commercial Code, it being the assumption that the depositor in the operation of his own facilities is reasonably careful, that he suffers "shrinkage" of a certain percent, and that the reasonably careful warehouseman should not be asked to assume liability should he fail to do better than the depositor.

The following article meets the above described situation:

In further consideration of the rates herein, and in keeping with the definitions of warehouseman's legal liability contained herein and in Article 7-204 of the Uniform Commercial Code, Depositor agrees to a shrinkage allowance of ___% of the value of the goods stored for which, in the case of loss or damage to goods or mysterious disappearance, however, caused, Warehouseman will not be liable.

Waiver of Subrogation

Most manufacturers, as a matter of sound business practice, carry "output insurance" on their goods (raw products through delivery of the finished product) wherever they may be—in their own facilities, in transportation, or in public warehousing. Practically all warehousemen carry warehouseman's legal liability insurance to cover losses which may result from their failure to exercise the reasonable care required by law. The goods in the hands of the warehouse-

PRACTICAL HANDBOOK OF WAREHOUSING

man are, therefore, the object of double insurance. If there is a major loss, the depositor's output insurance covers, because payment is not dependent upon the circumstances involved. Since no insurance company likes to pay a loss, it is not uncommon for the depositor's insuror to gamble that case of negligence can be made against the warehouseman, retain counsel on a contingent fee basis, and (in the name of the depositor) sue the warehouseman. This is referred to as "subrogation." Subrogation is particularly appealing to the depositor's insurance company because the premium rates on output insurance do not include any reduction factor for subrogation and, if the courts find the warehouseman negligent, the warehouseman's insurance company pays for the loss, resulting in a boon to the depositor's insuror.

The "double" insurance and expenses of litigation constitute an additional cost of doing business that, while hidden in the warehouseman's insurance premiums, must be reflected in charges for his services. In recognition of this, some depositors have provided the warehouseman with a "waiver of subrogation" (which is permitted under manufacturers' output policies) which waiver prevents depositor's insurance from seeking recovery on a loss paid under the policy. The waiver does not include any dollar loss that is less than the deductible in the depositor's policy.

One known waiver of subrogation clause, reflecting the above considerations, is as follows:

In further consideration of the rates herein, Depositor, for itself and its insurers, waives all claims against Warehouseman for loss and damage to goods, however caused, to the extent that such claims exceed $____ or the amount of the deductible on insurance carried by Depositor applicable to loss and damage to such goods, whichever is less.

Hazardous Materials

Increased regulations and new laws designed to protect the public have placed new responsibilities upon warehousemen. The regulations of the U.S. Department of Transportation concerning the shipping requirements for hazardous substances are the most extensive. To conform with these regulations a warehouseman must be furnished with necessary instructions and information. For warehouse agreements which will cover such substances the following language may be helpful:

When Depositor requests that Warehouseman ship goods of the Depositor which are considered hazardous materials under the regulations of the U.S. Department of Transportation, Depositor agrees to advise Warehouseman that such goods are hazardous under the regulations and to furnish Warehouseman with all the correct and necessary information and instructions to permit Warehouseman to prepare the shipment, and necessary shipping papers and certifications, to conform with the regulations, and Depositor appoints Warehouseman as its agent for the purpose of preparing the shipment and signing the certifications and shipping papers covering the shipment.

Index

A

Abels, Bruce, 212
accountability, 83-90
Ackerman Company, 107, 116, 213, 323, 326, 332, 335, 407, 420
act of God, 455
address designation, 171
Adolph Coors Company, 532, 533, 534
Advanced Handling Systems, 499
Affiliated Warehouse Companies, Inc., 215
air freight, 199
aisle planning, 173, 174, 488
alarms, 241
Alderson, Wroe, 197, 198
Allen, James L., 561
Allen, Jim, 558
Allied Warehousing Services, Inc., 466
American Institute of Baking, 255
American National Standards Institute (ANSI), 526
American Public Warehouse Register, 469
American Society for Testing and Materials (ASTM), 529
American Warehousemen's Association (AWA), 5, 6, 10, 45, 53, 84, 216, 265, 266, 267, 356, 414, 573
Ammons, W.S., 147
Apple, James M., 534
Arbuckle, William, 51
architecture, 149

audits
  performance, 113
Australia, 526
Autocon, 508
automated storage and retrieval systems, 517, 524
automatic guided vehicles, 478, 498, 507
automation, 513-24

B

backhaul, 23, 93, 374
bailee-for-hire, 265
bailment, 37
bar-code, 15, 61, 271, 585
bar-code scanners, 557, 558, 586, 587
bay size, 150
Benton, Todd, 384
Bergen Brunswig Drug Corp., 189
bill of lading, 45, 47, 50, 86, 115, 455
Black Box Catalog, 558
Blanding, Warren, 369
Bolger Group, The, 65
Bolger, Daniel, 65
Bowersox, Donald J., 16, 35
box cars, 63, 73, 80
boxcars, 15
brokers, 75, 76
building codes, 468
burden of proof, 456
Burns, David F., 74

C

California
  San Francisco, 4
Carlson, Roger W., 37
carousels, 352, 483, 493
carton clamps, 59
Cascade Corp., 296
Center for Values Research, 302, 339
charges
  demurrage, 5
  storage, 5
Chilton, 469
Civil War, 4
claims, 90
Clark Equipment Company, 295, 530
Clark Lift, 508
clerical activities, 549-554, 570
Clifton Budd Barlach and DeMaria, 302
Coca-Cola, 197, 200
Cohan, Leon, 252, 453, 459
Colombia, 6
common carriage, 6
common carriers, 47, 80
communications, 43-54, 15, 91, 99, 100, 376, 472
computers, 555-584, 8, 14, 494, 508, 521
  software, 567
  training, 582
Congress, U.S., 5, 6
consignment, 199
consolidation, 16, 17, 21, 54, 67, 69, 79, 371
construction, 147, 160
containerization, 67, 68
contracts, 265-270
conveyors, 8, 495
Cooperative Food Distributors Association (CFDA), 52
Coopers & Lybrand, 390
Coopersmith, Jeffrey A., 471
cost of living, 17
costs, 409-420, 8, 13, 14, 21, 26, 30, 38, 53, 56, 60, 61, 64, 70, 91, 92, 96, 149, 150, 258, 402, 435, 483, 510, 513, 514, 532, 533, 554, 557, 573
  equipment, 520
  floor, 518
  land, 518
  maintenance, 521
Council of Logistics Management (CLM), 9, 10
count on receipt, 453
Creel, Albert D., 297, 384
Crosby, Philip B., 378
Crossan and Nance, 414
cube utilization, 57, 73, 170, 172, 350, 381, 486, 498
customer service, 369-380, 22, 39, 91, 259, 369, 550, 567
cycle counting, 278

D

Dahm, George, 315
damage, 535-548, 47, 56, 63, 83, 257
  concealed, 537
decertification, 303, 306
Deloitte, Haskins & Sells, 136
DeMaria, Alfred T., 302
Department of Defense (DOD),

INDEX

117, 356, 414
Department of Transportation (DOT), 458, 467
deregulation, 14, 16, 68, 69, 79
design, 93
Directel, 471
discipline, 326
distribution, 7, 9, 21, 32, 83, 469
Distribution Centers, Inc., 113, 271
Distribution Warehouse Cost Digest, 539
Dixon, John, 144
docks, 154
Dockter, James E., 471
Doggett, John, 574
Drake Sheahan/Stewart Dougall, Inc., 353
Drucker, Peter, 9
Du Pont, 117
DuPont Model, 32, 33

E

East Coast, 17
Eastman Kodak, 345
Economic Recovery Tax Act of 1981, 136
Edwards, Bruce A., 271
Egypt, 3
Electronic Data Interchange (EDI), 52, 53, 54, 556, 560, 567, 568, 572
electronic data processing, 549, 553
electronic identification, 585-594
employees, 287-342, 29, 101, 182, 246, 258, 426, 444
    discharging, 311
    dismissing, 427
    interviewing, 330

    training, 287-300
energy, 381
England
    Cranfield, 524
Environmental Protection Agency (EPA), 466, 467
environmental studies, 148
Equal Emloyment Opportunity Commission (EEOC), 328
equipment, 485-512
    costs, 520, 521
equipment utilization, 186
ergonomics, 515
error reduction, 435-450
Europe, 4
Ewers, T.A., 493, 508

F

F.O.B. (terms of sale), 83
facsimile, 44
Factory Mutual Companies, 468
Fax, 44
fiberboard, 58
fidelity bonds, 89
finances, 127
fire, 227
    prevention, 163
    protection, 62, 86, 114, 151, 152, 211, 228, 237, 382, 485, 519
    regulations, 163
first aid, 114
first-in-first-out (FIFO), 476, 569
fixed slot system, 170
floating slot system, 170
floors, 144, 156, 518
Florida, 345

631

flow charts, 177, 482
Food and Drug Administration (FDA), 254, 465
Food Marketing Institute (FMI), 52
Food, Drug and Cosmetic Act, 254
forklifts, 499-512
forms, 43, 51, 74, 440
Fred Pryor Seminars Inc., 302
free time, 5
freight forwarders, 68, 75
Fulfillment Unlimited, 471
Fundamentals of Materials Handling, 291

G

Gabel, Jo Ellen, 319, 321
Gagnon, Gene, 186
General Systems Consulting Group, 315
Genesis, 3
Georgia
   Atlanta, 213
Gilmore, Russell A. III, 508, 593
government, 148, 155
   regulation, 211
Grant Thornton, 132
Grocery Manufacturers of America (GMA), 52, 527
Gulf Coast, 233

H

Hale, Bernard J., 189
Hall, Craig T., 51
handbook
   personnel, 287

Handbook of American Business History, 4
Hazardous Materials Storage: The Essential Differences, 466
Hazmat World, 470
heating, 153
Hepburn Act of 1906, 5
Heskett, James L., 198
Highway Users Foundation, 132
honeycombing, 167
housekeeping, 255, 465
Hughes, Charles, 302
Human Systems, Inc., 319, 321
Hunt Personnel, Ltd., 427
Hupp, Burr, 452

I

IBM, 560
identification, 55-64
   product, 60
Illinois
   Chicago, 125, 345
Improving Existing Warehouse Space Utilization, 490
Indiana
   Indianapolis, 406
industrial revenue bonds, 39
industrial revolution, 7
infestation, 88
inspection, 113
instruments
   negotiable, 4
insurance, 6, 62, 64, 75, 88, 90, 93, 151, 267, 468
International Air Transport Association (IATA), 467
International Association of Refrig-

INDEX

erated Warehouses (IARW), 53, 462
International Civil Aviation Organization (ICAO), 467
International Maritime Organization (IMO), 467
Interstate Commerce Commission (ICC), 5, 6, 76
inventory, 271-286, 6, 13, 15, 19, 25, 49, 86, 90, 96, 100, 124, 376, 390
   control, 48
   invoice, 52

J

Jahnigen, Rick, 499
Japan, 319, 401
Johnson, Monfort A., 237
Joseph, 3
just-in-time, 401-408, 16, 20, 79, 213, 553, 556, 569

K

Kalmar A.C., 493
Kalmar AC Handling Systems, 508
kan ban, 401
Kuhl, Norman E., 532

L

La Londe, Bernard J., 14, 343
labor relations, 301-314, 25, 32, 34, 95, 101, 182, 192, 335
Landes, Norman E., 358
last-in-first-out (LIFO), 476, 569
Latin America, 349
Lauer, David P., 136
law
   common business, 83
   English common, 86
liability, 455
   product, 88
lie detector, 246
local area networks (LAN), 560
location
   warehouse, 123-134, 15, 17, 39, 65, 91, 216
locations
   stocking, 16
logistics, 9, 14, 16, 20
loss
   and damage, 455
   casualty, 227-240
Louisiana, 84, 265
   New Orleans, 4

M

MacEwan, Bob, 508
Maginot Line, 248
maintenance, 114
Management Tools Inc., 325
manual
   operations, 107, 120
March Cold Storage Company, 406
March Plan, 406
Margo, 186
Marketing Publications, Inc., 37, 74, 75, 94, 141, 144, 147, 177, 181, 189, 204, 227, 241, 271, 272, 297, 302, 312, 313, 315, 321, 347,

353, 354, 358, 360, 366, 369, 384, 390, 422, 430, 435, 508
Martell, John P., 75
Maryland
    Baltimore, 4, 349
Master Standard Data, 414
materials handling, 451-548
Materials Handling Systems Design, 534
McBride, Jim, 215
McGraw-Hill, 414
Mead Corporation, 529
Menzies, John T., 116, 555
Metz, Alex, 427
Mexico, 6
mezzanine, 172
Miami University (Ohio), 297
Michigan State University, 55
Middle Ages, 4
military
    American, 9
Mitch, David, 4
Motor Carrier Act of 1980, 75
Mulvany, Stephen J., 325
Murphey, Richard R., 265
Murphey, Young & Smith, 265

N

National American Wholesale Grocers Association (NAWGA), 52
National Association of Retail Grocers in the US (NARGUS), 52
National Council of Physical Distribution Management, 10
National Fire Protection Association (NFPA), 468, 470
National Food Brokers Association (NFBA), 52
National Industrial Transportation League, 10
National Labor Relations Act, 304
National Labor Relations Board (NLRB), 34, 191, 304
National Materials Handling Centre, 524
Nebraska, 17
Nelson, Richard E., 175, 359
Ness, Robert E., 420
net profit, 32
New Jersey, 213
New York
    New York, 4, 125
NFPA 30 Flammable and Combustible Liquids Code, 469
non-negotiable warehouse receipt, 46, 47
North America, 4
North American Heating and Air Conditioning Wholesalers Association (NHAW), 291
North Carolina
    Charlotte, 213

O

obsolescence, 159
Occupational Safety and Health Administration (OSHA), 466
Occupational Safety and Health Act (OSHA), 251
Ohio
    Cincinnati, 22
    Dayton, 53
Ohio Ready Mixed Concrete Association, 144

INDEX

Ohio State University, 343
OKI Systems, Inc., 499
Operational Training Guide, 462
operators
    warehouse, 64
order checking, 435
order picking, 475-484, 50, 61, 182, 435, 440, 492, 593
ordinary care, 89
Orient, 4
over, short & damage reports, 47, 89, 115

P

packaging, 55-64, 536
pallet jack, 532
pallets, 8, 51, 57, 60, 74, 453, 526
paper
    negotiable, 6
Pareto's Law, 278, 348, 396
Pareto, Vilfredo, 396
peer review, 326
Pennsylvania
    Pittsburgh, 17
Peoples, Paul L., 55
performance audits, 113-122
personnel, 182, 246, 258, 444
Petcock, Frank, 315
Peter Principle, 326, 340
Peterson/Puritan, Inc., 237
Pfizer, Inc., 175, 359
picketing, 34
piggyback, 67, 68, 74, 80
Pilnick, Saul, 319, 321
Polan, Lake III, 466
Polaroid, 459

Pollock, T.E., 213
polygraph, 246
portable data collection, 558
Portman Equipment Company, 499
Portman, William C., 499
postponement, 197-204, 124, 401
Prior, Robert L., 332
private trucking, 69
productivity, 343-450
profit, 37, 116
Promisel, Bob, 525, 594
Pryor, Fred, 302
Pul-Pac, 530, 532
purchase order, 52

Q

Quaker Oats Company, 51
quality circles, 319
quality control, 343-450, 51, 204, 475
Quality Is Free, 378

R

rack
    drive-in, 490
    flow, 481, 483
    gravity-flow, 491
    roller-mounted, 490
    storage, 29, 56, 62, 172, 174, 232, 245
    tier, 490
railroads, 4, 125, 156, 534
Ransom, William J., 338, 475
Ransom & Associates, 475
rates, 36, 68

Reagan Tax Plan of 1984, 136
reasonable care, 84, 87
receiving, 451-460
recooperage, 90
roads, 153
robotics, 516
Rogala, Richard E., 323
roofs, 141, 151, 160
Ruscilli Construction Company, 141, 147
Ruscilli, L. Jack, 147
Ryan, Tom, 462

S

safety, 251-264, 114, 295, 482
sanitation, 254-257, 114
scanners, 61
Schaefer, David L., 490
scheduling, 363-368
Scotland, 349
seals, 244, 247, 456
security, 241-250, 92, 114, 521
Semco, Sweet & Mayers, Inc., 435, 491
Semco, W.B. "Bud", 435, 491
Sheehan, William G., 37, 113, 271, 567
shipper's advance notice, 46
shippers load and count, 455
shipping, 451-460
Slidemasters Inc., 358
slip sheets, 502, 529
Snakard, J.W., 120
space utilization, 62, 167, 381, 410
Speh, Thomas W., 297
sprinkler systems, 152, 211, 228, 237, 351, 382, 468

stack height, 62, 172, 485
stacker cranes, 8, 483, 486, 490
stacking height, 56, 57
Standard Time Values for Public Warehouse Handling, 573
stock location, 182
stock-status report, 49
stockpiling, 19
Stone, James H., 74
storage-in-transit, 17, 80
strategy, 205
strikes, 25, 191, 301, 303
    longshoremen's, 193
    public warehouse, 305
    rail, 193
    transportation, 192, 305
substance abuse, 332
Sun Oil Co., 201
sunbelt, 17
Sunoco, 201
supervision, 183

T

T. Marzetti Company, 453, 459
tariffs, 5, 14, 17, 76, 80
tax credits, 140
taxes, 5, 39, 93, 127, 163, 514, 518
Temperature Controlled Warehousing
    The Essential Differences, 462
Tetz, John A., 390
Texas
    Dallas, 213, 339
    Dallas-Ft. Worth, 124
theft, 234, 243, 452, 472, 521
Thomas J. Lipton Inc., 120
Thomas, Bill, 499

INDEX

Thomas, Lee P., 94, 499, 517
tie and high, 51
Tierney, Theodore J., 301, 326
title to goods, 83, 86, 90, 199
TLC Warehousing Services, 51
TOFC, 67
Tompkins Associates, Inc., 278
Tompkins, James A., 278, 535
Tower Bornes Publishing Company, 470
Towle, William H., 265
Toyota, 402
traffic, 9
    rail, 17
trailers, 73
    size, 155, 162
training, 287-300, 115
Transmart Company, 75
transportation, 65-82, 4, 7, 92, 126
    rail, 63
Transportation Data Coordinating Committee (TDCC), 52, 53
trucks
    carton-clamp, 62, 532
    delivery, 85
    double-reach, 503, 504
    Drexel, 503, 505
    electric, 464
    forklift, 499-512, 8, 532
    hand, 8, 478
    lift, 8, 74, 172, 173, 251, 384, 464, 499, 503, 521
        order picking, 502
        private, 15
        single-reach, 503, 504
        turret, 503, 505
        Waldon, 503, 505

U

U.S. Bureau of National Affairs, 192
U.S. Department of Agriculture (USDA), 183, 414, 510, 539
U.S. Supreme Court, 254, 455
Uniform Code Council (UCC), 53
Uniform Commercial Code (UCC), 84, 87, 89, 265, 267, 268
Uniform Communications Standards (UCS), 52, 53
Uniform Warehouse Receipts Act, 265
unionization, 302
unit loads, 525-534, 8, 59, 60, 69, 502
United Refrigeration Services, Inc., 462
United States, 7, 14, 124, 152, 163, 526
University Mall Properties, 147
University of Maryland, 4
utilities, 125, 149

V

vandalism, 236
Vedder, Price, Kaufman & Kammholz, 301
Venice, 4, 6
Von Thunen, J.H., 124

W

Wales Industries, Inc., 463
Warehouse Information Network

637

Standards (WINS), 53
warehouse
　receipt, 86
　receiving tally, 47
warehouses
　bank, 7
　bonded, 5
　consolidation, 21
　contract, 23, 36
　distribution, 22
　field, 7
　multi-story, 7, 8
　private, 8, 15, 17, 23, 35, 74, 76, 251
　product mixing, 20
　public, 15, 16, 23, 34, 36, 47, 65, 74, 75, 215, 251
　terminal, 7
warehousing, 3-43
　construction, 135-158
　definition, 3
　financing, 135, 136
　freight-depot, 5
　fulfillment, 461, 470
　future planning, 159-166
　hazardous materials, 466
　hazardous products, 461
　leasing, 135, 136
　location, 123-134
　management, 43-122
　performance audits, 113-122

planning, 167-196
private, 7
public, 215-226, 6, 7
purchasing, 135
rehabilitation, 135
starting up, 91-112
temperature-controlled, 461, 462
third party, 6, 26
Warehousing Education and Research Council (WERC), 10, 16, 35, 37, 175, 212, 452
Washington, D.C., 52, 132
Way, Howard, 406
Weber, Joel, 462
Westburgh, Jesse, 463
White, Prof. John A., 535, 546
Whitten, David O., 4
Williams, John, 524
wind-load, 151
windstorm, 227, 233
Wolff, Joel C., 347, 353, 354
Worker's Compensation, 115
World War II, 8, 9, 17, 30, 62, 266, 319, 526, 549

Y

Yeomans, Morton T., 594

About the Author

Since 1957, **Kenneth B. Ackerman** has been involved in Warehousing and Logistics Management.

Before entering the consulting field, he was chairman of Distribution Centers, Inc., a public warehousing company which operates in major cities throughout the U.S. In 1980 Ackerman sold the company and joined the management consulting service division of Coopers and Lybrand. In 1981, he formed the Ackerman Company.

He is the author of numerous articles and two books dealing with warehousing and general management. He was general editor of *Warehousing and Physical Distribution Productivity Report,* a newsletter published by Marketing Publications Inc. He is now publisher and editor of *Warehousing Forum,* a subscription newsletter. An article, *Making Warehousing More Efficient,* co-authored with Bernard J. LaLonde, was published by Harvard Business Review.

Mr. Ackerman is a frequent speaker at meetings and seminars. He has been a guest lecturer at many universities and management seminars. He has lectured at business meetings in Asia, Europe, and South America.

He has the following educational and professional credentials: B.A., Princeton University, M.B.A., Harvard University. He is a past president of Council of Logistics Management (formerly

NCPDM). He is a founder of Warehousing Education and Research Council.

In 1977, he was awarded the John Drury Sheahan Award for distinguished contribution to physical distribution. He is a former chapter chairman of the Young Presidents Organization. He is a past president and honorary life member of Ohio Warehousemen's Association and a former board member of American Warehousemen's Association.